FOOD SCIENCE AND TECHNOLOGY

FOOD HYDROCOLLOIDS: CHARACTERISTICS, PROPERTIES AND STRUCTURES

FOOD SCIENCE AND TECHNOLOGY

Food Science and Technology: New Research
Lorenzo V. Greco and Marco N. Bruno (Editors)
2008. ISBN: 978-1-60456-715-1

Food Science and Technology: New Research
Lorenzo V. Greco and Marco N. Bruno (Editors)
2008. ISBN: 978-1-61668-106-7 (Online Book)

The Price of Food
Meredith N. Fisher (Editor)
2009. ISBN: 978-1-60692-440-2

Food Processing and Engineering Topics
Maria Elena Sosa-Morales and Jorge F. Velez-Ruiz (Editors)
2009. ISBN: 978-1-60741-788-0

Food Processing and Engineering Topics
Maria Elena Sosa-Morales and Jorge F. Velez-Ruiz (Editors)
2009. ISBN: 978-1-91998-422-8 (Online Book)

Traditional Chinese Foods: Production and Research Progress
Li Zaigui and Tan Hongzhuo
2009. ISBN: 978-1-60692-902-5

Traditional Chinese Foods: Production and Research Progress
Li Zaigui and Tan Hongzhuo
2009. ISBN: 978-1-61668-277-4 (Online Book)

Food Science Research and Technology
Isaak Hülsen and Egon Ohnesorge (Editors)
2009. ISBN: 978-1-60741-848-1

Food Science Research and Technology
Isaak Hülsen and Egon Ohnesorge (Editors)
2009. ISBN: 978-1-61668-853-0 (Online book)

Food Safety, Fresh Produce and FDA Oversight
Russell P. Barton (Editor)
2010. ISBN: 978-1-60692-788-5

Industrial Fermentation: Food Processes, Nutrient Sources and Production Strategies
Jürgen Krause and Oswald Fleischer (Editors)
2010. ISBN: 978-1-60876-550-8

Food Hydrocolloids: Characteristics, Properties and Structures
Clarence S. Hollingworth (Editor)
2010 .ISBN: 978-1-60876-222-4

Food, Diet and Health: Past, Present and Future Tendencies
Raquel Pinho Ferreira Guine (Editor)
2010. ISBN: 978-1-60876-012-1

Sweet Potato: Post Harvest Aspects in Food, Feed and Industry
Ramesh C. Ray and K.I. Tomlins (Editors)
2010. ISBN: 978-1-60876-343-6

Microbial Interaction in Fermented Beverages
Marta E. Farías and Oscar A. Sosa; Lucía M. Mendoza, and Pedro A. Aredes Fernández
2010. ISBN: 978-1-60876-785-4

The Olive: Its Processing and Waste Management
José S. Torrecilla
2010. ISBN: 978-1-60876-719-9

Liquid Chromatography for the Determination of Mycotoxins in Foods
R. Romero-González, J. L. Martínez Vidal, and A. Garrido Frenich
2010. ISBN: 978-1-60876-882-0

Advances in Food Science and Technology
A. K. Haghi (Editor)
2010. ISBN: 978-1-61668-415-0

Applications of Natural Products in Food
Supayang Piyawan Voravuthikunchai and Wumi Ifesan
2010. ISBN: 978-1-60876-998-8

Applications of Natural Products in Food
Supayang Piyawan Voravuthikunchai and Wumi Ifesan
2010. ISBN: 978-1-61668-370-2 (Online Book)

Extraction of Metabolities from Plant Tissues
A. Mohdaly , and I. Smetanska
2010. ISBN: 978-1-61668-252-1

Extraction of Metabolities from Plant Tissues
A. Mohdaly , and I. Smetanska
2010. ISBN: 978-1-61668-616-1 (Online Book)

Food Hydrocolloid Edible Films and Coatings
O. Skurtys, C. Acevedo, F. Pedreschi, J. Enronoe, F. Osorio, and J.M. Aguilera
2010. ISBN: 978-1-61668-269-9

Food Hydrocolloid Edible Films and Coatings
O. Skurtys, C. Acevedo, F. Pedreschi, J. Enronoe, F. Osorio, and J.M. Aguilera
2010. ISBN: 978-1-61668-740-3 (Online Book)

Fermented Milk Products
R. Ahmed Abdelrahman, A.M Adel and I. Smetanska
2010. ISBN: 978-1-61668-299-6

Fermented Milk Products
R. Ahmed Abdelrahman, A.M Adel and I. Smetanska
2010. ISBN: 978-1-61668-741-0 (Online Book)

FOOD SCIENCE AND TECHNOLOGY

FOOD HYDROCOLLOIDS: CHARACTERISTICS, PROPERTIES AND STRUCTURES

CLARENCE S. HOLLINGWORTH
EDITOR

Nova Science Publishers, Inc.
New York

LIBRARY OF CONGRESS CATALOGING-IN-PUBLICATION DATA

Available upon request.

ISBN: 978-1-60876-222-4

Published by Nova Science Publishers, Inc. ✢ New York

CONTENTS

PREFACE

A hydrocolloid is defined as a colloid system wherein the colloid particles are dispersed in water. A hydrocolloid has colloid particles spread throughout water and depending on the quantity of water available that can take place in different states, e.g., gel or sol (liquid). Hydrocolloids can be either irreversible (single-state) or reversible. For example, agar, a reversible hydrocolloid of seaweed extract, can exist in a gel and sol state, and alternate between states with the addition or elimination of heat. Many hydrocolloids are derived from natural sources. Agar-agar and carrageenan are extracted from seaweed; gelatin is produced by hydrolysis of proteins of bovine and fish origins, and pectin is extracted from citrus peel and apple pomace. Gelatin desserts like jelly or Jell-O are made from gelatin powder, another effective hydrocolloid. Hydrocolloids are employed in food mainly to influence texture or viscosity. This new and important book gathers the latest research from around the globe in the study of food hydrocolloids and highlights such topics as: collagen and gelatin extracted from skate skin, sucrose pectin interaction, utiliztion of glucomannans for health and others.

Chapter 1 - Gelatin and collagen are different forms of the same macromolecule. Gelatin is obtained by the partial hydrolysis of collagen contained in the skin, the connective tissue and bones of animals. Collagen is a cheap and resourceful meat and fish by-product whose main product is a gelatin that is used extensively as a food additive to increase the texture, water-holding capacity and stability of several food products. The method of manufacture greatly affects the physicochemical and functional properties of the gelatin and collagen. Collagen must be pretreated to convert it into a form suitable for gelatin extraction. The degree of conversion of collagen into gelatin is related to the severity of both the pretreatment and the extraction processes, which depends on pH, temperature, and extraction time. The inconsistency in physicochemical and functional properties can be cause by differences in gelatin or collagen sources and processing conditions. Skate skin, which is presently considered to be a waste product, could be effectively used as a natural gelatin and collagen sources. Therefore, this review chapter focuses on the utilization of gelatin and collagen extracted from skate skin to evaluate its potential use as food ingredients and the efficacy of the collagen as a coating to enhance the shelf-life of food products.

Chapter 2 - In this chapter book, we present and discuss some aspects of hydrocolloid edible films and coating. The film-forming materials is presented and the physical properties (*e.g.* thickness, mechanical, optical and thermodynamic properties, water and gases barrier) of the film and coating are examined in detail. The main techniques used to form edible coatings as spray systems or immersion procedures are discussed as well as the solvent casting and the extrusion processes for the edible films. The main disadvantage of these techniques is the lost of quality of the edible coatings and films since there is no control over the shape, size and size distribution of the dispersed elements (*e.g.* additives, ingredients, etc.) in the support structure matrix is poor. Thus, microfluidic applications are presented and discussed to solve some of these problems and to improve film qualities.

Chapter 3 - Fresh (F) and frozen/thawed (F/T) mashed potatoes made with mixtures of sodium caseinate (SC) and xanthan gum (XG) were assessed for rheological properties as evaluated with two kinds of deformation tests—large deformation [instrumental texture profile analysis (ITPA) and cone penetration (CP)] and small deformation (oscillatory measurements)—colour, drip loss (DL), pH, and two different sensory analyses. The influence of SC and XG concentrations on rheological properties and quality characteristics was studied by means of response surface methodology (RSM) using a central composite rotatable experimental design with SC concentrations ranging between 2.5 and 7.5 g kg^{-1} and XG concentrations ranging between 0.5 and 2.5 g kg^{-1}, as independent variables. The effects were strongly dependent on the levels of SC and XG added. The effect of XG concentration on the oscillatory measurements was more significant, although hydrophobic interactions between SC and XG may well have influenced the viscoelastic behaviour of the products. F mashed potatoes with added biopolymer mixtures behaved like a weaker gel structure; this fact is attributed to repelling forces between negatively charged phosphate groups on potato starch granules and negative charges on XG molecules. Processing significantly reduced the effect of adding SC and XG mixtures on the gel strength and the viscoelastic behaviour of the mashed potatoes due to high local starch concentrations and retrogradation of both amylose and amylopectin. Addition of biopolymer mixtures modified the pH of the product, mainly due to the presence of SC. DL values revealed that addition of XG at concentrations of 1.5 g kg^{-1} or higher, either individually or mixed with SC, were effective in stabilizing mashed potatoes against syneresis. Scores for overall acceptability (OA) were higher in the F products than in their F/T counterparts, and when SC/XG mixtures (each biopolymer at 1.5 g kg^{-1}) were included in the formulation, the product exhibited very acceptable sensory quality thanks to their creamy mouthfeel conferred by this mixture.

Chapter 4 - Sodium caseinate has been used to obtain edible films with interesting properties due to their high nutritional quality, excellent sensory properties and potential to adequately protect food products from their surrounding environment. Caseinates provide good mechanical properties and are effective barriers to aroma compounds and gases such as oxygen and carbon dioxide. However, the highly hydrophilic nature of sodium caseinate films limits their ability to provide a significant moisture barrier. In this sense, the incorporation of lipids to the film matrix can improve the caseinate functional properties.

In this chapter, the effect of the incorporation of lipid mixtures (oleic acid: OA and beeswax: BW) on different properties of the caseinate based films has been discussed. The kind and amount of plasticizer have been analyzed in order to obtain films with adequate flexibility. The protein:lipid ratio and the OA:BW ratio in the lipid mixture were varied to find the best film formulation for a determined use. The effect of moisture content and temperature on the film properties was also studied. Likewise the influence of calcium on some film properties was analyzed.

Barrier properties to water and aroma compounds which are the main justification for the film application were discussed. Tensile properties were commented on as a function of the above mentioned variables. Optical properties (gloss and transparency) which are decisive to define the final appearance of the coated products were also described. All these aspects were related to the film's microstructure which was analysed through Scanning Electron Microscopy (SEM) and Atomic Force Microscopy (AFM).

The protein:glycerol ratio which gives the best film properties was 1:0.30. The lipid ratios in the protein matrix which made the film preparation possible ranged between 0.15 and 0.50. Oleic acid had a plasticizing effect on the films, increasing their flexibility and stretchability, and reducing water vapor permeability with respect to sodium caseinate films. Aroma barrier efficiency is dependent on oleic acid:beeswax ratio and physico-chemical properties of aroma compounds. A close correlation was established between optical and microstructural parameters of the films. The addition of calcium caseinate to sodium caseinate-oleic acid-beeswax films modified the mechanical, barrier and optical properties of the films. Calcium ions promote inter- and intramolecular cross-linking, increasing brittleness and limiting elongation and biopolymer segmental mobility.

Chapter 5 - Many solutions of polysaccharides and proteins exist as a solution at high temperature and become a gel at low temperature, and they are widely used in food industries to make a gel. Rheological properties drastically change when the solution becomes a gel. This change is based on the structural change of the polysaccharides and proteins, such as that from random coil to helix structure. The structural change causes the change of tumbling motion and translational motion in the solutions. NMR is a powerful tool for investigating the molecular motion. Relaxation times for the macromolecules reflect the molecular motion, and relaxation times for water reflect the motion of water molecules and also macromolecules through the chemical exchanging between water proton and labile proton on the macromolecules. The diffusion coefficients of molecules in the hydrocolloid system give the information about the mobility of molecules and the structure of the gel network.

Chapter 6 - There is now new evidence concerning the participation of plant-derived polysaccharides that may play an important role in reactive oxygen species in the etiology and physiopathology of human diseases. Ethanol-extract and sulfated polysaccharides of sweet cassava and seaweed, respectively, were tested for their structural, antiproliferative, and antioxidative properties. Animal cell cultures, including hepatoma HepG2, glioma C6, and Chang liver cell lines, were used as models to evaluate the antiproliferative, DPPH, and hydrogen peroxide radical scavenging properties of the extracts. The antioxidative properties of seaweed were screened and investigated for its hepatoprotective activity in rat. Sweet cassava polysaccharide is mainly composed of the major neutral sugars sucrose, fructose, glucose, galactose, and arabinose. ^{1}H NMR and ^{13}CNMR showed that the cassava polysaccharides was composed of 1, 3-linked glucan chains with 1, 6-linked galactan side chains attached to position 6. Both sweet cassava polysaccharides and a seaweed diet were independently found to significantly increase UGT1A1 and UGT1A6 (phase II enzymes) mRNA levels and to decrease CYP1A1 and CYP1A2 (Phase I) mRNA levels in chemically-injured rat liver microsomes. *In vitro* and intracellular antioxidative enzyme activity assays demonstrated and confirmed the potential of cassava root and seaweed extracts and were attributed to the structural properties of the polysaccharide compounds. A mechanism relating the antioxidative properties of the polysaccharide glucan chains to stimulate surface cellular receptors and induce antioxidant response enzymes is proposed in this study, as the underlying factor for the immune-stimulating properties of polysaccharides.

Chapter 7 - Present knowledge on the chemistry, structure and architecture, and the functionality and health benefits of four commonly used seed galactomannans, namely, guar gum, locust bean gum, tara gum and fenugreek gum is reviewed to provide directions for research needed to improve the utilization of these gums as functional ingredients in the food industry.

Chapter 8 - Pectin is a very important polysaccharide in food technology, due to its ability to thicken or gel in aqueous solution under specific conditions. One of the most common uses of pectin is in making jellies or jams, whichs requires the addition of a large amount of sucrose to acidic solutions. The role of sucrose in promoting pectin gelation has been ascribed to the strengthening of hydrophobic interactions consistently with its well-known stabilizing effect on protein structure. On the other hand, more specific effects on dimension and stiffness of pectin chains have been suggested by computational work. Here we present measurements of rheology, static and dynamic light scattering on samples of pectin containing different amounts of sucrose, ranging from few percent up to close to the solubility limit of sucrose. This corresponds to spanning from low viscous liquid samples to strong gels. Results show that below the threshold value of 55% w/w, sucrose acts as a critical parameter for the sol/gel transition by increasing the strength of excluded volume and hydrodynamic interactions between polymer chains. A further increase of sucrose above the critical concentration yields gels with a higher viscoelastic component corresponding to a higher amount of frozen structural inhomogeneities.

Chapter 9 - Glucomannans are a source of soluble dietary fibre. In nature they are distributed in both softwood tissues such as roots or corms and the hemicellulose of hardwood plants where they serve as energy reserves and in some cases structural roles. As glucomannans contain components which may have a positive effect on health, for instance by selectively stimulating the gut-friendly bacteria and inhibiting pathogens, they can be utilised to serve as valuable ‘functional foods’. Outside the gut, the impact of ingested glucomannans for broader health benefits include cholesterol lowering, reducing the risk of chronic constipation, diabetes management, cancer prevention, treating diverticular disease, stimulating the immune system and contributing towards weight reduction.

Chapter 10 - The food industry is constantly searching for new food components to improve the texture and mouth-feel of food products; moreover, health-promoting capacity of food ingredients is also desirable.

In recent years, exopolysaccharide (EPS)-producing lactic acid bacteria (LAB) have gained attention due to their major role in functional foods and for improving the sensorial properties of fermented products. Certain LAB excrete a wide diversity of polysaccharides with respect to their monomer composition, molecular weight (MW), structure, and functionalities. These extracellular components or EPS can impart functional properties to food and additionally, may confer beneficial health-promoting properties. The role of EPS from LAB in the dairy food industry (i.e. fermented milks and cheeses) is well documented. Also, the addition of EPS-producing LAB to cereal products or their use as food additives have been successfully applied thus broadening their field of application.

Kefiran, a high-MW EPS produced by microorganisms present in the grains of the fermented milk kefir, has displayed novel applications in the food industry. Kefiran, which has a Newtonian behaviour in diluted solutions but pseudoplastic at high concentrations, improves the viscosity and viscoelastic properties of acid milk gels. This EPS is also able to form gels at low temperatures in addition to brittle and rigid films. However, glycerol addition confers it extremely high elongation values, allowing comparable flexibilities to those of synthetic material. Moreover, several health-promoting properties of kefiran such as immunomodulation or epithelium protection have been reported. Furthermore, kefiran has the ability to antagonize biological activity of *Bacillus cereus* on Caco-2 cells thus providing new perspectives for the role of this bacterial EPS in functional foods.

Chapter 11 - Antimicrobial Edible films were prepared from natural fiber of pectin and other food hydrocolloids for food packaging or wrapping by extrusion followed by compression or blown film method. Microscopic analysis revealed a well mixed integrated structure of extruded pellets and an even distribution of the synthetic hydrocolloid in the biopolymers. The resultant composite films possess the mechanical properties that are comparable to films cast from most natural hydrocolloids that consumed as foods or components in processed foods. The inclusion of poly(ethylene oxide) alters the textures of the resultant composite films and therefore, demonstrating a new technique for the modification of film properties. The composite films were produced in mild processing conditions, thus, the films are able to protect the bioactivity of the incorporated nisin, as shown by the inhibition of *Listeria monocytogenes* bacterial growth by a liquid incubation method.

In: Food Hydrocolloids: Characteristics, Properties... ISBN: 978-1-60876-222-4
Editor: C. S. Hollingworth, pp. 1-39

Chapter 1

UTILIZATION OF COLLAGEN AND GELATIN EXTRACTED FROM SKATE SKIN

Jong-Bang Eun[1], Jin han Shon[2], Ji hyoun Eo[1], and Sung Hun[1]
[1] Department of Food Science and Technology, Chonnam National University, 300 Yongbong-dong Buk-gu, Gwangju 500-757, Korea
[2] Functional Food Research Center, Chonnam National University, 300 Yongbong-dong Buk-gu, Gwangju 500-757, Korea

ABSTRACT

Gelatin and collagen are different forms of the same macromolecule. Gelatin is obtained by the partial hydrolysis of collagen contained in the skin, the connective tissue and bones of animals. Collagen is a cheap and resourceful meat and fish by-product whose main product is a gelatin that is used extensively as a food additive to increase the texture, water-holding capacity and stability of several food products. The method of manufacture greatly affects the physicochemical and functional properties of the gelatin and collagen. Collagen must be pretreated to convert it into a form suitable for gelatin extraction. The degree of conversion of collagen into gelatin is related to the severity of both the pretreatment and the extraction processes, which depends on pH, temperature, and extraction time. The inconsistency in physicochemical and functional properties can be cause by differences in gelatin or collagen sources and processing conditions. Skate skin, which is presently considered to be a waste product, could be effectively used as a natural gelatin and collagen sources. Therefore, this review chapter focuses on the utilization of gelatin and collagen extracted from skate skin to evaluate its potential use as food ingredients and the efficacy of the collagen as a coating to enhance the shelf-life of food products.

1. INTRODUCTION

Collagen and gelatin are a fibrous protein with a broad range of global industrial applications. Gelatin is obtained by the partial hydrolysis of collagen contained in the skin, connective tissue, and bones generated as waste during animal slaughtering and processing [1]. It improves the functional properties of food products and may also be used as an antioxidative and antimicrobial peptide. Gelatin is also used as a food additive to increase protein content, texture, and stability of food products [2]., and is more widely used in food, as a pharmaceutical and photographic agent, and in cosmetic materials [3]. Bovine and porcine wastes are the most frequent sources of high quality gelatin and collagen. Other sources of gelatin and collagen such as fish bones and skins are becoming increasingly relevant [4]. However, collagen and gelatin peptide from bovine and porcine materials can be a food allergen. As an approach to develop new collagen and gelatin materials having modified allergenicity, it was considered that aquatic animal collagens could be an alternative and source of collagen and gelatin obtained in an environmentally-friendly manner. Gelatin and collagen from mammals have been studied for their antioxidative and antimicrobial properties, and edible films are available. Synthetic antioxidants and antimicrobial additives are commonly used to preserve food products, although they are limited in their application as food additives because of potential health problems [5]. Recent edible film research has focused on casein, collagen, chitosan, corn zein, gelatin, calcium alginate, wheat gluten, soy protein, whey protein, and methyl cellulose-based films [6-10]. These edible films can be used as carriers for antioxidative and antimicrobials. Sufficient information for potential antioxidative and antimicrobial peptides and films in fish skin gelatin and collagen is not available, although researchers are now increasingly turning their attention to fish gelatin and collagen.

The amount of byproduct produced during fish processing can be high as 75% of the total weight of the fish, a very high percentage. Some of the byproducts are used in animal feed, but much of it is discarded and is a source of environmental contamination [4]. The amount of hazardous waste produced from fish processing has tended to increase annually.

Skate species are deep sea-dwelling benthic animals. Skate (*Raja Kenojei*) is a popular food item in South Korea. Great amounts of skate skin are disposed of as waste during processing. Utilization of this by-product would provide a ready and inexpensive source of material. In this regard, skate skin has been recognized as potential food components because of its excellent nutritional values and functional properties, ability to form gels and films, and because of its antioxidative and antimicrobial properties. The purposes of this review chapter are to summarize optimization the extraction of gelatin and collagen from skate skin, evaluate the compositional and functional characteristics of the extracted materials, identify antimicrobial peptide(s) from skate skin, and to evaluate the effect of collagen on antioxidative activity in an oxidation catalyzed model system compared with standard synthetic and natural antioxidants and the efficacy of the extracted collagen as a coating to enhance the shelf-life of pork loin.

2. EFFECT OF PROCESSING CONDITIONS ON THE PHYSICOCHEMICAL AND FUNCTIONAL PROPERTIES OF GELATIN FROM SKATE (*RAJA KENOJEI*) SKIN

2.1. Physicochemical properties of gelatin from skate skin

2.1.1. Proximate composition

General extraction procedures of gelatin from skate skin are described in Figure. 1.

The mean values of gelatin of moisture, protein, fat, and ash content (%) were 4.52%, 92.3%, 0.35%, and 1.42%, respectively [11]. The low fat and ash contents suggest that the extraction process is more effective.

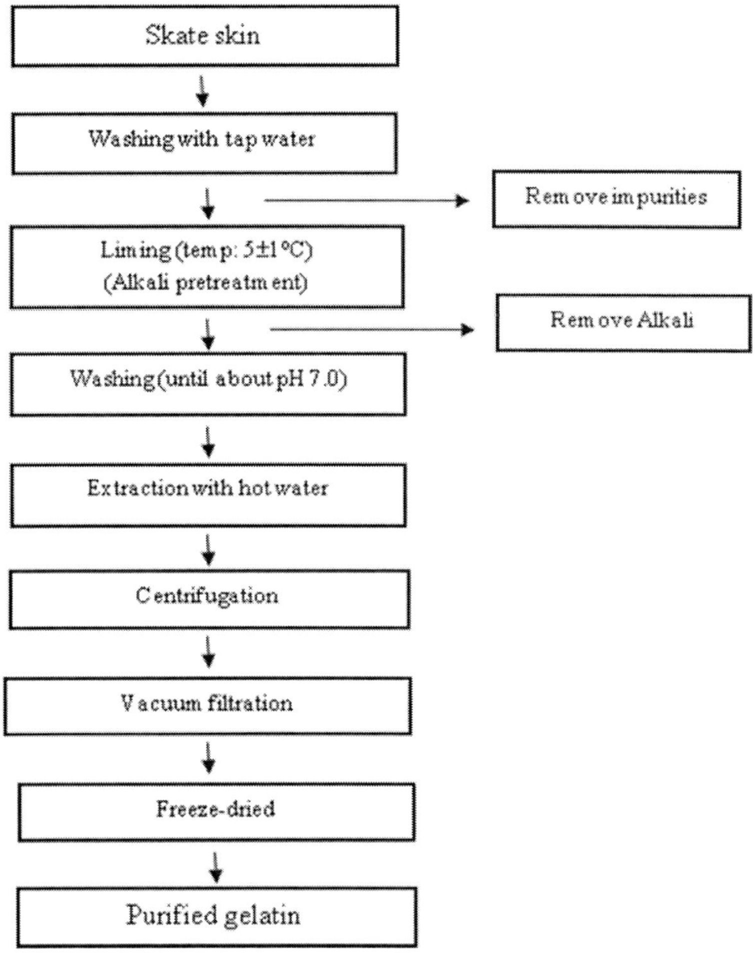

Fig. 1. Preparation of gelatin from skate skin

2.1.2. Effects of liming conditions on the extraction of gelatin

Gelatin is typically obtained by hydrolyzing the collagen present in the bones and skin generated as waste during animal slaughtering and processing. Therefore, the source, age of the animal, and type of collagen all influence the properties of the gelatins [12]. Gelatin and collagen are different forms of the same macromolecule, so that the gelation of gelatin could be described as a partial reformation of the collagen structure [13]. The method of manufacture greatly affects the physicochemical properties of the gelatin. The physicochemical properties of a gelatin depend on the species and tissues from which it is extracted, and by the severity of the manufacturing method [12]. Collagen must be pretreated to convert it into a form suitable for gelatin extraction, which is normally done with >45°C water. The degree of conversion of collagen into gelatin is related to the severity of both the pretreatment and the extraction processes, which depends on pH, temperature, and extraction time. A mild acid pretreatment is usually used for collagenous material containing a low concentration of intra- and inter-chain nonreducible crosslinks, that is, highly soluble collagens as is the case with fish skins [13, 14-16]. Extraction of fish gelatin has been reported for cod skin [17], tilapia skin [18], shark, lungfish and carp skin [19], as well as conger eel and arrow squid [20]. The quality of a fish gelatin is determined mainly by its bloom strength and heat stability (melting and gelling temperatures). The quality of a food grade gelatin, therefore, depends to a large extent on its rheological properties (essentially gel strength and viscosity), but is also determined by other characteristics, particularly color, transparency, flavor, and easy dissolution [1].

In general, calcium hydroxide ($Ca(OH)_2$) and sodium hydroxide ($NaOH$) solution was used to remove impurities (proteoglycan, blood, mucins, sugar, fat etc.) and change collagen to the optimal type for gelatin extraction [12]. According to [11], the extraction yields of gelatin from skate skin showed 11.5%, 12.1%, 16.8%, and 14.5% at respective $Ca(OH)_2$ concentrations of 0.5%, 1.0%, 1.5%, and 2.0%. The highest gelatin extraction yield from skate skins observed 16.8% at a 1.5% $Ca(OH)_2$ concentration [11]. The extraction yield of gelatin from skate skin was somewhat higher compared with results from Tilapia *spp.* [3] and cod skin [17], but less than mammalian gelatins [19]. Previous report indicated that the yield of gelatin from Dover sole (*Solea vulgaris*) decreased when pressurization was applied in the extraction step [21].

Viscosity is the second most important commercial physical property of a gelatin [19]. Viscosity of gelatins from skate skins showed 18.7, 19.0, 22.3 and 20.5 cps for respective $Ca(OH)_2$ concentrations of 0.5%, 1.0%, 1.5%, and 2.0% [11]. The viscosity of gelatins from skate skins at various $Ca(OH)_2$ concentrations was small. Viscosity is partially controlled by molecular weight and molecular size distribution [22]. For many gelatins, minimum viscosity occurs at pH 6.0–8.0 [23]. Gelatins have a minimum viscosity at their isoelectric point (pI) [19]. Thus, the viscosity of skate skin gelatins might be improved by maintaining pH of 2.0–3.0.

The Bloom value or jelly strength is associated with the mechanical elasticity of the gel, and allows gelatins to be classified. Gel strength of gelatins from skate skins varied (43.8, 63.5, 76.6 and 71.8 Bloom) with different $Ca(OH)_2$ concentrations (0.5%, 1.0%, 1.5%, and 2.0%), respectively [11]. The Bloom value was highest at a $Ca(OH)_2$ concentration of 1.5%, which was less compared with other marine fish and mammalian gelatins [24, 17, 25, 3]. A Bloom value between 50 and 300 allows the properties of the gels at the product storage temperature to be defined. Previous study reported that gel strength is dependent on the pI and it can be controlled by adjusting the pH [17]. More compact and stiffer gels can be formed by adjusting the pH of the gelatin at or near its pI [3]. The stability of collagen and gelatins is proportional to the total imino acid [proline (Pro) + hydroxyproline (Hyp)] and glycine (Gly) content [26]. The ultimate strength of the gel is related to its imino acid and Gly contents, and gelatins derived from fish collagens are weak with low melting points, compared with mammalian gelatins. The latter have a higher proportion of Pro, Hyp, and Gly [11]. The low gel strength of skate skin gelatins may be due to the lower amino acid (i.e., Pro and Gly) concentrations [27] compared with other fish and mammals. Gel strength of fish gelatins are less stable with poor physiological properties compared with mammalian gelatins [28], due to lower concentrations of proline [13]. Gelatin gels are formed by the entanglement of molecular chains due hydrogen, ionic, and hydrophobic bonds between the chains or the segments. These gels display a reversible sol-gel transition through change in temperature, solvent composition, and pH, as a few example parameters [29].

The color of the gelatin varied with concentration of extraction solution. Extraction liming (Ca $(OH)_2$) concentrations of 0.5%, 1.0%, 1.5%, and 2.0% produced Hunter color *L* value of 26.0, 36.4, 37.1, and 32.9, respectively, Hunter color *a* value of 0.03, 0.12, 0.17, and 0.46, respectively, representing a progressive increase, and the Hunter color *b* value was 0.53, 0.18, 0.25, and 0.87, respectively [11]. The Hunter color *L* and *b* values varied with extraction liming concentration. The skate skin gelatins had a snowy white color, which is a positive attribute. A difference in the color values at different extraction solution is due to the protein denaturation and structural change by hydrolysis. The color of the gelatin does not influence other functional properties [30].

2.1.3. Effects of pH on the extraction of gelatin

Few studies have been carried out to investigate the effects of pH on the extraction of gelatin from fish sources. Previous study has been evaluated the extraction yields of gelatin from skate skin at different pH values (4.0–9.0) of 1.5% Ca $(OH)_2$. The gelatin yield was 16.7%, 14.1%, 13.7%, 13.4%, 13.9%, and 12.5% at pH 4.0, 5.0, 6.0, 7.0, 8.0, and 9.0, respectively [11]. The gelatin yield was the highest at pH 4.0 compared with other pH values, which differed from [24] who reported that yields increased at pH 6.0–7.0.

Viscosity and gel strength of gelatins varies with pH of extraction solution. Viscosity of gelatin from skate skin showed 19.5, 20.3, 24.3, 22.4, 17.8, and 20.3 cps at pH 4.0, 5.0, 6.0, 7.0, 8.0, and 9.0, respectively [11]. Gel strength from skate skin showed 62.3, 68.6, 75.1, 72.4, 66.2, and 66.2 cps for pH 4.0, 5.0, 6.0, 7.0, 8.0 and 9.0, respectively [11]. These results showed that both viscosity and gel strength were highest at pH 6.0–7.0. This result might be because the molecular weight decreased slowly in the neutral region compared with alkali and acid conditions [18]. Gelatin polymers are closer to being neutrally charged and form more compact and stiffer gels when the pH of the fermented skate skin gelatin is near it pI [17]. These results disagree with those of previous study [23], who reported that minimum gelatin viscosity occurred at pH 6.0–8.0. Gelatin is characterized by its pH and its pI. Viscosity is lowest around the pI and increases when the pH moves away from this zone. The reason for this lies in the repulsion due to charges which are identical all along the chain when the pH is reasonably distant from pI.

The Hunter color *L*, *a* and *b* values of gelatins varied with pH values of extraction solution. The Hunter color *L* value showed 40.5, 39.6, 42.9, 43.8, 41.6, and 39.9, respectively, at an extraction pH of 4.0, 5.0, 6.0, 7.0, 8.0, and 9.0, the Hunter color *a* value showed 0.36, 0.33, 0.21, 0.23, 0.08, and 0.21, respectively, and the Hunter color *b* value showed 0.43, 0.36, 0.37, 0.58, 0.39, and 0.48, respectively [11]. A difference in the color values at different extraction solution is due to the protein denaturation and structural change by hydrolysis

2.1.4. Effects of temperature on the extraction of gelatin

Previous reports indicated that the yield, viscosity, gel strength, and color values of gelatin were affected by temperature of extraction. The extraction yields of gelatin from skate skin at various temperatures showed 5.25%, 13.7%, 15.3%, and 18.4% for temperatures of 40°C, 50°C, 60°C, and 70°C, respectively [11]. The higher the temperature, the higher levels of gelatin extracted. Viscosity of gelatin from skate skin at various temperatures showed 22.5, 20.5, 18.6, and 19.1 cps, respectively, and gel strength was 74.4, 67.2, 61.9 and 54.5 cps, respectively, for extraction temperatures of 40°C, 50°C, 60°C, and 70°C [11]. Gel strength decreased as temperature increased. In general, gel strength varies inversely with temperature. Both viscosity and gel strength of gelatin from skate skin were highest at an extraction temperature of 40°C compared with 50°C, 60°C, and 70°C. These results agree with previous study [19], who reported that at temperatures higher than 50°C, the gel forming ability and the physical properties decreased due to breakage of hydrogen bonds and free amino acid hydroxyl groups. Physical properties decreased at extraction temperatures greater than 40°C due to hydrolysis of cross-linkages in collagen and other proteins [24]. Extraction temperatures of gelatin from skate skin at 40°C, 50°C, 60°C, and 70°C produced respective Hunter color L values of 27.5, 43.7, 42.1, and 30.9, respective Hunter color a values of 0.39, 0.84, 0.82, and 1.40, and respective Hunter color b values of 1.35, 2.14, 2.82, and 3.64. The latter values increased as the extraction temperature increased [11]. A difference in the color values at different extraction solution is due to the protein denaturation and structural change by hydrolysis

2.1.5. Effects of time on the extraction of gelatin

The yield of collagen extraction was dependant upon the reaction time. Previous study found that the extraction yields of gelatin from skate skin at 3, 4, 5, and 6 h of extraction time were 9.82%, 13.6%, 15.5%, and 17.5%, respectively [11]. The extraction yields increased gradually as extraction time increased from 3–6 h to be maximum at 6 h. Yields of gelatin slightly increased with increasing extraction times [24]. During 24 h of protein extraction in HCl with pepsin treatment (3.3 mg/g raw material, about 33% of collagen was solublized, and after 72 h about twice more. Whereas in citric acid with the same concentration of enzyme the yield of collagen extracted after 72 h increased by about 65% in comparison with the 24 h digestion [31].

The inconsistency in physicochemical and functional properties can be cause by differences in gelatin sources and processing conditions. Viscosity of gelatin from skate skin showed 21.4, 23.7, 20.1, and 18.3 cps for extraction time of 3, 4, 5, and 6 h, respectively [11]. Gel strength was 67.3, 74.3, 70.7, and 53.9 cps for extraction time of 3, 4, 5, and 6 h, respectively [11]. Both viscosity and gel strength were highest at 4 h extraction time compared with the other extraction times.

Hunter color values of gelatin from skate skin varied with extraction time. At extraction times of 3, 4, 5, and 6 h, the Hunter color L value was 34.5, 41.7, 36.1, and 33.5, respectively, the Hunter color a value was 0.43, 0.94, 1.13, and 0.91, respectively, and the Hunter color b value was 2.67, 3.28, 3.65, and 3.89, respectively [11]. The Hunter color b value increased as extraction time increased. A difference in the color values at different extraction solution is due to the protein denaturation and structural change by hydrolysis

2.2. Functional properties of gelatin extracted from skate skin

As mentioned above, the inconsistency in functionality cause by differences processing conditions for the extraction of gelatin. Gelatin extracted from skate skins displayed gelling point, melting point, pI, and turbidity values of 16.1°C, 19.3°C, 6.45, and 6.98, respectively [11]. The gelling and melting points were lower compared with Alaska Pollack skin and commercial bovine skin [24], but were higher compared to cod skin [17]. Fish gelatins have lower melting and gelling points compared with mammalian gelatins and especially, fish gelatin extracted from cold water species, which represent the majority of the industrial fisheries [28, 14, 32]. The pI of gelatin derived from skate skin differed from previously-reported values of 4.8–5.0 for limed or alkaline processed commercial gelatins [11]. Gelatins produced at pH 6.0, which is near the pI, exhibits higher gel strengths. Gel strength is dependent on pI [17].

The turbidity of gelatin from skate skins was also lower compared with Alaska Pollack skin, but was much greater compared with commercial bovine skin [24]. The higher turbidity in skate gelatin reflects its poorer quality compared with commercial gelatins [33, 34]. Skate skin gelatins need to be processed further to lower the turbidity. Turbidity values reflected concentration of residual lipid and other colloidal material that were present in collagen. The total fat content was 0.35%, as determined immediately after preparation of the gelatin from the fermented skate skin [11]. The turbidity was also due to protein-protein aggregation leading to the formation of particles larger than the wavelength of light. Previous report showed that turbidity was significantly reduced from 0.236 to about 0.098 for NaOH-treated collagen and 0.329 to about 0.126 for Ca $(OH)_2$-treated collagen by $CaCl_2$ pretreatment(1.2 g L^{-1} Ca^{2+}), which effectively removed colloidal material and calcium phosphate-protein complexes [35].

2.3. Amino acid composition

The stability of collagen and gelatins is proportional to the total imino acid [proline (Pro) + hydroxyproline (Hyp)] and glycine (Gly) content [26]. Imino acid and Gly contents of the fermented skate skin gelatin showed 17.3% and 20.5%, respectively [11]. Gelatins derived from fish collagens are weak with low melting points compared with mammalian gelatins. Mammalian gelatins have a higher proportion of Pro, Hyp, and Gly [12]. The low gel strength of the fermented skate skin gelatin may be due to the lower amino acid (i.e., Pro and Gly) concentrations [27]. Calf skin gelatins and warm water fish gelatins from tilapia, they contain approximately 23.2% and 18.9% imino acids [36]. This is higher compared with fermented skate skin gelatin.

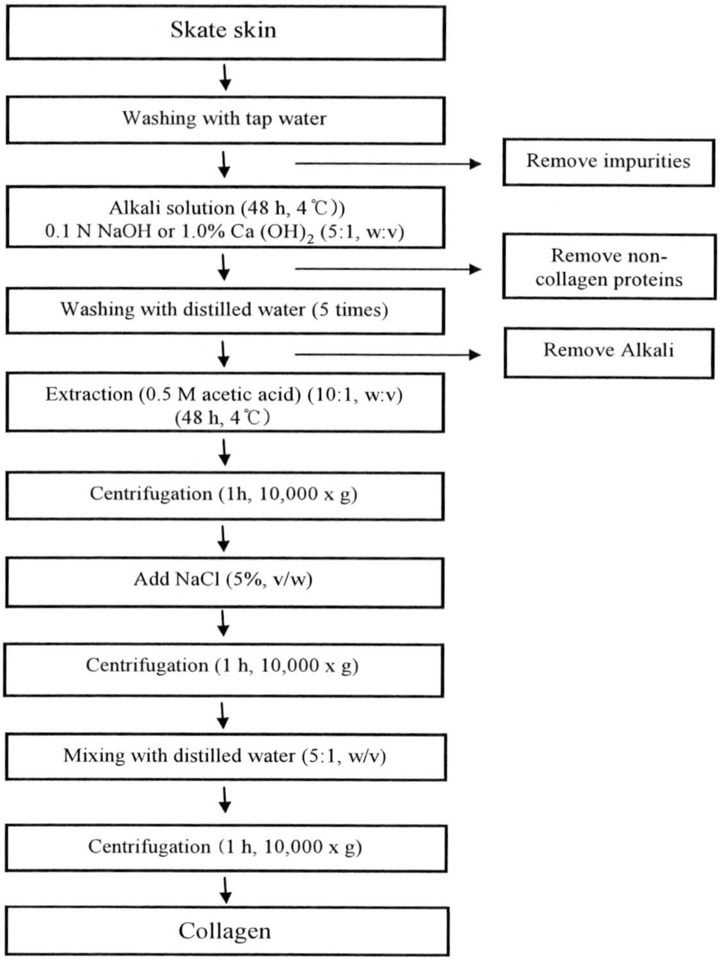

Fig. 2. Manufacture of collagen from skate skin.

3. Nutritional Composition and Microflora of Fresh and Fermented Skate Skin

3.1. Proximate compositions, VBN and pH

The nutritional values of fish skin, bone and fin are fairly high, but currently they are diverted to useful resources may be in fish meal manufacture. Skate skin, which is presently considered to be a waste product, could be effectively used as a natural gelatin and collagen sources. The nutritional composition and microflora of skate skin was determined to evaluate their potential use as possible gelatin food ingredients [11]. The mean values in fresh and fermented skate skin of moisture, protein, lipid, and ash content (%) were 75.9% and 74.5%, 22.7% and 21.8%, 0.51% and 0.74%, 0.62% and 0.93%, respectively [11]. The moisture and crude protein contents of fermented skin were lower compared with fresh skin. These values were similar to reported values for other marine fish species [24]. The crude ash and fat contents of fresh and fermented skate skin that affect a gelatin quality were lower than those of yellow-fin and Dover sole skins [37], and they are good sources of protein. Thus, there is a good potential for the use of fresh and fermented skate skin as a gelatin product. The degree of fermentation of skate skin was evaluated by volatile basic nitrogen (VBN) content and pH values [27]. Fermentation of skate skins accumulates trimethylamine (TMA), ammonia, and other organic bases [38]. The VBN and pH values in the fresh skin (14.1 mg% and 6.90, respectively) were lower compared with fermented skin (96.1 mg% and 7.80, respectively). The increase in VBN and pH values in the fermented skate skin due to accumulate trimethylamine (TMA), ammonia, and other organic bases, and to decrease in the protein concentration [38].

3.2. Mineral contents

The mineral contents (mg/100 g) of skate skin were determined to use as a gelatin and collagen sources [27]. Among the macro minerals, fermented skate skin had higher levels of potassium (104.6 mg/100 g), magnesium (38.8 mg/100 g), and phosphorus (105.1 mg/100 g) compared with fresh skin (53.5, 13.9, and 33.0 mg/100 g, respectively) [27]. In addition, potassium and phosphorus were the predominant minerals in fresh skin (53.5 and 33.0 mg/100 g, respectively) and fermented skin (104.6 and 105.1 mg/100 g, respectively) [27]. Macro minerals such as phosphorus are important for metabolic energy transfer and for formation of nucleic acids; neuromuscular control (magnesium); maintaining equilibrium of water, minerals, and acid-base balance; and contraction and relaxation of muscles (potassium) [39]. Trace minerals in fresh and fermented skate skins such as manganese (0.8 and 0.2 mg/100 g, respectively) and zinc (0.7 and 0.8 mg/100 g, respectively) are important as enzyme substrate activators. Iron (14.4 and 12.2 mg/100 g, respectively) and cupper (1.60 and 1.50 mg/100 g, respectively in fresh and fermented skate skins are the major trace element and consider an important trace mineral in food, especially for adult women [40].

3.3. Amino acid contents

A total of 17 amino acids in fresh and fermented skate skin were identified [27]. All amino acids were lower in fermented skate skin compared with fresh skate skin, due to conversion of amino acids to ammonia gas during the fermentation of the skate skin. The major amino acids of fresh and fermented skate skin were arginine, Pro, glutamic acid, but were deficient in sulphur-containing amino acids (methionine, cystine) and tyrosine. Except for threonine and isoleucine, other amino acids were lower than the recommended range [41]. However the Pro and arginine contents were high in the both fresh and fermented skin. The stability of collagen is proportional to the total collagen, and can be related to the pyrolidine content [12]. Thus, if skate skins with 12% Pro are used to produce the gelatins, it is expected that collagen stability would be enhanced. Skate skin contains approximately 19.8% of the total amino acid content as arginine. Arginine is widely recognized to have a beneficial effect on the immune system, and to improve of lymphocyte immune activity. Nitric oxide produced from arginine can maintain tissue integrity during lower inflammation and protects the liver during endotoxemia and chronic inflammation [42]. Arginine is also reported to act as an antioxidant in fat [43].

3.4. Free amino acid (FAA) contents

The FFA is responsible for the characteristics taste of seafoods in fresh and fermented skate skin. A total of 14 free amino acids in fresh and fermented skate skin were identified [27]. Histidine, glycine, alanine and glutamic acid were the major FAAs present in both fresh (1985.3, 1548.8, 1469.2, and 1006.6 mg/%, respectively) and fermented skin (2308.2, 1963.2, 1350.0, and 2771.4 mg/%, respectively). Serine, Pro, Gly, alanine and glutamic acids, which cause sweetness and savory taste of foods, were higher than 40% of total FAAs in the both fresh and fermented skins. The total content of FAAs in the fresh skin (9664.5 mg%) was lower compared with fermented skin (15134.0 mg%). Some kinds of microorganisms such as *Bacillus sp., Simonsiella sp., and Capnocytophage sp.* may hydrolyze protein during fermentation [44, 45]. Skate skins had a FAA pattern similar to the white muscle of milkfish, which also contained high levels of glycine and histidine, reported to function as a chemical buffer in migratory red-fleshed fish [46].

3.5. Fatty acid (FA) composition

The main FA in both fresh and fermented skate skin was palmitic acid (16:0) (16.7% and 20.4%, respectively) [27]. These values were not appreciably different compared with skin of three marine fish species (*Sardinella maderensis, Sardinella aurita and cephalopholis taeniops*) collected from the Senegalese coast [47]. Linoleic acid (18:2n-6) and linolenic acid (18:3n-3) contents in fresh (2.33% and 0.18%, respectively) and fermented skate skin (1.31% and 0.23%, respectively) were different. The arachidonic acid content of fresh skin (10.6%) was more than twice that of fermented skate skin (5.68%). Saturated FAs of fresh and fermented skate skins were 31.5% and 33.3%, respectively. The content of omega 3 polyunsaturated fatty acid (PUFA) was 22.2% in fresh skin and 24.6% in fermented skin. Major PUFA components were docosahexaenoic acid (DHA; 22:6, omega 3), arachidonic acid (20:4, omega 6) and eicosapentaenoic acid (EPA; 20:5, omega 3) (15.3%, 10.6%, and 3.36%, respectively, in fresh skin and 16.1%, 5.685, and 4.45%, respectively, in fermented skin). Both groups contained low concentrations of omega 6 PUFA. Omega 6 series in fish oils generally represents <5% of total FAs [48] and omega 3 FA concentrations are generally higher compared with omega 6 FA [49]. Saturated FAs have been implicated as an important dietary risk factor [50]. Previoue report indicated that the omega 3 and 6 FAs are reported to depress cholesterol metabolism, hence, skate skins containing approximately 41.4% PUFA may have human health benefits. Skate skin contains high amounts of polyunsaturated DHA and arachidonic acid, and should be further studied for its application in value-added products [51].

3.6. Microbiology

The identified microflora in fresh skate skin was *Bacillus sp., Photobacterium sp., Vibrio sp., Aeromonas sp., Simonsiella sp., Capnocytophage sp., and Proteus sp.* [27]. Microflora in fermented skate skin was *Bacillus sp., Staphylococcus sp., Photobacterium sp., Brevibacterium sp., Pseudomonas so., Enterobacter sp., Vibrio sp.* [27]. The predominant microorganisms were *Photobacterium sp., Vibrio sp.,* and spoilage bacteria, which may be related to ferment of skate skins [27]. Previous study reported that *Photobacterium sp., Pseudomonas sp.*, and *Vibrio sp.* are also found in sediments 810–4,000 m in deep in the ocean. The total plate counts for the fresh and fermented skate skin were 2.40×10^4 CFU/g and 7.70×10^7 CFU/g, respectively [52]. It may be indicated that fermentation increase the microbial population of fermented skin.

4. Antimicrobial Peptide Isolated from Skate Skin

4.1. Antimicrobial peptides in various sources

Antimicrobial peptides are widespread in nature and perform a multitude of offensive and defensive tasks in several organisms. These antimicrobial peptides have been isolated from various sources such as amphibians [53-55], insects [56, 57], mammals [58], plants [59] and fishes [60, 61]. Amphibians are especially rich in antimicrobial peptides, and many amphipathic α-helical antimicrobial peptides such as bombinins [62, 63], buforins [64], and dermaseptin [65]. Carboxy-terminal disulfide bond peptides such as esculentins and brevinins [66, 67] have been isolated from the glands of the skin, venom, and from the gastrointestinal tract. Most of the antimicrobial peptides kill target cells rapidly and specifically, and have advantages such that the peptides are immediately available from a reservoir in the cutaneous granular glands and these peptides usually have activity against a broad spectrum of microorganisms [68]. In addition to the microbial activity, antimicrobial peptides may also have other functions such as promotion of wound healing, stimulation of monocyte chemotaxis and inhibition of cytokine response [61]. So far, about 2,000 antimicrobial peptides have been identified. Generally, while all these antimicrobial skin peptides are cationic, it is already clear that large differences exist in features such as size, the presence of disulfide bonds, and structural motifs. To date, the majority of structural and biological studies on antimicrobial peptides have been limited to those from terrestrial biomes and only limited research has been conducted on antimicrobial peptides from aquatic organisms. Studies on antimicrobial peptides from aquatic organisms have been reported for shrimp [69], horseshoe crab [70], blue crab [71], tunicate hemocyte [72], mollusc [73] and a few fish (i.e., flounder [74], loach [75], rainbow trout [76], catfish [61], moses sole fish [77] and hagfish [60].

4.2. Antimicrobial activity of fermented skate skin extracts

Previous report studied that purification and partial characterization of the antimicrobial peptide from the fermented skate skin extracts [78]. They reported that skate skin extracts and the 60% MeOH fraction had strong inhibitory effect Gram-positive and Gram-negative bacteria and yeast despite the difference composition of their cell walls. It indicated that the skate skin extracts act in a non-selective manner in binding to lipid of cell membranes. Generally, antimicrobial activities against microorganisms have been suggested to depend on cell wall components. The cell wall components of Gram-negative bacteria differ from those of Gram-positive bacteria. Gram-negative bacteria have a smaller cell wall peptidoglaycan layer but possess an outer membrane in addition to the common cytoplasmic membrane. The cell wall of Gram-positive bacteria contains components including teichoic acid, teichuronic acid and peptidoglycan, while Gram-negative bacteria contain lipopolysaccharide and a lesser amount of peptidoglycan. Yeasts, one of the *Eumycetes* (true fungi), are composed of glucan, glucomannan, and ergosterol. Furthermore, the identical antimicrobial activity between the 60% MeOH fraction and the fresh and fermented skate skin extracts suggests that this antimicrobial peptide is fish origin.

Antimicrobial peptide, which was isolated from fermented skate skin extracts, has a molecular mass of 3058.75 Da [78], which is higher than that of the analogous peptides from aquatic organisms such as catfish (2000.4 Da) [61]) and hagfish (1279.5 Da) [60]), while lower than that of rainbow trout (6676.6 Da) [76].

Amino acid sequence of the antimicrobial peptide isolated from skate skin extract was determined [78]. Amino acid was determined to be: Gly-Lys-Gln-Tyr-Phe-Pro-Lys-Val-Gly-Gly-Arg-Leu-Ser-Gly-Lys-Ala-Pro-Leu-Ala-Ala-Lys-Thr-His-Arg-Arg-Leu-Lys-Pro-NH_2, consisting of 28 residues with three arginines and five lysines, which contributed to the overall cationic charge [78].

The synthetic antimicrobial peptide, which was isolated from skate skin extract, had an inhibitory effect against *B. subtilis, E. coli*, and *S. cerevisiae* [78]. This synthetic antimicrobial peptide showed more potent than a commercial buforin II or magainin against *E. coli* but less effect against *B. subtilis* and *S. cerevisiae.* The synthetic antimicrobial peptide was similar to peptides purified from *Xenopus laevis* [79], *Bufo bufo gagarizans* [64], and *Parasilurus asotus* [61].

Although the mechanism of action of the antimicrobial peptide is unclear, it has been suggested that the basic amino acid residues facilitate binding and transport across the Gram-negative outer membrane and Gram-positive cell wall membrane, and the hydrophobic residues are responsible for the disruption and permeation of the Gram-negative inner membrane and Gram-positive membrane that leads to cell death [80]. The large number of cationic residues (particularly lysine) in the peptides can interact with anionic bacterial membranes via electrostatic interaction and collapses the transmembrane electrochemical gradients in order to permeabilize the cell membrane. In addition, many antimicrobial peptides have a propensity to form amphipathic α-helical structures, and their mechanism of action involves a direct interaction with the fatty acyl chains in the cell membrane leading to disruption of normal membrane function responsible for osmotic balance [81]. Thus, antimicrobial activity of antimicrobial peptide isolated from skate skin extract is attributed to the presence of the numerous cationic lysine and arginine residues. The most potent antimicrobial peptides have been reported to have minimal inhibitory concentrations values in the range of 0.25–4 μg/mL [82].

5. Physicochemical and Functional Attributes of Collagen Powder extracted from Skate Skin

5.1. Extraction of collagen from skate skin

The functional properties of peptides are highly influenced by their molecular structure and weight, which are greatly affected by processing conditions. Enzymatic hydrolysis has become a valuable tool for modifying the functionality of proteins [83]. The inconsistency in physicochemical and functional properties can be cause by differences in collagen sources and processing conditions. Skate skin, which is presently considered to be a waste product, could be effectively used as a natural gelatin and collagen sources. Therefore, this review chapter focuses on the utilization of collagen powders extracted from skate skin to evaluate its potential use as possible food ingredients.

General extraction procedures of collagen from skate skin are described in Fig. 2. Calcium hydroxide ($Ca\ (OH)_2$) and sodium hydroxide (NaOH) solution was used to remove impurities and change collagen to the optimal type for gelatin extraction [12]. The method of manufacture greatly affects the physicochemical properties of the gelatin. Collagen must be pretreated to convert it into a form suitable for gelatin extraction, which is normally done with >45°C water. The degree of conversion of collagen into gelatin is related to the severity of both the pretreatment and the extraction processes, which depends on pH, temperature, and extraction time [13, 14-16].

5.2. Proximate analysis and pH

The moisture content of collagen powders extracted from skate skin were 3.68% for $Ca\ (OH)_2$-treated collagen powder and 5.01% for NaOH-treated collagen powder (Table 1) [35]. The moisture content was significantly higher in NaOH-treated collagen powder. The total ash content was higher in $Ca\ (OH)_2$-treated collagen powder (3.38%) compared to NaOH-treated collagen powder (2.18%). The total protein and fat content of collagen powders were 13.7% and 0.10% for $Ca\ (OH)_2$-treated collagen powder and 12.9% and 0.12% for NaOH-treated collagen powder (Table 1) [35].

The pH of collagen powders extracted from skate skin was 6.89 for $Ca\ (OH)_2$-treated collagen samples and 7.30 for NaOH-treated collagen samples (Table 1) [35]. There was a significant difference in pH between the differently pretreated samples. Organic acids in foods can influence flavor, color, stability, and food quality. Most common food acids are citric, malic, lactic, and tartaric acids.

5.3. Color measurements

The Hunter color values of collagen powders extracted from skate skin varied with pretreatment of collagen. The Hunter color *L* value, which reflects lightness, was affected by the extraction solution, being 91.7 and 88.4 for Ca $(OH)_2$ collagen samples and 7.30 for NaOH collagen samples [35]. The other relevant part of the Hunter color *a* value, which represents redness, was also affected by the extraction solution, being 0.42 and 0.92 for Ca $(OH)_2$-treated collagen samples and 7.30 for NaOH-treated collagen samples [35]. The Hunter color *b* value (yellowness) was 7.60 and 11.2 for Ca $(OH)_2$ collagen samples and 7.30 for NaOH collagen samples [35]. A difference in the color values at different extraction solution is due to the protein denaturation and structural change by hydrolysis [84]. The color of the gelatin does not influence other functional properties [30].

5.4. Consistency of suspensions of the collagen powders

Viscosity is important to determine physical property of proteins [19]. Consistency of suspensions (10%, w/v) of the collagen powder extracted from skate skin preparations in distilled water are determined [35]. Viscosity of collagen powders varied with various pH values. The viscosity was higher in NaOH collagen powder compared to Ca $(OH)_2$ collagen powder. The viscosity of both Ca $(OH)_2$ and NaOH collagen powder was highest at pH 7.0 (20.3 and 24.1, respectively) and lowest at pH 3.0 (11.3 and 13.6, respectively) [35]. This might be because the molecular weight decreased slowly in the neutral region compared with alkali and acid conditions [18]. Viscosity is partially controlled by molecular weight and molecular size distribution [22].

5.5. Gel strength (GS)

The Bloom value or jelly strength is associated with the mechanical elasticity of the gel. GS of the collagen powder from skate skin preparations is determined [35]. There was not a significant difference in GS between Ca $(OH)_2$- and NaOH-treated collagen samples. Many factors such as temperature, pH, and concentrations of protein, salt, calcium, and free sulfhydryls affect the formation and properties of protein gels [85].

5.6. Solubility (SO)

The SO is an important characteristic for the functional application of proteins in food systems [86]. The SO of collagen powders from skate skin was significantly higher in NaOH-treated collagen powder compared to Ca $(OH)_2$-treated collagen powder at all pHs [35]. The highest SO of both NaOH- and Ca $(OH)_2$-treated collagen powder was observed at pH 5.0. Generally, collagens are highly solubilized in the pH range of 2.0-5.0, with relative SO greater than 80% [87]. The variation in SO of collagens might be associate with a different isoelectric point (pI), which was altered by pepsin treatment. The SO of the collagen from the skin of Baltic cod (*Gadus morhua*) in acetic acid and lactic acid was similar with an average of 9%. However, collagen was dissolved about 60% in HCl solvent [31]. The SO of protein is affected by structural alterations that bring about intermolecular aggregation. The compositional differences between collagen powders that possibly can explain the difference in SO are lower bound fat and higher denaturation enthalpy, ash content, and Ca content [88]. Solubility of collagen from cod skin, after pepsin treatment increased from 55% to 90%. Collagen from cod skin was solubilized completely after enzymatic treatment in lactic acid, after 24 h [31]. High SO of proteins reflects the preponderance of protein-water interaction over protein-protein interaction. SO of Ca $(OH)_2$ collagen powders showed a lower overall SO than NaOH collagen powders, possibly because the higher ash content of Ca $(OH)_2$ collagen powders resulted in a higher degree of denaturation and insolubilization. Temperature and moisture affected water SO and absorption, which are indicators of degree of gelatinization [89]. The collagen SO of chicken and salmon increased with heating time and most changes occurred within the first 20 min [90].

5.7. Water holding capacity (WHC)

The WHC test measured the amount of water bound by the differently dried powders during their relatively short water contact time, as would be the case under manufacturing conditions in the food industry [91]. The WHC of NaOH collagen powder was significantly higher compared to Ca $(OH)_2$ collagen powder at all pH values [35]. However, WHC of both NaOH and Ca $(OH)_2$ collagen powder was not affected by pH. The WHC of Ca $(OH)_2$ collagen powder was lower than that of NaOH collagen powder, but water retention was still substantial (73.8%–77.9%) after 10 min centrifugation at 3,000 rpm. WHC can be affected by water that is physically entrapped within unfolding proteins and depends on the degree of denaturation [92]. The water holding properties of proteins are, therefore, conceivably related to protein swelling, which will influence SO, viscosity, and gelation. Generally, WHC of collagen powders decrease as bulk density increases [91].

5.8. Oil holding capacity (OHC)

The OHC is an important functional characteristic that influences the taste of a product [93]. NaOH-treated collagen powder had significantly higher OHC compared to Ca $(OH)_2$-treated collagen powder [35]. The OHC and WHC may help enhance sensory properties such as flavor retention and mouth-feel. The OHC is also significantly correlated with bulk density [94], an important characteristic that influences flow property and dispersibility.

5.9. Turbidity of collage powders

Turbidity of collagen dispersion at concentrations of 0.5%, 1.0%, 2.0%, 5.0%, and 10.0% (w/v) in distilled water is dertermined [35]. Ca $(OH)_2$-treated collagen powder gave lower turbidity compared to the NaOH-treated collagen powder indicative of greater dispersibility. Turbidity increased with concentration for all collagen dispersions [35]. Turbidity values reflected concentration of residual lipid and other colloidal material that were present in collagen. The total fat content of Ca $(OH)_2$-treated collagen powder and NaOH-treated collagen powder was 0.10% and 0.12%, respectively, as determined immediately after preparation of the lyophilized powders from freshly separated collagen. The turbidity was also due to protein-protein aggregation leading to the formation of particles larger than the wavelength of light [93]. The difference in turbidity of 10% NaOH- and Ca $(OH)_2$-treated collagen powder, and collagen powder pretreated with $CaCl_2$ (1.2 g L^{-1} Ca^{2+}) was also determined. Turbidity was significantly reduced from 0.236 to about 0.098 for NaOH-treated samples and 0.329 to about 0.126 for Ca $(OH)_2$-treated samples by $CaCl_2$ pretreatment, which effectively removed colloidal material and calcium phosphate-protein complexes [35].

5.10. Thermostability (TS)

TS is a desirable attribute for food proteins, especially for high protein beverages that require heat treatment [93]. TS of Ca $(OH)_2$ collagen powder was significantly higher than the NaOH collagen powder at all pH levels (3.0, 4.5, 7.0, and 9.0) and heating times studied [35]. TS of collagen powders was highest after 1 min of heating at all pH levels and heating times. However, TS decreased precipitously after further heating for 20 min. For Ca $(OH)_2$ collagen powder, TS was highest at a pH level of 3.0 at all heating times studied indicating that Ca $(OH)_2$ collagen powder was more resistant to precipitation on extended heating. This heat resistance property would be useful in food items such as health and sports drinks, which are acidic in nature and must be pasteurized, and in other foods that have to be retort-stable.

5.11. Emulsifying properties of collagen powders

Emulsion activity (EA) is based on the relationship between turbidity and surface area [95] and is associated with the ability of the protein to absorb, spread, and stabilize the interface formed. EA was highest at pH 9.0, followed by 5.0, and by 3.0 [35]. The collagen powders had the lowest EA at pH 7.0, which corresponds to the isoelectric region of collagen, and increased at pH values above or below the pI, perhaps due to increased protein-protein interaction resulting in low surface hydrophobicity and decreased net charge and protein solubility. EA was significantly different in collagen samples at different pHs. NaOH-treated collagen powder had a higher EA compared to Ca $(OH)_2$-treated collagen powder at all pH levels [35]. The EA and emulsion stability (ES) of proteins correlated linearly with surface hydrophobicity.

Emulsion stability (ES) reflects the ability of the proteins to impart strength to emulsion for resistance to stress [88]. ES of collagen powders was highest at pH 9.0 and lowest at the pI (pH 7.0). At acidic and alkaline pHs, the collagen is more soluble due to charge repulsion in the absence of sufficient counter ions, which effectively increases the number of protein molecules or aggregates availably to coat the interface. Emulsion globules coalesce due to Brownian motion unless contact between them is prevented by an opposition stronger than the kinetic energy. Opposition sufficient to produce stability includes electrostatic repulsion and degree of hydration, orientation, flexibility, and intertwining of amphiphiles at the interface [96]. NaOH-treated collagen powder had higher ES compared to Ca $(OH)_2$-treated collagen powder in all cases except pH 7.0 [35]. The ES of proteins is related to a suitable balance between the hydrophilic and hydrophobic groups [97]. The stable emulsifying properties of collagens at acidic pH would be useful in industrial applications such as beverages and dairy products. ES of proteins is negatively correlated with bound fat and is positively correlated with protein content, denaturation enthalpy, and the product of protein content and solubility [88]. Generally, emulsifying properties of proteins are positively correlated with surface hydrophobicity [98].

The creaming stability (CS) of emulsions, expressed as stability rating (SR), for collagen powders at different pH values was determined [35]. NaOH-treated collagen powder showed higher CS compared to Ca $(OH)_2$-treated collagen powder. No significant difference in CS of the NaOH collagen sample was evident at all pH levels (3.0, 4.5, 7.0, and 9.0) [35]. CS of the Ca $(OH)_2$-treated collagen powder was lower at pH 5.0 and the alteration of pH from 5.0 to either 3.0, 7.0, or 9.0 increased the CS [35]. These results suggest that the CS of the emulsion strongly depends on the electrostatic nature of protein. At pH 7.0, which is close to collagen pI, the net protein charge is diminished and molecular repulsive forces are eliminated.

6. Antioxidative Activity of Collagen Powder Extracted from Skate Skin

6.1. Antioxidative activity of proteins and peptides

Synthetic antioxidants and antimicrobial additives such as butylated hydroxyanisole (BHA) and butylated hydroxytoluene (BHT), and propyl gallate have been commonly used to preserve food products by delaying lipid oxidation and preventing bacterial growth are factors that can have a significant contribution towards the extension of shelf-life [99]. However, these synthetic antioxidants and antimicrobial additives have been limited in their applications as food additives because of potential health problems [100, 5]. The past two decades have witnessed a growing interest in finding safe and natural antioxidants and antimicrobial additives that enhance the body's antioxidant defenses through dietary supplementation, and inhibit lipid oxidation in foods [101].

Several natural proteins have been reported as water soluble antioxidants due to their chelating effect on metal ions [99, 102, 103]. Antioxidative activity has also been identified in several protein hydrolysates such as those in collagen [99, 104-106], soybean [107], myofibrillar [108], and milk proteins [109], egg-white albumin [110], and soluble elastin [111]. A number of antioxidative peptides, usually composed of 3 to 16 amino acid residues, have been isolated from these hydrolysates. Among these peptides, some contain hydrophobic amino acids (Val or Leu) at the N-terminus, Pro, His, or Tyr in sequences [112, 113], and some containing mainly acidic acid residues (Glu, Asp) [108]. However, little is known about the structure of antioxidative peptides in the various proteins.

Collagen is a cheap and resourceful meat and fish by-product whose main product is a gelatin that is used extensively as a food additive to increase the texture, water-holding capacity and stability of several food products. Although several studies have been conducted to identify potential antioxidative peptides in fish skin gelatin [104, 106], there has yet to be any research on antioxidative peptide activity in skate skin collagen. The antioxidative property of proteins and peptides is highly influenced by the processing conditions. While the antioxidative activity of protein hydrolysates has been examined using a single enzyme, little is known about the antioxidative activity of protein hydrolysates with mixture of enzymes.

Among the amino acids in egg white protein hydrolysates, Trp, Tyr and Met showed the highest antioxidative activity, followed by Cys, His and Phe [114]. The rest of the amino acids (Arg, Asn, Gln, Asp, Pro, Ala, Val, Lys, Ile, Tre, Leu, Glu and Gly) did not exhibit antioxidative activity. Met and Cys may be important to the radical scavenging activity of the identified peptides because Met is proton to oxidation of the Met sulfoxide and Cys donates the sulfur hydrogen (Hernandez-Ledesma et al., 2005) [108]. The antioxidative peptides isolated from soybean hydrolysate [112, 107] and egg-white albumin hydrolysate [110] have been attributed to His due to the proton-donation ability of the His imidazole group. It has also reported that His and Pro play an important role in the antioxdative activity of peptides test, among which Pro-His-His was the most antioxdative activity [113]. Generally, the quenching of free radical has been attributed to the donation of hydrogen. Further, some amino acids such as His Leu, Tyr, Met enhance the scavenging activities of peptides.

Collagen has a specifically arranged amino acid sequence where glycine strictly represents every third amino acid residue. Gelatin-derived antioxidative peptides containing the Gly-Pro-Hyp sequence from marine fish and bovine skin were isolated [104, 105]. The strong radical scavenging peptide from Hoki fish skin (His-Gly-pro-Leu-Gly-pro-Leu) also produced Gly-Pro amino acid residues with the exception of His [106] and from porcine collagen (Gln-Gly-Ala-Arg) by enzymatic hydrolysis, which has high antioxidative activity without strong proton-donating amino acid residues in the sequence [99].

6.2. Antioxidative activity of collagen in oxidation catalyzed model system

The pH and color changes of stored (40°C) collagen dispersions were determined to spectrophotometically monitor development of Mallard reaction products (MRPs) [115]. Initial pH value was 6.99 and 7.31 at the initial storage time and dropped to a final pH of 6.89 and 7.11 at 120 h storage for Ca $(OH)_2$ and NaOH collagen dispersions, respectively. The pH of a system decreases as Maillard reaction progresses due to the disappearance of basic amino groups [116]. The pH drop is predominantly due to the disappearance of lysine as a result of MRP formation increases [117].

The colorimetric L value, which reflects lightness, changed from 98.0 and 90.9 to 90.0 and 80.0 for Ca $(OH)_2$ and NaOH collagen dispersions, respectively, during storage [115]. A decrease in the L value during storage represented the formation of a darker color in collagen dispersions due to the browning reaction. Collagen dispersions (5%, w/v) were yellowish in color at the initial storage time and gradually changed to brown after 120 h reaction time at 40°C [115]. Natural yellow color of collagen dispersion reflects the un-oxidized state and brown color indicates highly oxidized state. Browning of NaOH collagen dispersions was slower than that of Ca $(OH)_2$ collagen dispersion. This indicated that NaOH collagen was resistant to browning compared to that of Ca $(OH)_2$ collagen.

The spectrum between 420 nm and 490 nm is commonly used to monitor quantity of browning in collagen solution [118]. Values for 5% (w/v) collagen dispersions increased generally and in a linear fashion as reaction time increased [115]. A 430 nm was selected to evaluate the extent of browning in the collagen dispersion at different reaction times. The absorbance value of each collagen dispersion as related to reaction time was compared at 430 nm. Collagen dispersions of 20% (w/v) had highest absorbance followed by 15%, 10%, and 5% [115]. The absorbance values indicated browning increased as reaction time increased. Thus, browning in collagen was dependent on reaction time and on the concentration of reactants.

The ability of collagen to inhibit lipid oxidation was tested in a model system containing catalyzed oil-in-water emulsions [115]. The ability of NaOH collagen to limit TBARS formation varied at different pH ranges (3.0, 4.0, 6.0, 7.0, and 9.0) after storage for 120 h at 40°C. The inhibitive effect of NaOH collagen on TBARS formation was highest at pH 3.0 and lowest at pH 7.0, as reflected by TBARS, after storage for 120 h at 40°C [115].

Peroxide value (PV) formation during storage showed a similar pH dependant trend as for TBARS formation [115]. Peroxides (R-OOH) are the primary reaction products formed in the initial stages of oxidation; therefore, given an indication of the progress of lipid oxidation. On the other hand, TBARS are a measure of the concentration of relatively polar secondary reaction products (i.e. aldehydes) that take [119].

The effects of collagen powders and antioxidants on TBARS formation and PV were compared up to 120 h storage under the same conditions using the catalyzed emulsion model [115]. Inhibition of TBARS formation by both Ca $(OH)_2$ and NaOH collagen powder was significantly lower at 20% (w/v) concentration compared to the standard and natural antioxidants. In thc case of PV formation, after 120 h of storage Ca $(OH)_2$ collagen powder (20%) was consistently better than all other antioxidants tested and was statistically comparable to ascorbic acid, tocopherol, BHT, and BHA. Only BHT was statistically more effective than NaOH collagen powder. Collectively, the data demonstrate that collagen powder is an effective natural antioxidant

7. Antioxidative and Antimicrobial Coating Containing Collagen Powder from Skate Skin

7.1. Application of Edible Films or Coatings

Edible films or coatings can improve the quality of fresh, frozen, and processed meat, poultry, and seafood products by retarding moisture loss, reducing lipid oxidation and discoloration, enhancing product appearance in retail packages by eliminating dripping, and sealing in volatile flavors, functioning as carriers of food additives such as antimicrobial and antioxidant agent, and reducing oil uptake. Edible films or coatings from polysaccharides, proteins, and lipids can extend the shelf-life of foods by functioning as solute, gas, and vapor [120].

7.1.1. Protein-based coatings or films

Formation and properties of films or coatings from animal and plant proteins such as collagen, gelatin, milk proteins, wheat gluten, soy protein, corn zein, and peanut protein, have been studied [121, 122]. Overall, in a similar manner to polysaccharide films, protein films exhibit relatively high water vapor permeability values. The limited resistance of protein films or coatings to water vapor transmission is attributed to the substantial inherent hydrophilicity of proteins and to the significant amounts of hydrophilic plasticizer, such as glycerin and sorbitol incorporated into films or coatings to impart adequate flexibility. The sorbitol or glycerol in the solution plays the role of a plasticizer agent, which prevents breaking of the film [123-127]. During the preparation of the coating material, the mixture was heated at 90°C for 30 min resulting in the formation of intermolecular disulfide and hydrophobic bonds [128]. This process is necessary to obtain a flexible coating solution with good mechanical properties [129, 130]. Previous research reported that heat-cured (80℃ for 24; 90℃ for 12 h) WPI-based films plasticized with glycerol and containing candelilla wax had mechanical properties similar to collage films [131]. A WPI-based film that can withstand meat-processing conditions may provide the meat industry with an alternative to collagen films.

On the other hand, the good oxygen barrier properties in low relative humidity environments of films or coatings from collagen [132], wheat gluten [133-135], corn zein [133-135], soy protein [136], and whey protein [137] have been studied. Previous reports have found antioxidative activity for sour whey coating of fresh-cut vegetables and fruit [124], cut low-fat sausages [125], cut beef steaks [126], and soy protein isolate coating of Korean cow beef [127].

A problem associate with protein-based coatings or films on raw meat, poultry, and seafood is the expected susceptibility of proteins to proteolytic enzymes present in these foods. Food allergens have been identified among the protein fractions of milk, egg white, peanuts, soybeans, and rice [138]. Lactose intolerance exhibited by lactase deficient individuals is another common dietary intolerance [139].

7.1.2. Collagen

Collagen casings are the most widely used edible sausage casings as alternatives to natural casings. Production of collagen sausage casings from the regenerated corium layer of food-grade beef hides is a well established technology and has been studied. Furthermore, use of proteins other than collagen, such as wheat gluten, corn zein, soy protein, peanut protein, and feather keratin, in manufacturing of sausage casing has been suggested [140, 141].

Collagen-based edible coatings also have been proposed for use on meat products other than sausage. A method where ground collagen was mixed with an aqueous mixture of lactic acid and glyceraldehyde, heated to about 75℃, and neutralized to pH 7.0 to make a coating for hamburgers capable of withstanding cooking temperatures without melting [142]. Previous reported that both refrigerated and frozen/thawed round beef steaks wrapped on Coffi collagen film prior to standard retail packaging (permeable film overwrap) or vacuum packaging exhibited significantly less fluid exudate than unwrapped control [143]. Moreover, based on thiobarbituric acid value and on Hunter color value and sensory color evaluation, the collagen films had no significant effect on meat oxidation and color [143]. Beef cubes wrapped in collagen-based films and stored at -18℃ for 20 weeks were not significantly different than plastic wrapped control in terms of oxidation, color, microbial growth, and sensory [144].

7.1.3. Gelatin

Film-forming applications of gelatin, which is derived from partial hydrolysis of collagen, in pharmaceutical and food industries include microencapsulation of ingredients and manufacture of tablet and capsule coatings [120]. Preservation of meat and other foods with gelatin coatings was proposed in patent disclosures.

Similar to other types of edible films, gelatin coatings have shown potential as carriers of antioxidants. Turkey steaks sprayed with an aqueous gelatin suspension of various antioxidants developed less lipid oxidation (lower peroxide value by 60-90%) in skin and meat fat than uncoated controls during frozen storage at -12℃ for 6 months [145]. An adherent gelatin coating applying on sausages followed by a cellulosic casing was described by [146]. Acidic aqueous solutions of gelatin and a meta phosphate polymer, such as sodium hexameta phosphate as well as aqueous solutions of metal gelatinates were applied on processed meats (e.g. sausages, Canadian bacon, boned hams). Moisture loss from frozen or refrigerated primal meat cuts was reduced by coating with hot melts consisting of gelatin, a polyhydric alcohol (e.g. propylene glycol, glycerin, sorbitol), and small amounts of water (<50 g/kg). Smoked-cured chicken meat treated with an ethylene glycol/gelatin (4:1,w/w) coating was reduced moisture loss by 15 to 20% than uncoated meat after 7 d at 27 to 31℃ [147]. Gelatin coatings formed on surface of battered and breaded fish, poultry, and meat pieces by applying aqueous gelatin (100 to 350 g/kg) solutions at 18 to 38℃ decreased oil absorption considerably (by 25 to 50%) during subsequent frying [148].

7.1.4. Milk proteins

Edible films or coating from casein, whey protein, and total milk proteins have been discussed [120, 149-151]. Previous studies have combined caseinate with stearic acid for coating peeled carrots [152] and with acetylated monoglycerides for coating zucchini [153]. Whey protein-based coatings have been tested on breakfast cereal, raisins, frozen peas, and cheese pieces and as oxygen barriers on peanuts [154-156]. Also food additives have been microencapsulated in whey protein or in whey protein/carbohydrate wall systems [157-162]. Whey protein-based coatings prevented oxidative browning of fresh-cut apples and potatoes [163, 124], cut low-fat sausage [125], cut beef steak [126], and soy protein isolate of Korean cow beef [127].

Applications of coating that combine acetylated monoglycerides with caseinates [164] or with whey protein isolate [165] on frozen king salmon pieces were discussed earlier. Whey protein isolate coatings alone, followed by an antioxidant (ascorbic acid and citric acid) spray did not affect moisture loss rate but delayed onset of lipid oxidation and reduced peroxide values in frozen King salmon samples [165].

Over the years, various applications of whey protein-based films have been proposed. One potential application is their use in meat processing as an alternative to currently used commercially available collagen casings and films [166, 167, 10]. Milk proteins are particularly suitable in manufacturing sausage casing because milk ingredients are already incorporated into many of the sausage formulation. Heat-cured whey protein isolate (WPI)-based films provide the benefit of a natural pigmented appearance similar to caramel or smoked colored casing, without the addition of food coloring. Previous reports demonstrated that Hunter *b* values (yellowness) increased by heat curing of these films [168, 10]. These findings further indicate the potential for whey protein-based films to be used as sausage casings. Previous research reported that heat-cured (80℃ for 24; 90℃ for 12 h) WPI-based films plasticized with glycerol and containing candelilla wax had mechanical properties similar to collage films [130, 131]. A WPI-based film that can withstand meat-processing conditions may provide the meat industry with an alternative to collagen films.

7.2. Antioxidative and Antimicrobial Edible Coating Containing Collagen Powder Extracted from Skate Skin on Cut Pork Loin

7.2.1. pH values

The pH values for control and collagen-coated pork loin over the 5 d of refrigerated storage under aerobic package were measured [169]. The initial pH increased from 5.85 to 7.02 for the control pork loins and from 4.84–5.38 to 5.32–6.63 for coated pork loins, respectively, during refrigerated storage time of 5 days [169]. The pH values of the coated fresh pork loin were lower than the control and were quite stable over 72 h of storage. The initial pH values of pork loins ranged from 4.84–5.85. This may be partially due to the difference in the pH of the coating solutions. Data indicated that pH values increased with increased storage time in aerobic packaged pork loins. The pH values of minced beef decrease in vacuum packaging but increase in aerobic packaging with increased storage time at 10°C.

The reduction of pH may be partially due to the production of lactic acid by contaminating lactic acid bacteria, the activity of lactobacilli and the dissolution of CO_2 into meat tissue.

7.2.2. Hunter color values

Hunter color values for control and collagen coated pork loin were determined during refrigerated storage time of 5 days [169]. The colorimetric *L* value, which reflects lightness, changed from 62.8 to 54.6 for control loins and from and 56.4 to 50.0 for NaOH collagen coated-loin during refrigerated storage. A decrease in the *L* value during storage was manifest as formation of a darker color, which is due to the browning reaction and the drying of pork loins in the air flow [170]. Lightness refers to the relation between reflected and absorbed light, without rcgard to a specific wavelength.

The Hunter a value of pork loins tended to increase at 48 h of refrigerated storage but decreased thereafter. This is partially due to the denaturation of nitrosomyoglobin. Lipid oxidation may initiate the oxidation of myoglobin to metmyoglobin, and change the meat color from red to an unattractive brown. The rate of meat discoloration is closely related to the rate of myoglobin oxidation induced by lipid oxidation [171].

The Hunter color *b* value (yellowness) increased in both control and treated pork loin during storage [169]. This indicated that pork loins changed from oxymyoglobin to metmyoglobin and consequent decreased oxymyoglobin which contributes yellowness color. The increase in yellowness may be partially due to rancidity during storage. No differences in yellowness were observed between the control loin and loins containing antioxidants. Color is probably the most important attribute that influences consumer decision-making regarding acceptability of fresh meat and cured meat products for consumption. The rapid color change was mainly due to increased lipid oxidation associated with unsaturated fatty acids. Another relationship between lipid oxidation and metmyoglobin formation is that increases in redox potential during lipid oxidation tend to oxidize myoglobin [172]. Ascorbic acid, phosphates, nitrites, and nitrates have been used to enhance meat color quality under exposed retail display conditions [173]. Previous reports indicated that soy-protein coatings with calcium salts better than control in maintaining the color of fresh-cut fruits and vegetables [174] and Korean cow beef [127] and sour whey protein coating of fresh-cut fruit and vegetables [124], cut low-fat sausage [125], and cut beef steaks [126]. Based on the previous studies, the color changes of the meat products during storage depend on the various product types and processing conditions of the products. Redness or yellowness results from differences in absorption of radiant energy at various wavelengths. Nitrite is the agent added to meats and meat products to produce a pink color. Nitrosylhemochrome cured meat pigment is formed when nitric oxide and myoglobin complex are heated. The most significant commercial problem associated with meat color is rapid fading in air and light, which can be delayed by vacuum-packaging. Light-induced dissociation of nitric oxide from heme is followed by oxidation of heme and nitric oxide.

Previous report indicated that calcium chloride, a divalent salt, is incorporated into the coating solution to improve the texture and color of food products. Collagen, whey-protein, and soy protein edible coatings with calcium chloride better than control in maintaining the color of cut beef steaks, and controlled darkening and oxidative browning of fresh-cut fruit and vegetables [124-127, 169]. Based on the previous studies, the color changes of the meat products during storage depended on the various product types and processing conditions of the products.

7.3. Antioxidative activity

The TBARS values for control and collagen-coated pork loin over the 5 d of refrigerated storage under aerobic package were measured [169]. Uncoated control pork loin had higher TBARS formation than antioxidant-containing loins over the 120 h storage period. TBARS formation of control loins significantly increased during storage time, indicating that lipid oxidation had occurred in pork loin during refrigerated storage under aerobic packaging. The initial TBARS formation of uncoated control pork loin increased from 0.17 to 0.57 mg MDA/kg at 72 h of storage time and then increased continuously to 0.68 mg MDA/kg at 120 h of storage time. Previous research reported that malondialdehyde (MDA) concentrations higher than 0.5 mg/kg are the threshold for the perception of rancidity [175]. Uncoated pork loin would, therefore, be perceived as rancid after 72 h of storage. However, Ca $(OH)_2$ and NaOH coated-pork loin increased TBARS formation from 0.175 and 0.168 to 0.512 and 0.503 mg MDA/kg, respectively, at 96 h of storage time. The MDA content of collagen-coated pork loins did not exceed 0.5 mg/kg until 96 h of refrigerated storage under aerobic packaging.

TBA values, which represent the malondialdehyde concentration in pork loin. TBA values in meat are well correlated with oxidative rancidity. However, MAD, highly reactive secondary product of lipid oxidation, can react with other meat components, such as amino acids and amines, forming complexes that are not detected by TBA analysis. Therefore, competition for MAD in a system containing protein could result in reduced color development and incomplete quantification of MAD.

Peroxide values (PV) for the control and collagen-coated loins increased continuously up to 120 h of refrigerated storage [169]. Uncoated control pork loin had higher PV formation than pork loins containing antioxidants during storage in excess of 120 h. At day 120 of storage, PV formation was reduced by 28.5% and 34.6% for Ca $(OH)_2$ and NaOH coated-pork loin, respectively, compared to uncoated control pork loins. PV formation was 1.623, 0.885, and 0.859 meq/ kg fat for control and Ca $(OH)_2$ and NaOH coated-pork loin, respectively, respectively, at 120 h of refrigerated storage periods. Free radicals in meat and meat products react with the free FAs released via lipolysis, and these reactions result in the formation of hydroperoxide.

Carboxymethyl cellulose was incorporated into coating solution, and the carboxylic groups, which act as a chelating agent under certain conditions, may cause the antioxidative activity [176]. There is considerable interest in edible films and coatings because of their potential to improve the quality and shelf-life of foods, and to reduce complexity and cost of packaging systems [149]. Edible films and coatings preserve fresh and frozen meat products and have been summarized in previous studies [177]. Antimicrobial packaging can be a promising tool for protecting ready-to-eat meat products from *L. monocytogenes* contamination [178]. Over the years, edible films or coatings that have been investigated include casein, collagen, chitosan, corn zein, gelatin, soy protein, wheat gluten, whey protein, calcium alginate, and methyl cellulose-based films [6-10, 124-127]. These edible films or coatings can also be used as carriers for antimicrobials and antioxidants. Applications of natural products with both antioxidants and antimicrobial additives in meat products may be necessary and useful to prolong the shelf-life of beef due to health concern of consumers [179]. Therefore, delaying lipid oxidation and preventing bacterial growth can contribute to the extension of shelf-life.

Other natural antioxidant coatings or films containing radical scavenging effects have been investigated in relation to the inhibition of lipid oxidation. Previous reports show that nonhydrolyzed SPI, regardless of heat treatment, possessed appreciable antioxidative capability due to the residual antioxidative phenolic compounds in SPI [107]. The SPI contain 1 to 1.5 mg phenolic compounds per g, including isoflavonoids that could stabilize free radicals [180]. Phenolic compounds are direct scavengers of reactive oxygen species and have the ability to chelate metals, such as iron, with their *o*-diphenol groups [181]. Previous reports have found antioxidative activity for sour whey coating of fresh-cut vegetables and fruit [124-126, 102]. This could be due to the presence of FAs (about 5.4% in sour whey powder) that significantly improve the moisture barrier properties and significantly reduce browning in fresh-cut fruit and vegetables. Nonhydrolyzed SPI inhibits lipid oxidation, retards rancidity odor development in ground beef [182], and decreases TBARS formation in fresh pork patties during storage [179]. Previous study also demonstrated that soy protein coating effectively retards lipid oxidation in precooked beef patties [183]. This may be attributed to the effective oxygen barrier properties of soy protein films or coatings [184]. The antioxidative activity contained in soy protein preparations has been well-established [185, 186], in that soy protein in ground beef produces progressive decreases in TBARS values in the cooked beef. During the preparation of the coating material, the mixture was heated at 90°C for 30 min resulting in the formation of intermolecular disulfide and hydrophobic bonds, and producing MRP that may possess antioxidant activity [128]. The soluble protein and peptide rich collagen favors the formation of Maillard reaction products (MRPs) which involves reaction of carbonyl groups of amino groups of proteins and, more readily, with the free amino acids and peptides.

Hydrolyzates react much more rapidly to yield MRPs [187] and MRPs have good free-radical scavenging capability [188]. This could explain why the collagen powder is comparable, if not better, than many commonly used food antioxidants [115] and how it reduces retail storage related oxidative degradation of cut pork loins [169].

7.4. Percent moisture loss (PML)

Since the contribution of volatiles and other compounds to the exposed storage-related weight loss is negligible, it is assumed that the weight loss is due to the evaporation of water. The storage time had a significant effect on MC, while the interaction of treatment and storage time was also significant [169]. The mean initial MC was 74.2%, 74.1%, 73.4%, 72.6%, 71.4%, 74.5%, and 72.5% for control, Ca $(OH)_2$, NaOH, BHA, BHT, ascorbic acid, and tocopherol coated pork loin, respectively. The initial MC decreased linearly to a final MC of 62.3%, 67.1%, 65.9%, 59.8%, 61.0%, 65.7%, and 64.8%, in the same respective order. This reflected a 16.1%, 9.38%, and 10.1% loss of moisture for control, Ca $(OH)_2$, and NaOH-coated pork loins, respectively, after 5 days of refrigerated storage. Higher released water is considered to represent a lower water holding capacity. Moisture changes can have adverse effects on texture, wholesomeness, and acceptability.

Moisture loss was due to evaporation of moisture during refrigerated storage. PML was greatly affected by the change of moisture contents in pork loins during storage. The storage time had a significant effect on PML, which uniformly increased with increased storage period. The control was less protected, with a PML value of 16.1% compared to 9.38% and 10.1% for Ca $(OH)_2$- and NaOH-coated pork loins. Thus, collagen-coated pork loins had less moisture loss than the uncoated control. There was a significant effect in PML of Ca $(OH)_2$- and NaOH-coated pork loins compared to the control pork loins after 2 days of storage. A significant reduction of PML by 42.7% and 38.1% over the control was achieved by Ca $(OH)_2$ and NaOH coatings, respectively, at the end of 5 days. This may be attributed to the good moisture-barrier properties of collagen coating. BHT, ascorbic acid, and tocopherol coatings also showed reduced PML of 9.9%, 27.6%, and 35.0%, respectively. However, BHA coating did not effectively reduce PML.

The types of macromolecules typically used in edible films or coatings are hydrocolloids and lipids. Hydrocolloids are either proteins, such as corn, soy, wheat, and milk proteins, or polysaccharides, such as pectin, starch, and cellulose derivatives [189]. However, hydrocolloids are sensitive to water and show high permeation rate to water vapor because of their hydrophilic nature. The more hydrophobic material, such as lipids include fatty acids, fatty alcohols or materials consisting of a combination of these, like acetoglycerides and waxes, are incorporated into the film or coating, the slower the moisture loss in film coated samples. It is conceivable that the sour whey powder coating was effective as a moisture barrier because of its hydrophobic nature [128].

Edible films or coatings produced from natural proteins provide generally good oxygen barrier but show high permeability to water [190], while lipid films or coatings show good water barriers and high oxygen permeability. By combining a hydrocolloid and a lipid to produce a composite edible film or coating, it is possible to decrease the water vapor permeability, and at the same time, keep the low oxygen barrier values.

Because protein-based compounds are immiscible, they have to be emulsified in the coating to be effective. Therefore, good emulsification properties of the proteins would be required to prevent fat loss and ineffectiveness as a result of phasing off into a planar interface. During the preparation of the coating solution, the mixture was heated at 90°C for 30 min, resulting in the formation of oil-in-water emulsion and producing disulfide bonds and greater surface hydrophobicity [128]. This process is known to improve the moisture barrier capability of protein-based edible films [174]. The emulsion activity and emulsion stability of proteins correlates linearly with surface hydrophobicity [128]. This could explain why the collagen powder is comparable, if not better, than many commonly used food antioxidants [115], and how it reduces retail storage related moisture loss of cut pork loins [169]. Previous reports demonstrated that soy protein coating reduces the moisture loss in precooked beef patties [182] and Korean cow beef [127], and fresh-cut vegetables and fruit [124], cut low-fat sausages [125] and in cut beef steaks [126] with whey protein coating.

Several calcium salts can be employed to gel coating formation. The $CaCl_2$ incorporated into the coating solution improved the moisture-barrier properties of collagen films or coatings. Previous reports have shown that whey protein edible films containing calcium salts have the potential to decrease water vapor permeability in fresh-cut vegetables and fruit [124], cut low-fat sausage [125], cut beef steak [126], and Korean cow beef with SPI coating [127]. Moisture loss is also a critical factor affecting the quality and shelf life of meats and meat products.

7.5. Microbial changes

The effect of collagen-based coating on microbial growth in pork loins was determined over a 96 h storage period at 4°C under aerobic packaging [169]. The total bacteria counts (TPC) were approximately 3.69, 3.81, 3.65, 3.74, 3.45, 3.97, and 3.93 log colony forming units (CFU)/g for control, $Ca(OH)_2$, NaOH, BHA, BHT, ascorbic acid, and tocopherol coated pork loins, respectively, at the beginning of the storage period. A significant constant increase in the TPC was observed with increased storage time for all pork loin samples during refrigerated storage. The TPC increased progressively to reach 8.55, 6.95, 6.83, 5.80, 5.96, 6.46, and 6.02 log CFU/g for uncoated control, $Ca(OH)_2$, NaOH, BHA, BHT, ascorbic acid, and tocopherol coated pork loins, respectively, at the end of the storage at 4°C. Approximately 96 h were required to reach total bacterial counts of 7.0 log CFU/g in control pork loins. Significant differences were observed between control and collagen-coated pork loins. However, TPCs were lower in pork loin cuts containing antioxidants than the control ($p < 0.05$). These results indicate that the collagen and antioxidant coatings on pork loins effectively inhibited the growth of TPC. A previous report indicated that off-odor and slime formation commences when total viable counts on meat tissue exceed log 10^7/g [191]. Therefore, pork loins may develop off-flavors on spoilage with increased storage and when the collagen coating and antioxidants do not inhibit microbial growth.

The effect of collagen coating on *Enterbacteriaceae* in pork loins over the 96 h storage at 4°C under aerobic package was measured [169]. The storage time had a significant effect on the growth of *Enterbacteriaceae*, with populations increasing from 3.16, 3.21, 3.31, 2.65, 2.59, 2.76, and 2.81 log CFU/g to 7.99, 6.14, 5.79, 4.73, 5.68, 5.92, and 4.84 log CFU/g for control, Ca $(OH)_2$, NaOH, BHA, BHT, ascorbic acid, and tocopherol coated pork loins, respectively, at the 5 d of the storage time. The slight differences in the initial populations of

Enterbacteriaceae might be due to differences in the attachment of bacteria to pork loin surfaces.

The effect of collagen-based edible coatings on lactic acid bacteria in pork loins was determined over a 96 h storage period at 4°C under aerobic packaging [169]. The microbial counts for lactic acid bacteria in pork loins also showed a similar trend to those of TPC and *Enterbacteriaceae*. Collagen coatings also reduced lactic acid bacteria growth compared to control after 96 h of storage at 4°C in aerobic packaging. Presently, collagen coating reduced the TPC count, *Enterbacteriaceae,* and lactic acid bacteria growth compared to uncoated control when the pork loins were stored at 4°C in aerobic packaging. Lactic acid bacteria comprise the major group of spoilage organisms in vacuum-packages meat products.

The various microbial reduction strategies used for processed meats and meat products typically select for specific groups of organisms, some of which can proliferate and cause spoilage during storage. Processed meats and meat products generally spoil from the growth of facultative Gram-positive bacteria (lactic acid bacteria) in vacuum-packaged products at refrigerated temperatures.

Edible films or coatings that have been utilized include casein, collagen, chitosan, gelatin, soy protein, whey protein, calcium alginate, and methyl cellulose-based films [8, 124-127]. These edible films or coatings can also be used as carriers for antimicrobials and antioxidants. Several studies has been extensively demonstrated that antimicrobial edible films or coatings in a variety of food systems. Previous reports showed that organic acids with calcium alginate can reduce the growth of *Listeria monocytogenes*, *Salmonella typhimurium*, and *Escherichia coli* O157:H7 on beef carcass tissue [192, 193]. WPI-based edible films containing 0.5%–1.5% sorbic acid (SA) or *p*-aminobenzoic acid (PABA) inhibit *L. monocytogenes*, *S. typhimurium*, and *E. coli* O157:H7 on trypticase soy agar [166]. WPI films containing SA or PABA can also reduce *L. monocytogenes*, *S. typhimurium*, and *E. coli* O157:H7 on bologna and summer sausages after 21 days of storage at 4°C [167]. In addition, soy protein edible films containing organic acids and nisin possess antimicrobial activity, and reduce *L. monocytogenes*, *S. gaminara*, and *E. coli* O157:H7 [194]. A previous report demonstrated that some plant proteins have antimicrobial properties; SPI hydrolysates (15%) incorporated into frankfurters can extend the shelf life of products stored aerobically at 25°C from 10 to 42 days [195]. The present study investigated coating of pork loins that relied entirely on the antimicrobial properties of collagen powder to enhance shelf-life. Additionally, collagen powders were not hydrolyzed and were incorporated at only 5% of final formulation.

CONCLUSIONS

Gelatins from skate skin have similar physicochemical properties, but higher yields compared with gelatins produced from other marine fish species. Additional research is needed to improve the overall quality of gelatins produced from skate skins. The skate skins have protein concentrations similar to other marine fish, but have relatively low fat and ash contents. Skate skins contain comparatively high omega 3 and omega 6 FA contents compared with other marine fish skins. Skate skin can be used for gelatin production. Skate skin also contains a substantial content of arginine. Antimicrobial peptide purified from fermented skate skin has high molecular weight compared with other marine fish skin and similar antimicrobial activity to several known peptides. These results indicate that skate skin is potentially antimicrobial, and this activity may be improved through suitable modification of the peptide structure by chemical synthesis. This review chapter suggested that collagen powder, which is presently considered to be a waste product, could be used effectively as natural food ingredients and natural antioxidative and antimicrobial coating of food products to enhance shelf-life foods.

References

[1] Gomez-Estaca, J., Montero, P., Fernandez-Martin, F., & Gomez-Guillen, M.C. (2009). Physico-chemical and film-forming properties of bovine-hide and tuna-skin gelatin: A comparative study. *J. Food Eng, 90,* 480-486.

[2] Montero, P., & Gomez-Guillen, M.C. (2000). Extracting conditions for Megrim (*Lepidorhombus boscii*) skin collagen affect functional properties of the resulting gelatin. *J. Food Sci, 65 (3),* 434-438.

[3] Jamilah, B., & Harvinder, K.G. (2001). Properties of gelatins from skins of fish-black tilapia (*Oreochromis mossambicus*) and red tilapia (*Oreochoomis nilotica*). *Food Chem, 77,* 81-84.

[4] Shahidi, F. (1994). Seafood processing by-products. In: Shahidi, F., & Botta, J.R. (Eds.), *Seafood chemistry, processing, technology and quality.* (pp, 320-334). Glasgow: Black Academic and professional

[5] Becker, G.L. (1993). Preserving food and health: antioxidants make functional, nutritious preservatives. *Food Proc, 12,* 54-56.

[6] Gennadios, A., & Weller, C.L. (1991). Edible films and coatings from wheat and corn proteins. *Food Technol, 44,* 66.

[7] Guilbert, S., Cuq, B., & Gontard, N. (1997). Recent innovation in edible and/or biodegradable păckaging materials. *Food Addit Contam, 14, 741-751.*

[8] Were, L., Hettiarachchy, N. S., & Coleman, M. (1999). Properties of cysteine-added soy protein-wheat gluten films. *J. Food Sci, 64*, 514-518.

[9] Coma, V., Martial-Gros, A, Garreau, S, Copinet, A, Salin, F., & Deschamps, A. (2002). Edible antimicrobial films based on chitosan matrix. *J. Food Sci, 67,* 1162-1169.

[10] Cagri, A., Ustunol, Z., Osburn, W., & Ryser, E. T. (2003). Inhibition of *Listeria monocytogenes* on hot dog using antimicrobial whey protein-based edible casings. *J. Food Sci, 68*, 291-8.

[11] Cho, S. H., Jahncke, M. L., Chin, K-B., & Eun, J-B. (2006). The effect of processing conditions on the properties of gelatin from skate (*Raja Kenojei*) skins. *Food Hydrocolloids, 20,* 810-816.

[12] Johnston-Banks, F. A. (1990). Gelatine. In *Food Gels*; P, Harris (ED.). (pp. 258-259) Inc: London and New York: Elsevier Applied Science Publishing Co.

[13] Ledward, D. A. (1986). In Functional properties of food macromolecules; Mitchell, J. R., Ledward, D. A., Eds.; *Gelation of gelatin.* (pp. 171-201). London and New York: Elsevier Applied Science Publishing Co.

[14] Norland, R. E. (1990). In Advances in fisheries and technology and biotechnology for increased profitability; Voight, M. N., Botta, J. K. Eds.; *Fish gelatin, technical aspects.* (pp. 325-332) Technomic Publishing Co.

[15] Montero, P., Borderias, J., Turnay, J., & Leyzarbe, M.A. (1990). Characterization of hake (*Merluccius merluccius* L) and trout (*Salmo irideus* Gibb.) collagen. *J. Agric. Food Chem, 38,* 604-609.
[16] Montero, P., Alvarez, C., Marti, M.A., & Borderias, AJ. (1995). Plaice skin collagen extraction and functional properties. *J. Food Sci, 60,* 1-3.
[17] Gudmunsson, M., & Hafsteinsson, H. (1997). Gelatin from cod skin as affected by chemical treatments. *J. Food Sci, 62(1)*, 37-39.
[18] Grossman, S., & Bergman, M. (1992). Process for the production of gelatin from fish skin. United States Patent, 5, 093. 474.
[19] Ward, A. G., & Courts, A. (1997). The Science and Technology of Gelatin. London: Academic press.
[20] Kim, J. S., & Cho, S. Y. (1996). Screening for raw material of modified gelatin in marine animal skins caught in coastal offshore water in Korea. *Agric. Chem. Biotechnol, 39,* 134-139.
[21] Gomez-Guillen, M. C., Gimenez, B., & Montero, P. (2005). Extraction of gelatin from fish skins by high pressure treatment. *Food Hydrocolloids, 19,* 923-928.
[22] Sperling, L. H. (1985). Introduction to physical polymer science.: New York: John Wiley and Sons.
[23] Stainsby, G. (1952). Viscosity of diluted gelatin solutions. *Nature, 169,* 662-665.
[24] Kim, S. K., Byun, H. G., & Lee, E. H. (1994). Optimum extraction conditions of gelatin from fish skins and its physical properties. *J. Korean Ind. Eng. Chem, 5(3),* 547-559.
[25] Arnesen, J. A., & Gildberg, A. (2002). Preparation and characterization of gelatin from the skin of harp seal (*Phoca groendlandica*). *Biores. Tech, 82,* 191-194.
[26] Lehninger, A. L., Nelson, D. L., & Cox, M. M. (1993). *Three-Dimensional Structure of Proteins*. In Principles of Biochemistry; 2nd ed.;, (pp. 160-197) Inc: New York: Worth Publishers
[27] Cho, S. H., Jahncke, M. L., & Eun, J-B. (2004). Nutritional composition and microflora of the fresh and fermented skate (*Raja Kenojei*) skins. *Int. J. Food Sci & Nutr, 55,* 45-51.
[28] Leuenberger, B. H. (1991). Investigation of viscosity and gelatin properties of different mammalian and fish gelatins. *Food Hydrocolloids, 5,* 353-361.
[29] Keiji, Y., Mariko, T., Daiki, H., Takahito, E., Yoshihiro, N., Yasuhiro, I., & Kunio, S. (2000). Physical properties of shark gelatin compared with pig gelatin. *J. Agric. Food Chem, 48,* 2023-2027.
[30] Ockerman, H.W., & Hansen, C.L, (1988). Glue and gelatin, in *Animal by-product processing,* England, Chichester, Ellis Horwood Ltd.
[31] Skierka, E., & Sadowska, M. (2007). The influence of different acids and pepsin on the extractability of collagen from the skin of Baltic cod (Gadus morhua). *Food Chemistry, 105(3),* 1302-1306.
[32] Gimenez, B., Gomez-Guillen, M.C., & Montero, P. (2005). Storage of dried fish skins on quality characteristics of extracted gelatin. *Food Hydrocolloids, 19,* 958-963.
[33] Montero, P., Fernandez-Diaz, M. D., & Gomez-Guillen, M. C. (2002). Characterization of gelatin gels induced by high pressure. *Food Hydrocolloids, 16,* 197- 205.
[34] Yang, S. T., Ryu, B. H., Moon, Y. H., Kim, D. S., & Kim, H. S. (1994). Effects of protein concentration, heat treatment and pH on the functional properties of gelatin. Bull. Kyoung Sung Nat. Univ., 15(4), 79-88.

[35] Eo, J-H., Shon, J., & Eun, J-B. (2009a). Physicochemical, functional,and antioxidative properties of collagen extracted from skate (*raja kenojei*) skin. *Food Sci. Biotechnol.* Submitted.
[36] Piez, K. A., & Gross, J. G. (1960). The amino acid composition of some fish collagens: The relation between composition and structure. *J. Biol Chem, 235(4),* 995-998.
[37] Kim, J. S., Kim, J. G., Cho, S.Y., Ha, J. H., & Lee, E. H. (1993). Characteristics of the yellowfin sole and Dover sole skins as processing material of gelatin. *J. Korean Agric. Chem. Soc, 36(4),* 290-295.
[38] Park, Y. H. (1989). *Susan Sikpum Gagonghak*, pp. 89-116. Taegu, South Korea: Hyong-Seol Press.
[39] Han, S.B., Choi, E. H., Chung, H. W., Park, H. K., Hong, J. H., Kim, D. S., Kim, K. J., Park, J. S., Lee, E. J., Lee, J. S., Song, K. H., & Kim, H. Y. (2001). Analysis of mineral contents of edible medicinal plants in Korea. *Food Sci. Biotechnol, 10(3),* 225-230.
[40] Carpenter, J. R., Goodman-Lowe, G. D., Atkinson, S., & Ako, H. (1999). Nutrient, fatty acid, amino acid and mineral analysis of natural prey of the Hawaiian monk seal, Monachus shauinsl andi. *Comp. Biochem. Physiol. 123A,* 137-146.
[41] Perumal, S., Karuppanan, V., & Karnam, J. (1996). Chemical composition and protein quality of the little-known Legume, Velvet bean. *J. Agric. Food Chem. 44,* 2636-2641.
[42] Adawi, D., Kasravi, F. B., Molin, G., & Jeppsson, B. (1996). Oral arginine supplementation in acute liver injury. *Nutr. Res. 12,* 529-533.
[43] Cho, H. S. (1999). Antioxidant effect of tryptophan and arginine on soybean oil rancidity. *J. Korean Dietary Culture, 14 (1),* 21-28.
[44] Gregory, D. A., Kuhn, D.A., Daly, K. R., & Flygenring, F. (1985). Statistical association of dietary components with Simonsiella resding in normal human mouths. *Appl. Environ. Microbiol, 50(3),* 704-705.
[45] Ochiai, K., Senpuku, H., & Kurita-Ochiai, T. (1998). Purification of immunosuppressive factor from *Capnocytophaga ochracea. J. Med. Microbiol, 47(12),* 1087-1095.
[46] Shiau, C. Y., Pong, Y. J., Chiou, T. K., & Chai, T. J. (1997). Effect of growth on the levels of free histidine and amino acids in whtite muscle of milkfish (*Chanos chanos*). *J. Agric. Food Chem, 45,* 2103-2106.
[47] Njinkone, G. B., Mirall, J., & Gaydou, A. S. (2002). Lipids and fatty acids in muscle, liver and skin of three edible fish from the *Senegalese* coast. *Comp. Biochem. Physiol, 131B,* 395-402.
[48] Ratnayake, M. W. N., Olson, B., & Ackman, R. G. (1989). Novel branched-chain fatty acids in certain fish oils. *Lipids, 24,* 630-637.
[49] Green, D. H. S., & Selivonchick, D. P. (1987). Lipid metabolism in fish. *Progr. Lipid Res, 26,* 53-85.
[50] Bender, A. (1992). Meat and meat products in human nutrition in developing countries. *FAO Food and Nutr. 53.* FAO, Rome, Italy.
[51] Park, B. S., Lee, N. H., & Lee, Y. C. (1992). Effect of dietary omega 3 PUFA and omega 6 PUFA on redistribution of cholesterol from blood to tissue in rat. *J. Kor. Anim. Nutr. Feed, 16(4),* 176-182.
[52] Lee, W. J., & Ohwada, K. (1995). Studies on the ecological characteristics of marine bacteria isolated from deep sea. *J. Korean Fish. Soc, 28(4),* 401-411.

[53] Rinaldi, A.C. (2002). Antimicrobial peptides from amphibian skin: An expanding scenario: Commentary. *Curr. Opin. Chem. Biol, 6,* 799-804.
[54] Simmaco, M., Mignogna, G., & Barra, D. (1998). Novel antimicrobial peptides from skin secretion of the European frog, *Rana esculenta. Biopolymers, 47,* 435-450.
[55] Soravia, E., Martini, G., & Zasloff, M. (1988). Antimicrobial properties of peptides from *Xenopus* granular gland secretions, Elsevier Science Publishers B. V. (Biomedical Division). 28, 337-340.
[56] Steiner, H. (1982). Secondary structure of the *Cecropins*: Antibacterial peptides from the moth *Hyalphora cecropia.* FEBS Lett, 137, 283-287.
[57] Ho, C. L., & Hwang, L. L. (1991). Structure and biological activities of a new mastoparan isolated from the venom of the *Hornet vespa basallis. J. Biochem, 274,* 453-456.
[58] Lehrer, R. I., Lichtenstein, A. K., & Ganz, T. D. (1985). Natural peptide antibiotics from human neutrophils. *J. Clin. Invest, 76,* 1427-1435.
[59] Cammue, B. P., De Dolle, M. F., Terras, F. R., Proost, P., Van Damme, J., Rees, S. B., Vanderleyelen, J., & Broekaert, W. F. (1992). Analysis of two novel classes of plant antifungal proteins from radish (*Raphanus Sativus L.*) seeds. *J. Biol. Chem*, 267, 2228-2233.
[60] Hwang, E. Y., Kim, C. H., Kim, E. J., Chung, J. K., Ryu, H. S.,& Park, N. G. (1999). Purification and charaterization of a novel antimicrobial peptide from the skin of the hagfish, *Eptatretus burgeri. J. Food Sci. Nutri.* 4, 28-32.
[61] Park, I. Y., Park, C. B., Kim, M. S., & Kim, S. C. (1998). Parasin I, an antimicrobial peptide derived from histone H2A in the catfish, *Parasilurus asotus*, *FEBS Lett, 437,* 258-262.
[62] Gibson, B. W., Tong, D. Z., Mandrell, R., Kelly, M., & Spindel, F. R. (1991). Bombinin-like peptides with antimicrobial activity from skin secretions of the Asian toad, *Bombina orientalis*, *J. Biol. Chem, 266 (1991),* 23103-23111.
[63] Simmaco, M., Barra, D., Chiarini, F., Noviello, L., Melchiorri, P., Kreil, G., Richter, K. (1991). A family of bombinin-related peptides from the skin of *Bombina variegate. Eur. J. Biochem, 199,* 217-222.
[64] Park, C. B., Kim, M. S., & Kim, S. C. (1996). A novel antimicrobial peptide from Bufo bufo gargarizans. Biochem. Biophys. *Res. Commum, 218,* 408-413.
[65] Mor, A. H., Van, N., Antoine, D., Daniele, M. S., & Pierre, N. (1991). Isolation, amino acid sequence and synthesis of dermaseptin, a novel antimicrobial peptide of amphibian skin. *Biochemistry, 30,* 8824-8830.
[66] Simmaco, M., Mignogna, G., Barra, D., & Bossa, F. (1993). Novel antimicrobial peptides from skin secretion of the European frog, *Rana esculenta. Trends Biotechnol, 324,* 159-161.
[67] Simmaco, M., Mignogna, G., Barra, D., & Bossa, F. (1994). Antimicrobial peptides from skin secretions of Rana esculenta: Molecular cloning of cDNAs encoding esculentin and brevinins and isolation of new active peptides. *J. Biol. Chem, 269,* 11956-11961.
[68] Barra, D., & Simmaco, M. (1995). Amphibian skin, a promising resource for antimicrobial peptides. *Trends Biotechnol, 13,* 206-209.
[69] Bachere, E., Destoumieux, D., & Bulet, P. (2000). Penaeidins, antimicrobial peptides of shrimp: a comparison with other effectors of innate immunity, *FEBS Lett, 191,* 71-88.

[70] Osaki, T., Omotezako, M., Nagayama, R., Hirata, M., Lwanaga, S., Kasahara, J., Hattori, S., Ito, I., Sugiyama, H., & and Kawabata, S. I. (1999). Horseshoe crab hemocyte-derived antimicrobial polypeptides, tachystatins, with sequence similarity to spider neutrotoxins. *J. Biol. Chem, 274,* 26172-26178.

[71] Khoo, L., Robinette, D. W., & Noga, E. J. (1999). Callinectin, an antibacterial peptide from blue crab, *Callinectes sapidus*, hemocytes. *Mar. Biotechnol, 1,* 44-51.

[72] Lee, I. H., Zhao, C., Cho, Y., Harwing, S. S., Cooper, E. L., & Lehrer, R. I. (1997). Clavanins, α-helical antimicrobial peptides from tunicate hemocytes. *FEBS Lett, 400,* 158-162.

[73] Mitta, G., Hubert, F., Noel, T, & Roch, P. (1999). Myticin, a novel cysteine-rich antimicrobial peptide isolated from hemocytes and plasma of the mussel *Mytilus galloprovincialis*. *Eur. J. Biochem, 265,* 71-78.

[74] Alexander, M. C., Peddric, W., & Gill, D. (1997). Isolation and characterization of Pleurocidin, an antimicrobial peptide in the skin secretion of flounder. *J. Biol. Chem, 272,* 12008-12013.

[75] Park, C. B., Lee, J. H., Park, I. Y., Kim, M. S., & Kim, S. C. (1997). A novel antimicrobial peptides from the loach, *Misgurnus anguillican*. *FEBS Lett, 411,* 173-178.

[76] Fernades, J. M. O., & Smith, V. J. (2002). A novel antimicrobial function for a ribosomal peptide from rainbow trout skin. *Biochem Biophys Res Commun, 296,* 167-171.

[77] Tompson, S. A., Tachibana, K., Nakanishi, K., & Kubota, I. (1986). Melittin-like peptides from the shark-repelling defense secretion of the sole *Pardachirus pavoninus*. *Science, 233,* 341-343.

[78] Cho, S-H., Lee, B-D., An, H., & Eun, J-B. (2005). Kenojeinin I, antimicrobial peptide isolated from the skin of the fermented skate, *raja kenojei*. *Peptides, 26,* 581-587.

[79] Zasloff, M. (1987). Maginins, a class of antimicrobial peptide from *Xenopus* skin, in: isolation characterization of two active forms and potential cDNA sequence of a precursor, Proc. Natl. Acad. Sci., USA, 1987, pp. 5449-5453.

[80] Hancock, R. E., Falla, T., & Brown, M. (1995). Cationic bactericidal peptides. *Adv. Microb. Physiol, 37,* 135-175.

[81] Cruciani, R. A., Barker, J. L., Durell, S. R., Raghunathan, G., Guy, H., Zasloff, M., & Stanley, F. (1992). Magainin 2, a natural antibiotic from frog skin, forms ion channels in lipid bilayer membrane. *Eur. F. Pharmacol, 226,* 287-296.

[82] Hancock, R. E. (1997). Peptide antibiotics. *Lancet, 349,* 418-422.

[83] Korhonen, H., Pihlanto-Leppala, A., & Tupasela, T. (1998). Impact of processing on bioactive proteins and peptides. *Trends in food science and technology,* 9, 307-319.

[84] Cole, C. G. B., & Roberts, J. J. (1997). Gelatin color measurement. *Meat Sci., 45,* 23-31.

[85] Mangino, M. E. (1984). Physicochemical aspects of whey protein functionality. *J. Dairy Sci, 67,* 2711-2722.

[86] De Wit, J. N. & Klarenbeek, G. (1984). Effects of various heat treatments on structure and solubility of whey proteins. *J. Dairy Sci,* 67, 2701-2710.

[87] Nalinanon, S., Benjakul, S., Visessanguan, W., & Kishimura, H. (2007). Use of pepsin for collagen extraction from the skin of bigeye snapper (*Priacanthus tayenus*). *Food Chemistry, 104(2),* 593-601.

[88] Patel, M. T., & Kilara, A. (1990). Studies on whey protein concentrates. 2. Foaming and emulsifying properties and their relationship with physicochemical properties. *J. Dairy Sci., 73,* 2731-2740.

[89] Kirby, A. R., Ollett, A. L., Parker, R., & Smith, A. C. (1989). An experimental study of screw configuration effects in the twin-screw extrusion-cooking of maize grits. *J. Food Eng, 8,* 247-272.

[90] Kong, F., Tan, J., Lin, M., & Rasco, B. (2008). Thermal effect on chicken and salmon muscles: Tenderness, cook loss, area shrinkage, collagen solubility and microstructure. *Swiss Soc Food Sci Technol, 41,* 1210-1222.

[91] Jelen, P., Kalab, M., & Greig, R. I. W. (1979). Water-holding capacity and microstructure of head-coagulated whey protein powders. *Milchwissenschaft, 34,* 351-356.

[92] Fiora, F. A., Pilosof, A. M. R., & Bartholomai, G. B. (1990). Physicochemical properties of soybean proteins related to flow, viscoelastic, mechanical and water-holding characteristics of gels. *J. Food Sci, 55,* 133-136.

[93] Haque, Z. U., & Antilla, P. (1993). Milk peptides: effect on the enzymatic hydrolysis of sodium caseinate. *Agric Sci in Finland, 2,* 371-377.

[94] Jelen, P. (1973). Whipping studies with partially delactosed cheese whey. *J. Dairy Sci, 56(12),* 1505-1511.

[95] Pearce, K. N., & Kinsella, J. E. (1978). Emulsifying properties of proteins: evaluation of a turbidimetric technique. *J. Agric Food Chem,* 26, 716-723.

[96] Chansiri, G., Lyons, R T., Patel, M. V., & Hem, S.L. (1999). Effect of Surface Charge on the Stability of Oil/Water Emulsions during steam sterilization. *J. Pharm Sci, 88,* 454-458.

[97] Aoki, H., Taneyama, O., & Inami, M. (1980). Emulsion properties of soy protein characteristics of 7S and 11S proteins. *J. Food* Sci, 45, 534-538, 546.

[98] Yildirim, M., Hettiarachchy, N. S., & Kalapathy, U. (1996). Properties of biopolymers from cross-linking whey protein isolate and soybean 11S globulin. *J. Food Sci, 61(6),* 1129-1134.

[99] Li, B., Feng, C., Wang, X., Ji, B., & Wu, Y. (2007). Isolation and identification of antioxidative peptides from porcine collagen hydrolysate by consecutive chromatography and electrospray ionization-mass spectrometry. *Food Chemistry, 102,* 1135-1143.

[100] Branen, A.L. (1975). Toxicology and biochemistry of butylated hydroxyanisole and butylated hydroxytoluene. *J. Ame Oil Chem' Soc, 52,* 93-63.

[101] Finkel, T., & Holbrook. (2000). Oxidant, oxidative stress and the biology of aging. *Nature,* 408-293.

[102] Shon, J., & Haque, Z. U. (2007a). Antioxidative ability of native and thermized sour whey in oxidation-catalysed model systems. *Int. J. Dairy Technol, 60,* 143-148.

[103] Tong, L. M., S. Sasaki, D. J. McClements, and E. A. Decker. (2000). Antioxidant activity of whey in a salmon oil emulsion. *J. Food Sci.,* 65(8), 1325-1329.

[104] Kim, S., Kim, Y., Byun, H., Nam, K., Joo, D., & Shahidi, F. (2001a). Isolation and characterization of antioxidative peptides from gelatin hydrolysate of Alaska Pollack skin. *J. Agric Food Chem, 49,* 1984-1989.

[105] Kim, S.K., Kim, Y.T., Byun, H. G., Park, P.J., & Ito, H. (2001b). Purification and characterization of antioxidative peptides from bovine skin. *J. Biochem Mol Biol, 34,* 219-224.

[106] Mendis, E., Rajapakse, N., Kim, S. (2005). Antioxidative properties of a radical-scavenging peptide purified from enzymatically prepared fish skin gelatin hydrolysate. *J. Agric Food Chem, 53,* 581-587.

[107] Pena-Ramos, E. A., & Xiong, Y. L. (2002). Antioxidant activity of soy protein hydrolysates in a liposomal system. *J. Food Sci. 67,* 2952-2956.

[108] Saiga, A., Tanabe, S., & Nishimura, T. (2003). Antioxidant activity of peptides obtained from porcine myofibrillar proteins by protease treatment. *J. Agric Food Chem, 51,* 3661-3667.

[109] Hernandez-Ledesma, B., Davalos, A., Bartolome, B., & Amigo, L. (2005). Preparation of antioxidant enzymatic hydrolysate from α-lactalbumin and β-lactoglobulin identification of active peptides by HPLC-MS/MS. *J. Agric Food Chem, 53,* 588-593.

[110] Tsuge, N., Eikawa, Y., Nomura, T., Yamamoto, M., & Sugisawa, K. (1991). Antioxidant activity of peptides prepared by enzymatic hydrplysis of egg-white albumin. *Nippon Nogeikagaku Kaishi, 65,* 1635-1641.

[111] Hattori, M., Yamaji-Tsukamoto, K., Kumagai, H., Feng, Y., & Takahashi, K. (1998). Antioxidant activity of soluble elastin peptide. *Journal of Agriculture and Food Chemistry, 46,* 2167-2170.

[112] Chen, H. M., Muramoto, K., & Yamauchi, F. (1995). Structural analysis of antioxidative peptides from soybean beat-conglycinin. *J. Agric Food Chem, 43,* 574-578.

[113] Chen, H. M., Muramoto, K., Yamauchi, F., & Nokihara, K. (1996). Antioxidant activity of designed peptides based on the antioxidative peptide isolated from digests of a soybean protein. *J. Agric Food Chem, 44,* 2619-2623.

[114] Da'valos, A., Miguel, M., Bartolome, B., & Lo pez-Fandino, R. (2004). Antioxidant activity of peptides derived from egg white proteins by enzymatic hydrolysis. *J. Food Prot, 67,* 1939-1944.

[115] Eo, J-H., Shon, J., & Eun, J-B. (2009b). Antioxidative ability of collagen powder in oxidation-catalysed model systems. *Food Sci. Biotechnol.* Submitted.

[116] Browdy, A. A., & Harris, N. D. (1997). Whey improves oxidative stability of soybean oil. *J. Food Sci, 62(2),* 348-350, 376.

[117] Kanterewicz, R. J., & Chirife, J. (1986). Color changes and available lysine during storage of shelf-stable concentrated cheese whey. *J. Food Sci, 51,* 826-828.

[118] Petriella, C., Resnik, S. L., Lozano, R. D., & Chirife, J. (1985). Kinetics of deteriorative reactions in model food systems of high water activity: color changes due to nonenzymaic browning. *J. Food Sci, 50(3),* 622-626.

[119] Nawar, W. W. (1996). Lipids. In: Fennema, O.R, editor. Food Chemistry, 3rd ed. Marcel Dekker, Inc., New York. P 225-319.

[120] Gennadios, A., Hanna, M. A., & Kurth, L. B. (1997). Application of edible coatings on meats, poultry and seafoods: A review. *Swiss Soc. Food Sci Technol,* 30, 337-350.

[121] Gennadios, A., McHugh, T.H. Weller, C.L., & Krochta, J.M. (1994). Edible coatings and films based on proteins. In: Krochta, J.M., Baldwin, E.A., & Nisperos-

Carriedo, M. (Eds), *Edible coatings and Films to Improve Food Quality*. (pp. 201-277).Lancaster, PA: Thchnomic Publishing Company, Inc.

[122] Torres, J.A. (1994). Edible films and coatings from proteins. In: Hettiarachchy, N.S., & Ziegler, G.R. (Eds), *Protein Functionality in Food Systems*. New York: Marcel Dekker, Inc., pp. 467-507.

[123] Ressouany, M., Vachon, C., & Lacroix, M. (1998). Irradiation does and calcium effect of the chemical properties of cross-linked caseinate films. *J. Agric Food Chem, 46(4),* 1618-1623.

[124] Shon, J., & Haque, Z. U. (2007b). Efficacy of sour whey as a shelf-life enhancer: Use in antioxidative edible coatings of cut vegetables and fruit. *J. Food Qual, 30,* 581-93.

[125] Shon, J., & Chin, K-B. (2008). Effect of whey protein coating on quality attributes of low-fat, aerobically packaged sausage during refrigerated storage. *J. Food Sci, 73,* C469-C475.

[126] Haque, Z. U., Shon J., & Willians, B. (2009). Efficacy of sour whey as a shelf-life enhnacer: Use in antioxidative edible coatings of beef steak. *J. Food Qual, 32,* 381-397.

[127] Shon, J., Eo, J-H., & Eun, J-B. (2010). Effect of soy protein isolate coating on quality attributes of cut raw Han-Woo (Korean Cow) beef, aerobically packaged and held refrigerated. *J. Food Qual,* Accepted.

[128] Shon, J., &Haque, Z.U. (2007c). Functional attributes of native and thermized sour and sweet whey. *Int. J. Dairy Technol, 60,* 135-42.

[129] Le Tien, C., Letendre, M., Ispas-Szabo, P., Mateescu, M. A., Delmas-Patterson, G., Yu, H. L., & Lacroix, M. (2000). Development of biodegradable films from whey proteins by cross-linking and entrapment in cellulose. *J. Agric. Food Chem, 48,* 5566-5575.

[130] Vachon, C., Yu, H. L., Yefsah, R., Alain, R., St-Gelais, D., & Lacroix, M. (2000). Mechanical and structural properties of milk protein edible films cross-linked by heating and irradiation. *J. Agric. Food Chem, 48,* 559-566.

[131] Simelane S., & Ustunol, Z. (2005). Mechanical properties of heat-cured whey protein-based edible films compared with collagen casings under sausage manufacturing conditions. *J. Food Sci, 70 (2),* E131-134.

[132] Lieberman, E. R., & Gilbert, S. G. (1973). Gas permeation of collagen films as affected by cross-linkage, moisture, and plasticizer content. *J. Polym Sci, 41,* 33-43.

[133] Park, H. J., & Chinnan, M. S. (1995). Gas and water vapor barrier properties of edible films from protein and cellulosic materials. *J. Food Eng, 25,* 497-507.

[134] Aydt, T. P., & Weller, C. L., & Testin, R. F. (1991). Mechanical and barrier properties of edible corn and wheat protein films. *Transactions of the ASAE, 34,* 207-211.

[135] Gennadios, A., Weller, C. L., & Testin, R. F. (1993). Temperature effect on oxygen permeability of edible protein-based films. *J. Food Sci, 58,* 212-214, 219.

[136] Brandenburg, A. H., Weller, C. L., & Testin, R. F. (1993). Edible films and coatings from soy protein. *J. Food Sci, 58,* 1086-1089.

[137] McHugh, T. H., & Krochta, J. M. (1994). Sorbitol- vs glycerol-plasticized whey protein edible films: integrated oxygen permeability and tensile property evaluation. *J. Agric Food Chem, 42,* 841-845.

[138] Taylor, S.L. (1992). Chemistry and detection of food allergens. *Food Technol, 46 (5),* 146, 148-152.

[139] Houts, S.S. (1988). Lactose intolerance. *Food Technol, 42 (3),* 110-113.

[140] Hood, L. L. (1987). Collagen in sausage casings. In: Pearson, A. M., Dutson, T. R., & Bailey, A. J. (Eds), *Advances in Meat Research,* Vol. 4. (pp. 109-129) New York: Van Nostrand Reinhold Company.

[141] Rust, R. E. (1987). Sausage products. In: Price, J.F., & Schweigert, B. S. (Eds), *The Science of Meat and Meat Products.* (pp. 457-485). Westport, CT: Food and Nutrition Press, Inc.,

[142] Jones, H. W., & Whitmore, R. A. (1972). Collagen food coating composition and method of preparation. U.S. Patent3, 694, 234, September 26.

[143] Farouk, M. M., Price, J. F., & Salih, A. M. (1990). Effect of an edible collagen film overwrap on exudation and lipid oxidation in beef round steak. *J. Food Sci, 55,* 1510-1512, 1563.

[144] Rice, J. (1994). What's new in edible film?. *Food Proc, 55 (7),* 61-62.

[145] Klose, A. A., Mecchi, E. P., & Hanson, H. L. (1952). Use of antioxidants in the frozen storage of turkeys. *Food Technol, 6,* 308-311.

[146] Childs, W. H. (1957). Coated sausage and method of preparing same. *U.S. Patent* 2,811,453, October 29.

[147] Mooriani, M. N., Raia, K. C. M., Puttaraiappa, P., Khabade, V. S., Mahendrakar, N. S., & Mahade-Vaswamy, M. (1978). Studies on curing and smoking poultry meat. *Indian J. Poultry Sci, 13,* 52-57.

[148] Olson, S., & Zoss, R. (1985). Fried foods of reduced oil absorption and methods of preparation employing spray of film forming agent. *U.S. Patent* 4,511,583, April 16.

[149] McHugh, T. H., & Krochta, J. M. (1994). Milk protein-based edible films and coatings. *Food Technol.*, 48(1), 97-103.

[150] McHugh, T. H., & Krochta, J. M. (1994). Sorbitol- vs glycerol-plasticized whey protein edible films: integrated oxygen permeability and tensile property evaluation. *J. Agric. Food Chem,, 42,* 841-845.

[151] Chen, H. (1995). Functional properties and applications of edible films made of milk proteins. *J. Dairy Sci, 78,* 2563-2583.

[152] Avena-Bustillos, R. J., Cisneros-Zevallos, L. A., Krochta, J. M., & Saltveit, M. E. (1993). Optimization of edible coatings on minimally processed carrots using response surface methodology. *Transactions of the ASAE, 36,* 801-805.

[153] Avena-Bustillos, R. J., Krochta, J. M., & Saltveit, M. E. (1994). Optimization of edible coatings on formulations on zucchini to reduce water loss. *J. Food Eng, 21,* 197-214.

[154] Mate, J. I., & Krochta. J. M. (1996a). Comparison of oxygen and water vapor permeabilities of whey protein isolate and β-lactoglobulin edible films. *J. Agric. Food Chem, 44,* 3001-3004.

[155] Mate, J. I., & Krochta. J. M. (1996b). Whey protein coating effect on oxygen uptake of dry roasted peanuts. *J. Food Sci., 61,* 1202-1206, 1210.

[156] Mate, J. I., Frankel, E. N., & Krochta, J. M. (1996). Whey protein isolate edible coatings: effect on the rancidity process of dry roasted peanuts. *J. Agric. Food Chem., 44*, 1736-1740.

[157] Moreau, D. L., & Rosenberg, M. (1993). Microstructure and fat extractability in microcapsules based on whey proteins or mixtures of whey proteins and lactose. *Food Structure, 12,* 457-468.

[158] Rosenberg, M., & Young, S. L. (1993). Whey proteins as microencapsulating agents. Microencapsulating of anhydrous milkfat-structure evaluation. *Food Structure, 12,* 31-41.

[159] Young, S. L., Sarda, X., & Rosenberg, M. (1993). Microencapsulating properties of whey proteins. 1. Microencapsulating of anhydrous milk fat. *J. Dairy Sci, 76,* 2868-2877.

[160] Young, S. L., Sarda, X., & Rosenberg, M. (1993). Microencapsulating properties of whey proteins. 2. Combination of whey proteins with carbohydrates. *J. Dairy Sci, 76,* 2878-2885.

[161] Sheu, T. Y., & Rosenberg, M. (1995). Microencapsulation by spray drying ethyl caprylate in whey protein and carbohydrate wall system. *J. Food Sci, 60,* 98-103.

[162] Le Tien, C., Letendre, M., Ispas-Szabo, P., Mateescu, M. A., Delmas-Patterson, G., Yu, H. L., & Lacroix, M. (2000). Development of biodegradable films from whey proteins by cross-linking and entrapment in cellulose. *J. Agric. Food Chem, 48,* 5566-5575.

[163] Le Tien, C., Vachon, C., Mateescu, M. A., & Lacroix, M. (2001). Milk protein coatings prevent oxidative browning of apples and potatoes. *J. Food Sci, 66,* 512-516.

[164] Hirasa, K. (1991). Moisture loss and lipid oxidation infrozen fish- Effect of a casein-acetylated monoglyceride edible coating. M.S thesis, University of California-Davis, Davis, CA, USA.

[165] Stuchell, Y. M., & Krochta, J. M. (1995). Edible coatings on frozen King salmon: Effect of whey protein isolate and acetylated monoglycerides on moisture loss and lipid oxidation. *J. Food Sci, 60,* 28-31.

[166] Cagri A., Ustunol, Z., & Ryser, E. T. (2001). Antimicrobial, mechanical, and moisture barrier properties of low pH whey protein-based edible films containing p-aminobenzoic or sorbic acids. *J. Food Sci, 66*, 865-70.

[167] Cagri, A., Ustunol, Z., & Ryser, E. T. (2002). Inhibition of three pathogens on bologna and summer sausage using antimicrobial edible films. *J. Food Sci, 67,* 2317-2324.

[168] Kim, K-M., Weller, C. L., Hanna, M. A., & Gennadiios, A. (2002). Heat curing of soy protein films at selected temperatures and pressure. *Lebensm Wiss Technol, 35,* 140-145.

[169] Eo, J-H., Shon, J., & Eun, J-B. (2009c). Antioxidative and antimicrobial edible coatings of cut pork loin. *Food Sci. Biotechnol.* Submitted.

[170] Papastamatiou, F., Gerasopoulos, D., Siomos, A., & Bloukas, J. G. (2007). Effect of degree of cutting of leek on physicochemical characteristics of Greek traditional sausages. *Meat Sci, 75,* 648-54.

[171] Yin, M. C., & Faustman, C. (1993). The influence of temperature, pH, and phospholipid composition upon the stability of myoglobin and phospholipid: a liposome model. *J. Agric. Food Chem, 41,* 853-7.

[172] Hur, S. J., Ye, B. W., Lee, J. L., Ha, Y. L., Park, G. B., & Joo, S.T. (2004). Effects of conjugated linoleic acid on color and lipid oxidation of beef patties during cold storage. *Meat Sci, 66,* 771-5.

[173] Mercier, Y., Gatellier, P., Viay, M., Remignon, H., & Renerre, M. (1998). Effect of dietary fat and vitamin E on lipid and protein oxidation in turkey meat during storage. *Meat Sci, 48*, 301-17.

[174] Park, S. K., Rhee, C. O., Bae, D. H., & Hettiarachchy, N. S. (2001). Mechanical properties and water-vapor permeability of soy-protein films affected by calcium salts and glucono-delta-lactone. *J. Agric. Food Chem, 49*, 2308–2312.

[175] Gray, J. I., & Pearson, A. M. (1987). Rancidity and warmed-over flavor. In A.M. Pearson, & T. R. Dutson. Advances in meat research, Vol. 3. Restructured meat and poultry products (pp. 221-269). New York, NY: Van Nostrand Reinhold Co.

[176] Sapers, G. M. (1993). Browning of foods: Control by sulfites, antioxidants and other means. *Food Technol,* 47, 75-84.

[177] Krochta, J. M., & De Mulder-Johnson, C. (1997). Edible and biodegradable polymer films: Challenges and opportunities. *Food Technol, 51,* 61-74.

[178] Janes, M. E., Kooshesh S., & Johnson M. G. (2002). Control of *Listeria monocytogenes* on the surface of refrigerated, ready-to-eat chicken coated with edible zein film coatings containing nisin/or calcium propionate. *J. Food Sci. 67,* 2754-2757.

[179] McCarthy, T. L., Kerry, J. P., Kerry, J. F., Lynch, P. B., & Buckley, D. J. (2001). Evaluation of the antioxidant potential of natural food and plant extracts as compared with synthetic antioxidants and vitamin E in raw and cooked pork patties. *Meat Sci, 57*, 45-52

[180] Seo, A., & Morr, C.V. (1984). Improved high-performance liquid chromatographic analysis of phenolic acids and isoflavonoids from soybean protein products. *J. Agric. Food Chem, 32,* 530-533.

[181] Rojas, M. C., & Brewer, M. S. (2008). Effect of natural antioxidants on oxidative stability of frozen, vacuum-packaged beef and pork. *J. Food Qual, 31,* 173-188.

[182] Wu, S. Y., & Brewer, M. S. (1994). Soy protein isolate antioxidant effect on lipid peroxidation of ground beef and microsomal lipids. *J. Food Sci, 59,* 702-706.

[183] Wu, Y., Rhim, J. W., Weller, C. L., Hamouz, F.A., Cuppett, S., & Schnepf, M. (2000). Moisture loss and lipid oxidation for precooked beef patties stored in edible coatings and films. *J. Food Sci, 65,* 300-304.

[184] Brandenburg, A. H., Weller, C. L., & Testin, R. F. (1993). Edible films and coatings from soy protein. *J. Food Sci, 58,* 1086-1089.

[185] Bowers, J. A., & Engler, P. P. (1975). Freshly cooked and cooked, frozen reheated beef and beef-soy patties. *Food Sci, 40,* 624-625.

[186] Romjin, A., Cuppett, S. L., Zeece, M. G., Parkhurst, A. M., & Lee, M. L. (1991). Impact of soy protein isolates and specific fractions on rancidity development in a cooked, refrigerated beef system. *J. Food* Sci, 56, 188-190.

[187] De St. Louvent, E., Locquet, N., Stahl, P., & Birlouez-Aragon, I. (2004). Protein hydrolysis of infant formulas strongly activates the Maillard reaction. Pediatric Gastroenterology, Reports from the World Congress of Pediatric Gastroenterology, Hepatology and Nutrition, 2nd, Paris, France, July 3-7, 2004

[188] Jing, H., & Nakamura, S. (2005). Production and use of Maillard products as oxidative stress modulators. *J. Med. Food, 8(3),* 291-298.

[189] Kester, J. J., & Fennema, O. R. (1986). Edible films and coatings: a review. *Food Technol, 40,* 47-49.

[190]) Rayas, L. M., Hernandez, R. Z., & Ng, P. (1997). Development and characterization of biodegradable/edible wheat protein film. *J. Food Sci., 62(2),* 160-162, 189.

[191] Verma, S. P., & Sahoo, J. (2000). Improvement in the quality of ground chevron during refrigerated storage by tocopherol acetate pre-blending. *Meat Sci, 56,* 403-13.

[192] Siragusa, G. R., & Dickson, J. S. (1992). Inhibition of *Listeria monocytogenes* on beef tissue by application of organic acids immobilized in a calcium alginate gel. *J. Food Sci, 57(2),* 293-296.

[193] Siragusa, G. R., & Dickson, J. S. (1993). Inhibition of *Listeria monocytogenes, Salmonella Typhimurium* and *Escherichia coli O157:H7* on beef muscle tissue by lactic or acetic acid contained in calcium alginate gels. *J. Food Safety, 13(2),* 147-158.

[194] Eswaranandam, S., Hettiarachchy, N. S., & Johnson, M. G. (2004). Effects of citric, lactic, malic, and tartaric acids on antimicrobial activity of nisin-incorporated soy protein film against *Listeria monocytogenes, Escherichia coli* O157:H7, and *Salmonella gaminara. J. Food Sci, 69,* M79-84.

[195] Vallejo-Cordoba, B., Nakai, S., Powrie, W. D., & Beveridge, T. (1987). Extended shelf life of frankfurters and fish frankfurter-analogs with added soy protein hydrolysates. *J. Food Sci. 52,* 1133-1136.

In: Food Hydrocolloids: Characteristics, Properties... ISBN: 978-1-60876-222-4
Editor: C. S. Hollingworth, pp. 41-80

Chapter 2

FOOD HYDROCOLLOID EDIBLE FILMS AND COATINGS

***O. Skurtys*[1*]*, C. Acevedo*[1]*, F. Pedreschi*[2]*, J. Enrione*[1]*, F. Osorio*[1]* and J.M. Aguilera*[2]

[1]Department of Food Science and Technology, Universidad de Santiago de Chile
Av. Ecuador 3769. Santiago, Chile.
[2]Department of Chemical Engineering and Bioprocesses, Pontificia Universidad Católica de Chile
Av. Vicuña Mackenna 4860. Santiago. Chile

ABSTRACT

In this chapter book, we present and discuss some aspects of hydrocolloid edible films and coating. The film-forming materials is presented and the physical properties (*e.g.* thickness, mechanical, optical and thermodynamic properties, water and gases barrier) of the film and coating are examined in detail. The main techniques used to form edible coatings as spray systems or immersion procedures are discussed as well as the solvent casting and the extrusion processes for the edible films. The main disadvantage of these techniques is the lost of quality of the edible coatings and films since there is no control over the shape, size and size distribution of the dispersed elements (*e.g.* additives, ingredients, etc.) in the support structure matrix is poor. Thus, microfluidic applications are presented and discussed to solve some of these problems and to improve film qualities.

---

* **Corresponding author:** Dr. Olivier Skurtys
Department of Food Science and Technology, Universidad de Santiago de Chile
Address: Avenida Ecuador 3769, Santiago, Chile.
Phone : 56-2-718 45 26.
Fax : 56-2-776 40 80.
e-mail: olivier.skurtys@usach.cl

1. INTRODUCTION

An edible film is defined as a thin layer, which can be consumed, coated on a food or placed as barrier between the food and the surrounding environment. For the past 10 years, research on edible films and coatings in foods is driven by food engineers due to the high demand of consumers for longer shelf–life and better quality of fresh foods as well as of environmentally friendly packagings [1-4]. The most familiar example of edible packaging is sausage meat in casing that is not removed for cooking and eating. Such films can mechanically protect foods, prevent the contamination from microorganisms, prevent quality loss of foods due to mass transfer (*e.g.* moisture, gases, flavours, etc.). Indeed, edible films and coatings can be used as a vehicle for incorporating natural or chemical antimicrobial agents, antioxidants, enzymes or functional ingredients such as probiotics, minerals and vitamins [5-7]. The edible films are classified into three categories taking into account the nature of their components: hydrocolloids (containing proteins, polysaccharides or alginates), lipids (constituted by fatty acids, acylglycerols or waxes) and composites (made by combining substances from the two categories) [8].

Thus, the objective of this work is to present and discuss only some aspects of hydrocolloid edible films and coatings. In particular, the film-forming materials is presented and the physical properties (*e.g.* thickness, mechanical, optical and thermodynamic properties, water and gases barrier) of the film and coating are examined. The main techniques used to form edible coatings as spray systems or immersion procedures are detailed as well as the solvent casting and the extrusion processes for the edible films. The main disadvantage of these techniques is the lost of quality of the edible coatings and films since there is no control over the shape, size and size distribution of the dispersed elements (e.g. additives, ingredients, etc.) in the support structure matrix is poor. An other disadvantage of these techniques is that the thickness of the films is generally no constant and no controlled. Finally, microfluidic applications are presented and discussed to solve some of these problems and to improve film qualities.

2. FILM-FORMING MATERIALS

2.1. Hydrocolloidal Materials

Hydrocolloids are hydrophilic polymers, of vegetable, animal, microbial or synthetic origin, that generally contain many hydroxyl groups and may be polyelectrolytes (for example alginate, carrageenan, carboxymethylcellulose, gum arabic, pectin and xanthan gum). Nowadays, they are widely used as film-forming solution to perform and control the texture, flavour, and shelf-life of foods [9]. Hydrocolloidal materials, *i.e.* proteins and polysaccharides, used extensively for the formation of edible films and coatings are presented in Table 1. All hydrocolloids used are fully or partially soluble in water and are used principally to increase the viscosity of the continuous phase (aqueous phase) *i.e* as gelling agent or thickener [1, 8, 10-11]. They can also be used as emulsifier since their stabilizing effect on emulsions derives from an increase in viscosity of the aqueous phase of the edible film. The kinetic motion of the oil droplets is reduced, resulting in a lower rate of flocculation and coalescence in the film.

2.1.1. Polysaccharides-Based Films and Coating

Agar (E406)

Agar is a hydrophilic colloid consisting of a mixture of agarose and agaropectin that have the ability to form reversible gels simply by cooling a hot aqueous solution. Agar gel melts on heating and resets on cooling. Typical gel temperatures of agar for 1.5% solution are in the range 35-45 °C [12]. Because of its ability to form very hard gels at very low concentrations and the simplicity of the extraction process [13], agar has been used extensively as a gelling agent in the food industry. However, despite its biodegradability and its enormous gelling power, agar has been few used as edible film due to a poor aging. Indeed, both photodegradation and fluctuations in ambient temperature and humidity alter agar crystallinity, leading to formation of micro-fractures and polymer embrittlement [14-15]. However, it was reported that agar-based film displays a better moisture barrier properties than cassava starch film [16]. The influence of agar on the structure and the functional properties of emulsified edible films has been recently studied by Phan The *et al.* [17]. Gelled agar chains can stabilize film-forming emulsion to create a macronetwork. This macronetwork entraps flattened lipid particles improving barrier performance by increasing tortuosity. The mechanical properties of agar-hydrogenated vegetable oil emulsified films are comparable with some protein and Low-density polyethylene (LDPE) films.

Alginate (E400-E404)

Alginate is an indigestible biomaterial produced by brown seaweeds (Phaeophyceae, mainly Laminaria) therefore it may also be viewed as a source of dietary fibre. Alginate has a potential to form biopolymer film or coating component because of its unique colloidal properties, which include thickening, stabilizing, suspending, film forming, gel producing, and emulsion stabilizing [18-19]. An attractive feature of alginate solutions is the gelling capacity in presence of Calcium. Edible films prepared from alginate form strong films and exhibit poor water resistance because of their hydrophilic nature [10, 20-21]. The water permeability and mechanical attributes can be considered as moderate compared to synthetic films [22]. Alginate edible-films are appropriated to load additives and antibacterial compounds. Good results are obtained applying probiotics [23] and oil compunds such as garlic oil [24] and oregano oil [25]. A mixture of starch and alginate to form edible film has been studied by Wu *et al.* [26] achieving to improve the mechanical properties of film.

Carrageenans (E407)

Carrageenans are water-soluble polymers with a linear chain of partially sulphated galactans, which present high potentiality as film-forming material. These sulphated polysaccharides are extracted from the cell walls of various red seaweeds (Rhodophyceae). Different seaweeds produce different carrageenans. The positions and numbers of sulfate ester groups are important because they are, together with the anhydrogalactose bridge, responsible for carrageenan functionality and determine a classification in three major types:

κ, ι and λ. κ- and ι-carrageenans contain the 3,6-anhydro units and are used as gelling agents because of their property to produce thermoreversible gels on cooling below the critical temperature. λ-Carrageenan, with only sulfated galactose groups and no anhydrogalactose bridge, is a thickening polymer [27]. The use of carrageenan as edible films and coatings already covers various fields of the food industry such as application on fresh and frozen meat, poultry and fish to prevent superficial dehydration [28], sausage-casings [29], dry solids foods, oily foods [30], etc. Carrageenan film formation includes this gelation mechanism during moderate drying, leading to a three-dimensional network formed by polysaccharide double-helices and to a solid film after solvent evaporation [31-32]. Recently, carrageenan films were also found to be less opaque than those made of starch [33].

Cellulose Derivatives

Cellulose derivatives are polysaccharides composed of linear chains of β (1–4) glucosidic units with methyl, hydroxypropyl or carboxyl substituents. Only four cellulose derivative forms are used for edible coatings or films: Hydroxypropyl cellulose (E463; HPC), hydroxypropyl methylcellulose (E464; HPMC), Carboxymethylcellulose (E466; CMC) or Methyl cellulose (E461; MC). Cellulose derivatives exhibit thermo-gelation therefore when suspensions are heated they form a gel whereas they returns to originally consistency when cooled [34]. The films cast from aqueous solutions of MC, HPMC, HPC and CMC tend to have moderate strength, are resistant to oils and fats, and are flexible, transparent, flavourless, colourless, tasteless, water-soluble and moderate barriers to oxygen. MC is the most resistant to water and it is the lowest hydrophilic cellulose derivatives [10]. However, cellulose derivative films are poor water vapour barriers because of the inherent hydrophilic nature of polysaccharides and they possess poor mechanical properties [35-36]. A way to improve the moisture barrier would be the incorporation of hydrophobic compounds, such as fatty acids, into the cellulose ether matrix to develop a composite film [37]. As other example, Debeaufort *et al.* [38], developed bilayers by adding a mixture of lipids (paraffin oil, paraffin wax, or a mixture of hydrogenated palm oil and triolein) onto a methylcellulose layer.

Edible coatings made of CMC, MC, HPC, and HPMC have been applied to some fruits and vegetables for providing barriers to oxygen, oil, or moisture transfer [7, 39].

Chitosan

Chitosan (β-(1, 4)-2-amino-2-deoxy-D-glucopyranose), which is mainly made from crustacean shells, is the second most abundant natural and non-toxic polymer in nature after cellulose [40-42]. Chitosan shows antifungal and antibacterial properties, which are believed to be originated from its polycationic nature [41, 43-44]. However, a major drawback of chitosan is its poor solubility in neutral solutions. The required degree of deacetylation to obtain a soluble product must be 80–85% or higher [45]. Chitosan products are highly viscous, resembling natural gums [46]. Chitosan can form transparent films to enhance the quality and extend the storage life of food products [33, 47-50]. Pure chitosan films are generally cohesive, compact and the film surface has a smooth contour without pores or cracks [51]. Chitosan films such as many polysaccharide based films, tend to exhibit fat and oil resistance and selective permeability to gases but lack resistance to water transmission [52-53]. Chitosan-based edible films and coatings allowed to extend and preserve shelf life of fresh strawberries [54-55], of slices mango fruit [56]. Finally, there is an available review on chitosan films [57].

Gums

Gums in edible-forming preparation are used a for their texturizing capabilities. All gums are polysaccharides composed of sugars other than glucose. Gums are differentiate in three groups [9]: exudate gums (gum Arabic; mesquite gum), the extractive gums (come from endosperm of some legume seeds or extracted from the wood: guar gum) and the microbial fermentation gums (xanthan gum). In edible-forming preparations, guar gum (E 412) is used as a water binder, stabilizer and viscosity builder. Gum arabic (E 414), owing to its solubility in hot or cold water, is the least viscous of the hydrocolloid gums. Xanthan gum (E 415) is readily dispersed in water, hence high consistency is obtained rapidly in both hot and cold systems. A blend of guar gum, gum arabic and xanthan gum provided uniform coatings with good cling and improved adhesion in wet batters [58-60]. The coating based from mesquite gum preparation and applied to Persian limes provided a lowest physiological weight loss, a best dark shade green colour retention and an unaltered physicochemical parameters to Persian limes [61]. The mesquite gum forms films with excellent water vapor barrier properties when small amounts of lipids are added in their formulation [62]. Film-forming polymers based on gum arabic, showed to be an inhibitor of after-cooking darkening of potatoes [63].

Pectin

Pectin (E440) is a heterogeneous grouping of acidic structural polysaccharides, found in fruit and vegetables and mainly prepared from citrus peel and apple pomace. This complex anionic polysaccharide is composed of β-1,4-linked d-galacturonic acid residues, wherein the uronic acid carboxyls are either fully (HMP, high methoxy pectin) or partially (LMP, low methoxy pectin) methyl esterified. Commercial extraction causes extensive degradation of the neutral sugar-containing side chains. With Chitosan, HMP or LMP forms excellent films. Indeed, the cationic nature of chitosan offers the possibility to take advantage of the electrostatic interactions with anionic polyelectrolytes, such as pectin [64-65]. The effect of antimicrobial plant essential oils such as oregano, lemon grass, or cinnamon oil at different concentrations, on physical properties of apple puree-pectin based edible coatings was reported by Rojas-Grau *et al.* [66]. It is also interesting to point out the application of LMP based edible coatings as a pre-treatment in osmotic dehydration for obtaining a better dehydration efficiency [67-68]. Finally recently, an investigations demonstrated the prevention of crumb ageing of dietetic sucrose-free sponge cake when a pectin-containing edible film was used. This sponge cake had better preserved freshness, especially up to the fifth day of storage [69].

Starch

Starch is the major carbohydrate reserve in plant tubers and seed endosperm where it is found as granules, each typically containing several million amylopectin molecules accompanied by a much larger number of smaller amylose molecules [70]. Amylose is responsible for the film-forming capacity of starch [71]. The largest source of starch is corn (maize) with other commonly used sources being wheat, potato, tapioca and rice. Genetic modification of starch crops has recently led to the development of starches with improved and targeted functionality [72]. Starch is used to produce biodegradable films to partially or entirely replace plastic polymer [73]. The films are transparent or translucent, flavourless, tasteless and colourless [74]. However, starch film application are limited by poor mechanical strength and its efficient barrier against low polarity compound [10]. Many research reported that film forming conditions have an effect on crystallinity of the starch films and, therefore, their properties [75]. Films of high-amylose corn starch or potato starch was more stable during aging, lost little of their elongation and had not or a slight increasing in tensile strength [76]. Films from cassava starch had good flexibility and low water permeability, indicating potential application as edible films [77]. Plasticizer is generally required for starch-based edible films to overcome film brittleness. The most commonly used plasticizers for starch films are glycerol and sorbitol [78]. Recently, Hu *et al.* [79] have developed starch films from oxidized potato starch (OPS) with glycerol as a plasticizer at different contents. The OPS films were transparent and flexible with interesting mechanical properties.

2.1.2. Proteins-Based Films and Coatings

Animal (milk and whey proteins) and plant (such as soy and zein proteins) protein films exhibit better oxygen barrier, carbon dioxide barrier and mechanical properties than polysaccharide films. However, similar to polysaccharide films, due to their hydrophillic nature, protein films generally possess poor water vapor barrier property [80]. Studies have shown that the concentration, composition, size, and shape of plasticizers all affect the properties of protein films [81-85]. Proteins that are insoluble in water, such as corn zein and wheat gluten, produce insoluble coatings, whereas proteins that are soluble in water produce coatings of varying solubility, depending on the protein and the conditions of coating formation and treatment. Most of the protein in soybeans is insoluble in water but soluble in dilute neutral salt solutions therefore soy protein are not considered as a hydrocolloid.

Gelatin

Gelatin is prepared by the thermal denaturation of collagen, isolated from animal skin, bones and fish skins [86]. The physical properties of gelatins are related not only to the molecular weight distribution but also to the amino acid composition. Gelatin is readily soluble in water at temperatures above 40ºC, forming a viscous solution of random-coiled linear polypeptide chains. On cooling a gelatin solution around 20ºC, collagen like helices are formed, albeit not very long ones and including only part of the material: this is form a gel [70]. Mammalian gelatins commonly have better physical properties and thermostability than most fish gelatins [87], and this has been related mainly to their higher amino acid content [86]. The use of mammalian gelatin in the elaboration of edible film or coatings was very well studied until the sixties, which resulted in many patents, mainly in the pharmaceutical area. Consequently, these materials characteristics are not easily available in the literature [88]. The properties and film forming ability of gelatins are directly related to the molecular weight, i.e., the higher the average molecular weight, the better the quality of the film [89]. The molecular weight distribution depends mainly on the degree of collagen cross-linking and the extraction procedure [90]. However, in the years 2000, the gelatin films forming principally with fish gelatin have returned to the attention of researcher [91-94]. Films from tuna skin gelatin plasticized with glycerol presented lower water vapor permeability (WVP) compared to values reported for pigskin gelatin [95].

Whey Proteins

Table 1. Hydrocolloidal materials that have been studied extensively for the formation of edible films and coatings in foods

	Film Forming Material		Principal Function	References
	Agar (E406)		Gelling agent	[14-17]
	Alginate (E400-404)		Gelling agent	[10, 19-24]
	Carrageenan (E407)		Gelling agent	[26-29, 31-32]
		Carboxymethyl cellulose (E466)	Thickner	
		Hydroxypropyl cellulose (E463)	Thickner and emulsifier	
Polysaccharide	Cellulose Derivative	Hydroxypropyl methyl (E464)	Thickner	[7, 10, 36, 38-39, 107-112]
		Methyl cellulose (E461)	Thickner, emulsifier and gelling agent	
	Chitosan		Gelling agent Antimicrobials	[33, 47-50, 53-56, 113]
		Arabic gum (E414)	Emulsifier	
	Gum	Guar gum (E412)	Thickner	[58-63, 114]
		Xanthan gum (E415)	Thickner	
	Pectin (E440)		Gelling agent	[67-69, 115]
Protein	Starches		Thickner and gelling agent	[76-79, 116-117]
		Bovine gelatin	Gelling agent	[88, 92, 94]
	Gelatin	Fish gelatin	Gelling agent	[92, 95, 118-119]
		Pig gelatin	Gelling agent	[88, 92-93]
	Whey Protein			[96, 102-103, 105-106]

Whey protein isolate produces totally water-soluble coatings but heat denatured solutions of whey protein isolate produce coatings in which the protein is insoluble [96]. Industrially produced whey protein concentrate (WPC) have a protein content between 25 and 80% whereas the whey protein isolate (WPI) have protein content above 90% [97]. Whey proteins have been the subject of intense investigation over the past decade or so [98]. Whey proteins can produce transparent, flexible, colourless and flavourless films, with a poor moisture barrier [99-100]. Moreover protein-based films possess a good aroma barrier [101], and low oxygen permeability [102]. Whey protein films produced without addition of any plasticizers are very brittle therefore the addition of plasticizers provides flexibility to the films but also increases their water vapour permeability. Increased concentration of plasticizers in edible whey protein films decreased tensile strength and Young's modulus, but increased elongation [103]. The water vapour permeability can also be improved by the incorporation of hydrophobic materials such as lipids [104]. However, the incorporation of lipid materials into edible film formulations into films could adversely affect the sensory characteristics of films. The incorporation of antimicrobial agents into edible coating formulations is needed to obtain stronger inhibitory effect against microbial growth (See paragraph 2.3). Optimization of edible whey protein films containing sorbitol, beeswax and potassium sorbate concentrations for mechanical and optical properties was reported by Ozdemir and Floros [105]. Recently WPI and mesquite gum (MG) are showed totally compatible to form solutions and films. Indeed, the incorporation of different amounts of MG on the WPI allowed to obtain composite films with improved flexibility without increasing plasticizer content and consequently without affecting negatively others characteristics such as water vapour permeability [106].

2.2. Plasticizers

Edible films and coatings need to have good elasticity and flexibility, a low brittleness, a high toughness and to prevent cracking during handling and storage [120]. Therefore, plasticizers of low molecular weight (non volatile) are typically added to hydrocolloid film forming solutions to modify the flexibility of edible films. Plasticizers with characteristics such as small size, high polarity, more polar groups per molecule, and greater distance between polar groups within a molecule generally impart greater plasticizing effects on a polymeric system. Indeed, they act by increasing the free volume or in other word by decreasing intermolecular attractions between adjacent polymeric chains by reducing hydrogen bonding between polymers chains [121]. Generally, plasticizers are required for polysaccharides (or proteins) based edible films. Their amount added into hydrocolloid film-forming preparations varies between 10% and 60% by weight of the hydrocolloid. The most commonly used plasticizers are: polyols (propylene glycol [122], glycerol [121], sorbitol [123], polyethylene glycol [124]), oligosaccharides (sucrose [125]) and water. In fact, water is a ubiquitous natural diluent, which plasticizes and/or antiplasticizes some films depending on the amount sorbed onto the films matrix [126]. Thus, film moisture content, as affected by the relative humidity of the surrounding environment, largely affects film properties [127]. However, addition of plasticizers may cause significant changes in the barrier properties of the films, *e.g.* increase film permeability to gases (or aroma compounds), decrease the ability of the film to attract water or decrease the tensile strength [81, 128].

2.3. Antimicrobial, Antioxidant and Other Functional Agents

The new generation of edible films and coatings is being especially designed to increase theirs functionalities by incorporating natural or chemical antimicrobial agents, antioxidants, enzymes or functional ingredients such as probiotics, minerals and vitamins [5-7].

Antimicrobial and antioxidant coatings have advantages over direct applications of the antimicrobial or antioxidant agents because they can be designed to slow down the diffusion of the active compounds from the surface of the food. Edible coating can enhance the nutritional value of foods by carrying basic nutrients and / or nutraceuticals in its matrix. The sensory quality of coated products can be also improve if in the matrix flavour and pigment are added.

2.3.1. Antimicrobial Agents

Common chemical antimicrobial agents used in food systems, such as benzoic acid, propionic acid, sodium benzoate, sorbic acid, and potassium sorbate, may be incorporated into edible films and coatings to inhibit the outgrowth of both bacterial and fungal cells [129]. In fact, the antimicrobial compounds, when establish contact with food, inhibit the growth of microorganisms present in the surface [130]. However, due to the health concerns of consumers related to chemical preservatives, the demand for natural foods has spurred the search for natural biopreservatives in edible films-forming preparations. Chitosan, which has so far attracted numerous works regarding its antifungal and antibacterial properties is discussed in the above paragraph. The most frequently used biopreservatives for antimicrobial are lysozyme and nisin. These biopreservatives are shown bactericidal effect on Gram (+) bacteria but they can also become effective on Gram (-) bacteria if they are combined with chelating agents such as ethylene diamine tetraacetic acid (EDTA) [131]. Common other biopreservatives that may be used in edible films and coatings are bacteriocins, such as lacticin [132] and pediocin [133] and antimicrobial enzymes, such as chitinase and glucose oxidase [134]. Mitrakas et al. [135] are reported that during the storage at 5 °C, the use of calcium alginate film with antimicrobial as potassium sorbate results in a significant delay in microbial growth on the cylindrical samples of potato, which may help extend the shelf life of an osmo-dehydrated, refrigerated product. Recently, different studies have also been conducted on the use of the enzyme lactoperoxidase (LPS) in edible films and coatings [136]. LPS, that has antimicrobial and antioxidant properties, is found in the mammary, salivary, and lachrymal glands of mammals and in their respective secretions [137]. LPS shows bactericidal effect on Gram-negative bacteria and bacteriostatic effect on Gram-positive bacteria [138]. LPS also presents antifungal and antiviral activities [137]. Natural antimicrobial compounds have also been incorporated into protein or polysaccharide-based matrices: rosemary, and garlic essential oils [139] oregano, lemon grass, or cinnamon oil at different concentrations [66].

2.3.2. Antioxidant Agents

Generally, edible films and coatings have as additive lipids to reduce water vapour transfer due to its hydrophobic character. Therefore, incorporation of antioxidants in edible films-forming preparations to increase product shelf life by protecting foods against oxidative rancidity, degradation, and discoloration is became very popular [11]. The reader can also note that most antimicrobial compounds have antioxidant properties. Natural antioxidants such as phenolic compounds, vitamins E and C in place of synthetic antioxidants are extensively used in edible films. For example, the antioxidants citric and ascorbic acid were incorporated into methylcellulose-based edible coatings in order to control oxygen permeability and reduce Vitamin C losses in apricots during storage [140]. Xanthan gum coatings mixed with vitamin E enhanced nutritional quality and improved the surface colour of peeled baby carrots [59]. Carrageenan or whey protein coatings with added antibrowning agents, ascorbic acid or citric acid, effectively maintained colour of apple slices during storage and extended the shelf-life of minimally processed apple slices by 2 week when stored in packed trays [29]. Recently, banana slices coated with calcium chloride, ascorbic acid and cysteine prevented product weight loss and increase of polyphenol oxidase activity during the 5 days of storage [141]. Essential oils exhibit a wide range of biological effects, including antioxidant and antimicrobial properties. In particular they exhibit antibacterial activity against food borne pathogens. The phenolic components, such as carvacrol, camphor, eugenol, linalool and thymol are most active and appear to act principally as membrane permeabilisers [142]. Oregano or garlic essential oil added whey protein isolate edible films exhibited larger inhibitory zones on *Staphylococcus aureus*, *Salmonella enteritidis*, *Listeria monocytogenes*, *Escherichia coli* O157:H7 and *Lactobacillus plantarum* as compared to rosemary essential oil incorporated into film-forming preparations [139]. Furthermore, antimicrobial films prepared by incorporating oregano oil into sorbitol-plasticized whey protein isolate films was effective in increasing the beef's shelf life by a factor of 2, while minimising changes in colour [143]. Maizura *et al.* [144] reported that starch-alginate edible films containing lemon grass oil are effective in inhibiting the growth of Escherichia coli O157:H7 at all levels. However, generally all these studies showed their efficacy in vitro against various microorganisms but they were not tested with real foods. Moreover, there is a lack of available information about their possible impact on the aroma and flavour of the coated products.

2.3.3. Other Functional Agents

Edible films and coatings a can be used as a host for carring basic nutrients and/or nutraceuticals that are lacking or are present in only low quantity in fruits and vegetables. Recently, a study reported by Tapia *et al.* [23] showed that alginate- and gellan-based edible films seem to be efficient in supporting *Bifidobacterium lactis* Bb-12 on fresh-cut apple and papaya. Indeed, values greater than 10^6 CFU/g of *Bifidobacterium lactis* Bb-12 were maintained for 10 days during refrigerated storage of fresh-cut apple and papaya. However, the bifidus-containing films were more permeable to water vapour than the corresponding cast films. The use of edible films and coatings based on fructooligosaccharides (FOS) showed benefits of FOS like their specific physiological effect on human digestive track especially growth stimulation of beneficial bifidobacteria in the digestive tract, low calorific value and non-cariogenicity [145].

The development of nutritionally fortified edible films and coatings strongly depends on the type of carriers (film-forming materials) and the type and concentration of nutraceuticals added into the film-forming solutions [146]. Nutraceuticals as calcium, zinc, or vitamin E incorporated in a chitosan-based film matrix provided alternative ways to fortify fresh fruits and vegetables [146]. Vitamin E is believed to protect the human body against certain types of cancers, cardiovascular disease, cataracts, arthritis, diabetes, and Alzheimer's disease whereas calcium is a common intracellular messenger, a cofactor for extracellular enzymes and proteins, and also is essential for the development of bone and teeth [147]. The integration of calcium and vitamin E into edible coating around baby carrots improve the nutritional value of this food. This result has also been successfully applied on fresh and frozen strawberries and raspberries [54]. Flavour and pigments agents may also be added to edible coatings to improve the sensory quality of coated foods. However, there is a lack of available information on these applications in the literature.

3. Properties of Hydrocolloid Edible Films and Coatings

3.1. Thickness

The thickness of edible films and coatings is an important parameter since it directly affects the biological properties and the shelf life of the coated food. The effectiveness of edible films and coatings for protection of food depends primarily on controlling the spreading of the coating solutions, which affect the thickness of the film.

When a drop of a liquid A is placed on a solid B (for example fruit peel´s or a vegetable) the liquid can wetting totally or partially the solid B. The parameter that distinguishes this two states is the spreading coefficient [148]:

$$S_{A/B} = [E_{solid}]_{dry} - [E_{solid}]_{wet} = \gamma_{solid/air} - \gamma_{solid/liquid} - \gamma_{liquid/air} = \gamma_{liquid/air}(\cos(\theta) - 1) \quad (1)$$

If $S_{A/B} > 0$, the liquid drop spreads completely in order lower its energy, i.e. the contact angle $\theta = 0$ (See Figure 1.B).

If $S_{A/B} < 0$, the liquid drop does not spread but forms at equilibrium a spherical cap resting on the solid with a contact angle θ. For $\theta < 90°$ the liquid is said to be "mostly wetting" and for $\theta > 90°$ "mostly non-wetting" (See Figure 1.C). Furthermore, by projecting the equilibrium forces onto the solid plane, one obtains Young's relation: $\gamma_{liquid/air}\cos(\theta) = \gamma_{solid/air} - \gamma_{solid/liquid}$.

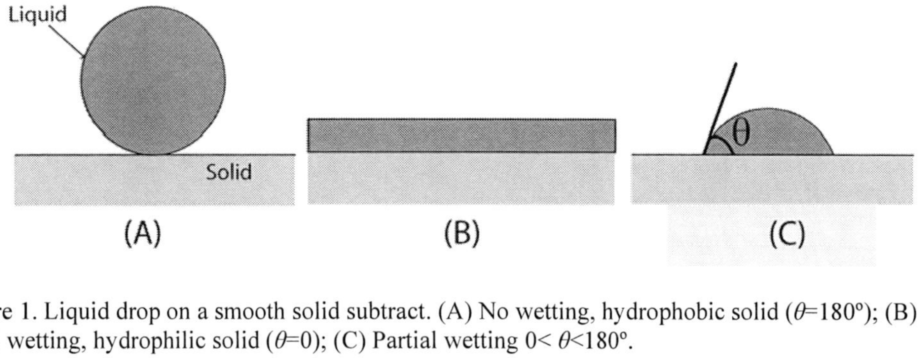

Figure 1. Liquid drop on a smooth solid subtract. (A) No wetting, hydrophobic solid (θ=180°); (B) Total wetting, hydrophilic solid (θ=0); (C) Partial wetting 0< θ<180°.

However, it is difficult to estimate $S_{A/B}$. Indeed, although the interfacial tension between liquid and air $\gamma_{liquid/air}$, can be measured accurately, there are no direct methods for measuring $\gamma_{solid/air}$ and $\gamma_{solid/liquid}$.

A commonly used approach is to express any surface tension γ (usually against air) as a sum of components due to dispersion forces (γ^d) and polar forces(γ^p) [149]: $\gamma=\gamma^d+\gamma^p$. For example, at 25°C an air-water interface has a $\gamma^p_{liquid/air}=50.4$ mJ.m^{-2} and a $\gamma^d_{liquid/air}=21.5$ mJ.m^{-2}.

The interfacial tension between two phases α and β is then expressed in terms of these two components for each phase:

$$\gamma_{\alpha\beta}=\gamma_\alpha+\gamma_\beta-2\sqrt{\gamma_\alpha^d\gamma_\beta^d}-2\sqrt{\gamma_\alpha^p\gamma_\beta^p} \tag{2}$$

At this point it is convenient to introduce two important works, namely, the ''work of cohesion'' and the ''work of adhesion''. The work of cohesion, W_c, is defined as the reversible work required to separate two surfaces of unit area of a single material with surface tension γ whereas W_a, the work of adhesion is defined as the reversible work required to separate unit area of interface between two different materials (See Figure 2). The works are given by:

$$W_c=2\gamma_{liquid/air} \tag{3}$$

$$W_a=\gamma_{solid/air}+\gamma_{liquid/air}-\gamma_{solid/liquid} \tag{4}$$

The balance between W_c and W_a allows to determine the spreading coefficient $S_{A/B}$, according to:

$$S_{A/B}=W_a-W_c \tag{5}$$

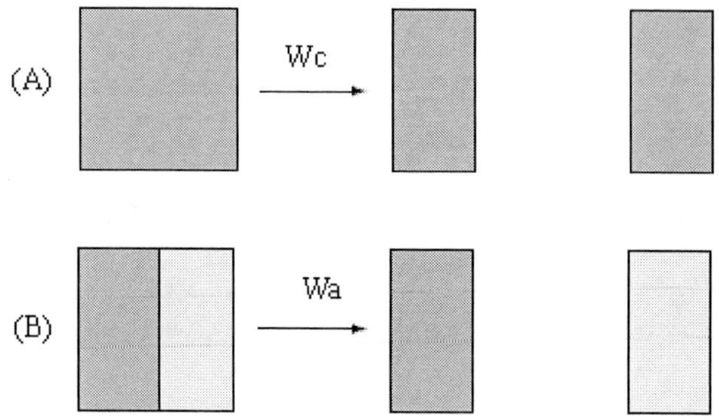

Figure 2. (A) Work of cohesion; (B) Work of adhesion.

Thus, using Eq. 2 the work of adhesion can be express:

$$W_a = W_a^d + W_a^p = 2\left(\sqrt{\gamma^d_{solid/air}\gamma^d_{liquid/air}} + \sqrt{\gamma^p_{solid/air}\gamma^p_{liquid/air}}\right) = \gamma_{liquid/air}\left(\cos(\theta)+1\right) \quad (6)$$

Rearranging Eq. 6, yields:

$$\frac{1+\cos(\theta)}{2}\frac{\gamma_{liquid/air}}{\sqrt{\gamma^d_{liquid/air}}} = \sqrt{\gamma^p_{solid/air}}\sqrt{\frac{\gamma^p_{liquid/air}}{\gamma^d_{liquid/air}}} + \sqrt{\gamma^d_{solid/air}} \quad (7)$$

It is possible to obtain $\gamma^p_{solid/air}$ and $\gamma^d_{solid/air}$ by plotting the experimental data in a graph of $\frac{1+\cos(\theta)}{2}\frac{\gamma_{liquid/air}}{\sqrt{\gamma^d_{liquid/air}}}$ vs $\sqrt{\frac{\gamma^p_{liquid/air}}{\gamma^d_{liquid/air}}}$. The reader can find various data $\gamma^p_{liquid/air}$ and $\gamma^d_{liquid/air}$ in [150]. Since $\gamma_{solid/air}$ is the sum of $\gamma^p_{solid/air}$ and $\gamma^d_{solid/air}$, the parameter $\gamma_{solid/liquid}$ can be determined by the Eq. 2. Thus, the important spreading coefficient $S_{A/B}$ can be calculated.

For a solution of pure starch or carrageenan coating a strawberry fruit S~-45mJ.m^{-2} whereas for a chitosan solution S~-38mJ.m^{-2} [33]. In all cases, the thickness of the edible film will be thick since the liquids do not spread.

Howvever, films and coatings must not exceed a critical thickness to no reduce drastically the internal O_2 concentration and to increase CO_2 concentration from anaerobic fermentation. Coating thickness depends on the physical properties of the solution (such as density, viscosity, and surface tension) but also of the method used in the film formation: dipping, spraying solvent casting, skimming, etc. [109]. The methods used for the film formation are discussed in the paragraph 4.

The method to measure the film or coating thickness can be classified in two groups: the contact method and the non-contact method. The non-contact method is more effective than the contact method since this technique is non-destructive for the sample. Among the contact methods, due to its simplicity, the direct metric measurement is most frequently used. The coating is peeled from the food surface and measured using a micrometer. Usually, a film has thickness inferior to 0.25mm or 0.001 inch. A thickness greater than 0.25mm is called a sheet.

Common non-contact methods to analyze coating thickness directly from the surface of coated foods include optical or scanning electron microscope, confocal Raman microspectrometry (CRM), Fourier transform (FT)-Raman spectrometer and surface enhanced Raman scattering (SERS) [151-154]. A representative example of the non-contact method is the optical method, which is based mainly on light interference [155]. The interference method can usually measure only a relative thickness by means of interference fringe. However, a method for an absolute measurement has recently been developed by means of multi-wave laser interference at many incident angle [153]. Two methods are proposed: one can be used when an index of refraction of the film, a wavelength λ; and two successive angles of incidence at which the sinusoidal light intensity has minimum values, are known (Method I), and another can be used without an index of film refraction when three successive angles of incidence and a wavelength are known (Method II). The smallest measurable thickness is 1.43 λ for Method I, and 2.5 λ for Method II. The largest measurable thickness is about 100 λ for both methods [153]. The simultaneously measure of the thickness and of the surface profile of single film as well as multiple-layer film can be obtained using the interferometry in combination with confocal microscopy [156-157]. However, when a food is coated by dipping, different theoretical approaches can be used to estimate film thickness from coating solution properties [158-159]. Cisneros-Zevallos and Krochta [109] used hydroxypropyl methylcellulose to coat Fuji apples by dipping. They reported that coating thickness varied with viscosity, concentration, density, and draining time of the biopolymer solution.

3.2. Mechanical Properties

Typically, the mechanical resistance of hydrocolloidal films was studied according to three parameters: tensile strength (*TS*), Young's modulus (*Y*) and percent of elongation at break (*E*). Tensile strength, elongation properties, Young's modulus are determined from stress–strain curves according to the norm of the American Society for Testing and Materials (ASTM).

In Figure 3, a strain-stress curve is presented. TS is calculated by dividing the maximum load on the film before failure (σ_Y) by the initial cross sectional area (*S*). *E* is expressed as the percentage of change in the original length of film before breaking and gives a measure of the extensibility of the films (ε_B). *Y* is experimentally determined from the slope of a stress-strain curve created during the tensile test ($Y = \frac{\sigma_L}{\varepsilon_L}$) which is a measure of film stiffness [160].

In the literature some results of mechanical properties (*TS*, *E*, *Y*) of biopolymer and synthetic films are reported [160-161]. However, due to potential difference in manipulation or testing procedures, it was very difficult to compare the results. Values of *TS* and *E* are reported in the Table 2 for various polymer films. For polysaccharides films, *T*S values vary between 10 and 100 MPa whereas the *E* values range between 1 and 80%. Generally, protein films have a *TS* lower while theirs *E* are higher [127]. Comparison with common synthetic polymers used in films-forming preparations, *TS* values approach those of polysaccharides films: 9 to 20 Mpa for low-density polyethylene (LDPE) and 10 to 60 MPa for high-density polyethylene (HDPE). However, LDPE and HDPE have higher *E* values (up to 1000%), which are greater than those of protein films. Thus, it is necessary to add plasticizers into polysaccharides or proteins film-forming preparations to increase *E* values. However, increasing concentration of these plasticizers resulted in decreased tensile strength (*TS*). For example, films produced with low molecular weight chitosan at 3% w/w in 1% acetic acid, with glycerol as a plasticizer at 0.25 and 0.50 mL/g of chitosan, were reported to have tensile strength (TS) of 15 to 35 MPa and percent elongation at break (%E) of 17 to 76 [162].

Lipids incorporated into film-forming preparation increase the hydrophobicity of edible films but also generally cause a weakening of film strength [37-163].

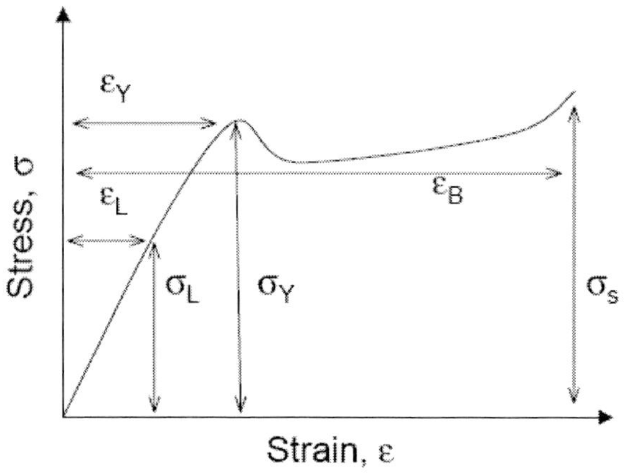

Figure 3: Stress–Strain behaviour of a hydrocolloid film. The explanation of all symbols are given in the text.

Table 2. Tensile strength (*TS*) and elongation at break (*E*) for various polymer films. All data are issue of [164] excepted a) [165], b) [166] and c) [2]

Polymer		Tensile Strength (TS) Mpa	Elongation at break (E) %
Cellulose	acetate	44 - 64	20 - 50
	Hydroxypropyl cellulose [c)]	14 - 15	30 - 205
	Methylcellulose [c)]	55 - 60	10 - 20
Collagen		1 - 70	10 - 70
Chitsan [a)]		10 - 100	20 - 80
Gelatin	Type A	25 - 85	7 - 22
	Type B	28 - 140	7 - 12
Starch	Waxy maize	34.9	1.7
	Tapioca	44	3.4
	Potato	44.2	3.1
	Wheat	46.3	2.9
	Corn (A)	46.4	2.5
	Corn (B)	46.7	3.2
Polyethylene	linear low density	9 - 20	100 - 1200
	linear high density	10 - 60	400 - 1800
zein [b)]		3 - 4	50 - 120

3.3. Barrier Properties

When an edible film is applied on the food, the surface gas concentration may change during storage. The chemical composition of the food surface is dynamic and these changes occur mainly due to food metabolism, microbial respiration, gas solubility and permeability of the edible film. Food and microbial metabolism are responsible for consumption of oxygen and CO_2 production. The microbial activity can affect the composition of oxygen and CO_2, however this is only significant when it has reached the end of shelf life and deterioration is evident [167]. Therefore, the efficiency of an edible coating strongly depends on its barrier properties to gas, water vapor, aroma and oil which in turn depends on the chemical composition and structure of the coating-forming polymers, the characteristics of the product, and the storage conditions.

3.3.1. Gas Barrier

Coatings are used to create a controlled or modified atmosphere inside the fruits and vegetables that will delay ripening and senescence in a manner similar to the most costly controlled or modified atmosphere storage. The transport of gas molecules in an edible film consists of 3 steps (See Figure 4): adsorption of the permeant onto the edible film surface, diffusion of the permeant from one side of the edible film to the other side, and then the desorption of the permeant from the edible film.

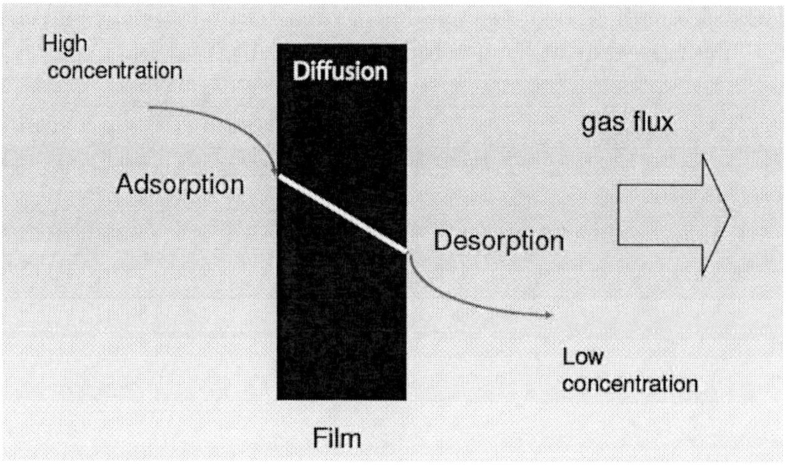

Figure 4. Transport of gas molecules in an edible film.

Diffusion of simple gases (e.g., 0_2, N_2, CO_2) through a film generally obeys Fick's law, i.e. [168]:

$$J=-D\frac{\partial C}{\partial x}=D\frac{C_1-C_2}{l} \tag{1}$$

Where: l is the thickness of the edible film, D the diffusion coefficient and C the concentration.

Moreover, the Henry's law says that the concentration of the gas can be expressed as:

$$C=\Sigma p \tag{2}$$

where Σ is the solubility coefficient of the gas in the membrane, and p is the pressure of the gas. The solubility, diffusivity, and kinetic diameter data of N_2, O_2 and CO_2 gas in water at 25°C are presented in the Table 3. The kinetic diameter of CO_2, O_2 and N_2 are very close. However CO_2 is much more permeable (product of diffusivity by the solubility coefficient) than the two other gases because of its much higher solubility in moist films due to presence of water in the films. Furthermore, the solubility and diffusivity of O_2 are larger than N_2. Therefore, in the case of a polar film matrix, these characteristics attribute to the permeability in the order of $CO_2 > O_2 > N_2$.

Table 3. Solubility, diffusivity, and kinetic diameter data of N_2, O_2 and CO_2 gas in water at 25°C [169]

Gas	Solubility Coefficient (mol.(L.atm)$^{-1}$)	Diffusivity 10^{-9}(m^2/s)	Kinetic diameter (10^{-10} m)
N_2	6.1×10^{-4}	2	3.65
O_2	1.3×10^{-3}	2.42	3.46
CO_2	3.4×10^{-2}	1.91	3.3

Furthermore, what happens inside the edible film is not relevant to the design, therefore the concept of permeability can be obtained by combining the Fick's first law (Eq. 1) with Henry's law (Eq.2):

$$J = D\Sigma \frac{p_1 - p_2}{l} \quad (3)$$

The product of diffusion coefficient and solubility coefficient is equal to the permeability coefficient, $\Pi = D\Sigma$, which characterizes the intrinsic permeability of the edible film.

For a i-multilayer edible-film, the flux J will be described by:

$$J = \frac{p_1 - p_2}{\sum_i \frac{l_i}{\Pi_i}} \quad (4)$$

It is important to point out that in respiring-foods the permeability of the edible-film is extremely relevant since foods consume O_2 to produce an equilibrium condition. Oxygen uptake by a food often results in deleterious reactions which affect its flavor, nutritional quality and acceptability. The oxygen uptake for the food (R) is usually estimated by a Michaellis Menten model [170]:

$$R = R_{max}\left(\frac{P_{O_2}}{K + P_{O_2}}\right) \tag{5}$$

Thus, it is possible predict easily the equilibrium partial pressure of oxygen when $R = J$ with the following equation:

$$V_F R_{max}\left(\frac{P_{O_2}}{K + P_{O_2}}\right) = \frac{A_F}{\sum_i \frac{l_i}{\Pi_{O_2,i}}}\left(p_{O_{2},AIR} - p_{O_2}\right) \tag{6}$$

where

V_F and A_F are volume of food and area of food, respectively.

In addition, this equation can be written with dimensionless numbers (α and β):

$$\alpha\left(\frac{\beta}{1+\beta}\right) = \left(\beta_{AIR} - \beta\right) \tag{7}$$

where

$$\alpha = \frac{V_F}{A_F} R_{max} \sum_i \frac{l_i}{\Pi_{O_2,i}}\ ,\quad \beta = \frac{P_{O_2}}{K} \text{ and } \beta_{AIR} = \frac{P_{O_{2,AIR}}}{K}$$

Therefore the partial pressure of oxygen in contact with the food will be minor that air (<0.21 atm):

$$\beta = \frac{1}{2}\left(\left(\beta_{AIR} - \alpha - 1\right) + \sqrt{\left(\beta_{AIR} - \alpha - 1\right)^2 + 4\beta_{AIR}}\right) \tag{8}$$

This dimensionless can be used to estimate the reduction in the oxygen uptake in comparison with a food respiring without edible-film (in contact with air):

$$\left(\frac{R}{R_{AIR}}\right) = \frac{R_{MAX}\left(\frac{p_{O2}}{K + p_{O2}}\right)}{R_{MAX}\left(\frac{p_{O2,AIR}}{K + p_{O2,AIR}}\right)} = \frac{\beta(1+\beta_{AIR})}{\beta_{AIR}(1+\beta)} \tag{9}$$

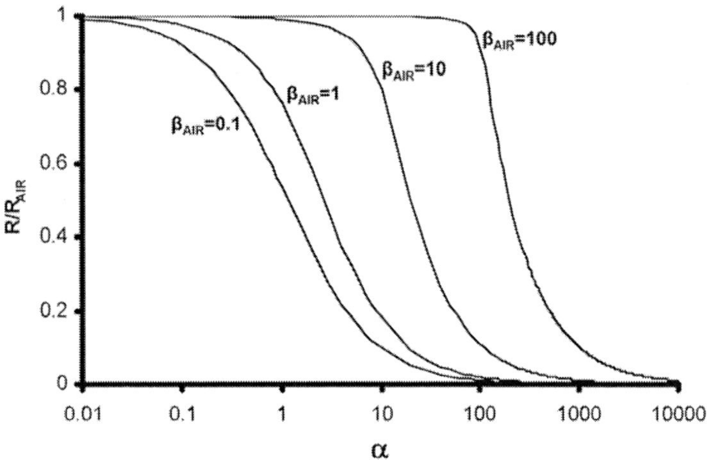

Figure 5 shows the relationship between the reduction of respiration (R/RAIR) and the edible-film to different □ and β values.

Figure 5. Reduction of respiration (R/R_{AIR}) and the edible-film to different α and β values.

It is important to point out that in respiring-foods foods consume O_2 to produce CO_2, which must be released to the environment to produce an equilibrium condition [171]. If the CO_2 concentration is higher, this gas could produce an inhibition on respiration. The uncompetitive enzymatic model is usually used to predict this effect:

$$R = R_{MAX}\left(\frac{p_{O2}}{K + p_{O2}\left(1 + p_{CO2}/K_I\right)}\right) \tag{10}$$

The CO_2 production is obtained with the respiration quotient (Q). Typically, Q varies between 0.8 and 1.5 [170]. Then, to predict the new equilibrium conditions is necessary to solve a 2 x 2 system equations:

$$V_F R_{\mathrm{MAX}}\left(\frac{p_{O2}}{K + p_{O2}(1 + p_{CO2}/K_I)}\right) = \frac{A_F}{\sum \frac{L_i}{\wp_{O2,i}}}(p_{\mathrm{O2,AIR}} - p_{\mathrm{O2}})$$

$$V_F \mathrm{Q} R_{\mathrm{MAX}}\left(\frac{p_{O2}}{K + p_{O2}(1 + p_{CO2}/K_I)}\right) = \frac{A_F}{\sum \frac{L_i}{\wp_{CO2,i}}}(p_{\mathrm{CO2}} - p_{\mathrm{CO2,AIR}})$$

3.3.2. Water Vapour Permeability (WVP)

In the case of water vapour diffusing through a polar film matrix, J will increase as Δp is moved up in the vapour pressure spectrum. This occurs because the adsorption isotherms for hydrophilic polymers are sigmoïdal, causing Σ to increase significantly when the vapour pressure rises. The greater amount of adsorbed water plasticizes more the film causing the diffusion coefficient D to increase. Hence, Fick's law is not strictly obeyed and D is not a constant for water vapour migration through hydrophilic film matrices [10].

The WVP is calculated by the following equation [172]:

$$WVP = \frac{\Delta m}{\Delta t}\frac{l}{A\Delta P} \tag{1}$$

The WVP correspond to the amount of water vapour (Δm) transferred through a film of area (A), of thickness l during a finite time (Δt). The ΔP is the vapour pressure difference across the film.

Published WVP values for edible films should be cautiously compared because of the difference in plasticizers, temperature, relative humidity gradient, etc. The increase of the film permeability with thickness indicates a water affinity of the film that could be attributed to hydrophilic compounds. Then it seems that the influence of thickness varies with the lipid nature or the film composition [37].

3.3.3. Aroma Barrier

An understanding of sorption between flavour compounds and edible film matrix requires knowledge of the chemical and physical structures of both the flavour compound and the polymer. [173]. Studies on the aroma permeability of hydrocoloid edible films are very scarce. The flavour diffusion coefficient D in edible films are in the order of 10^{-13} m^2 s^{-1} [1, 173]. Flavour absorption may affect the flavour of a food as well as the mechanical properties of the edible film, such as tensile and heat seal strength and permeability, or cause delamination of the polymeric structure [174]. Finally, an example of study aroma permeability was reported by Miller and Krochta [175]: a whey protein film was better than Vinylidene chloride copolymer (co.VDC) and was comparable to ethylene vinyl alcohol copolymer (EVOH) as barrier to limonene (citrus arome).

3.3.4. Oil Barrier

Film coatings may become a good alternative to reduce oil uptake during frying. Several hydrocolloids with thermal gelling or thickening properties, such as proteins or carbohydrates, were tested to reduce oil migration [176-178]. Hydrocolloid films are excepted to be oil-resistant due to their substantial hydrophilicity. García *et al.* [178] reported that, in deep-fat frying potato strips and dough discs, methylcellulose coatings are more effective in reducing oil uptake than hydroxypropylmethyl cellulose. Sorbitol addition was necessary to maintain coating integrity and improve barrier properties. Some potentiality and applications of edible coatings as an oil barrier are presented by García *et al.* [179].

3.4. Optical Properties: Gloss, Transparency and Colour

In the last years, research efforts have been made to improve mechanical and barrier properties of edible films and coatings but few studies have focussed on optical properties such as colour, gloss and transparency. Optical properties are characteristics of surfaces which are detected by human vision affecting some crucial aspects of food quality. The internal and surface film microstructure plays an important role in optical properties of the film. Indeed, the intensity of light reflected by the coated food is determined both by the light directly reflected at the interface between air and the coated food surface (specular reflection), and by both the light re-emitted out of the surface in all directions after penetrating into the coating of the food and scattering internally (indirect reflection). Specular reflection is related to the gloss of the coating. According to the ASTM standard D-523, it is measured at 20, 60 and 85° angles from the normal to the coating surface using a flat surface gloss meter. Polished black glass with a refractive index of 1.567 possesses a specular gloss of 100%. The gloss of edible coatings is affected by coating microstructure in particular, by the type and level of surfactant, the distribution and the size particle diameter of dispersed phase, the relative humidity, the storage time and the surface roughness [180-183].

The transparency of films was measured through the surface reflectance spectra in a spectrocolorimeter. A good method to determine the transparency is the Kubelka–Munk theory [184]. This theory models the reflected and transmitted spectrum of a coloured layer based on a material dependent scattering and absorption function. The assumptions of this theoretical model are:

(1) A translucent colourant layer on the top of an opaque background,
(2) Within the colourant layer both absorption and scattering occur,
(3) Light within the colourant layer is completely diffuse,
(4) Geometry should be diffuse illumination and diffuse collection.

The light flux passes through the layer and it is affected by the absorption coefficient *K* and the scattering coefficient *S*. The ratio known, as the Kubelka-Munk function, shows that the reflectance of an opaque layer depends only on the ratio of *K* to *S* and not on their individual values: $\frac{K}{S} = \frac{(1 - R_\infty)^2}{2R_\infty}$

where

R_∞ is the reflectance of an infinitely thick layer of the material: $R_\infty = a - b$. With $a = \frac{1}{2}\left(R + \frac{R_0 - R + R_g}{R_0 R_g}\right)$ and $b = \sqrt{a^2 - 1}$. R_0 is the reflectance of the layer with an ideal black background; R the reflectance of the sample layer backed by a known reflectance R_g.

The internal transmittance in the films is quantified by applying: $T_i = \sqrt{(a - R_0)^2 - b^2}$.

The color of edible films are evaluated by a colorimeter or a spectrophotometer. The parameters L (Luminosity), a (Red–Green) and b (Yellow–Blue) are the edible film colour values in the Cielab scale. A white disk (L_0 , a_0 and b_0) was used like standard colour. From these values, Chrome ($C=\sqrt{a^2+b^2}$), Hue ($H=\arctan\left(\frac{b}{a}\right)$), Whiteness Index ($WI=100-\sqrt{(100-L)^2+a^2+b^2}$) and difference of colour $\Delta E=\sqrt{(\Delta L)^2+(\Delta a)^2+(\Delta b)^2}$ can be calculated. Where $\Delta L=L-L_0$, $\Delta a=a-a_0$ and $\Delta b=b-b_0$.

3.5. Thermodynamic Parameters Associated to Sorption Phenomenon in Edible Films

The interactions of between the sorbant and solvent can define the efficiency of the edible film to control the gas permeability and therefore the stabilities of the food. The relationship between the total moisture content and the water activity (A_w), over a range of values, at constant temperature, yields the sorption isotherm when expressed graphically [185]. When obtained at different temperatures, the energy associated to moisture interactions at low water vapor pressures can be estimated. Various mathematical models have been developed to describe sorption isotherms. GAB equation [186-187] (Eq. 1) has proven to be one of the more versatile to represent sorption values over a wide range of A_w [188-191]. Furthermore its parameter can give information related to the sorption energies involved [198].

$$M=\frac{m_o CK\frac{p}{p_o}}{\left(1-K\frac{p}{p_o}\right)\left(1-K\frac{p}{p_o}+CK\frac{p}{p_o}\right)} \quad (1)$$

where

p/p_o = partial water vapor pressure in the food system, m_o = monolayer value (all sorption sites occupied by one molecule), C = constant (related to the net heat of sorption at the monolayer) K = constant (related to the heat of sorption at the multilayer).

The effect of temperature on sorption isotherms can be represented by the Clausius-Clayperon relation [192] (Eq. 2). Graphically, this equation produces the parameter q_{st}, net isosteric heat of sorption or the difference between the total heat of sorption (Q_{st}) and the heat of vaporization of water [193]. This parameter accounts for the excess in energy for removal of water [192].

$$q_{st} = -R\frac{\partial \ln(p/p_o)}{\partial(1/T)} \tag{2}$$

where

q_{st} = net isosteric heat of sorption ($Jg^{-1}mol^{-1}$), p/p_o = partial water vapor pressure in the food system, T= temperature (K) and R = universal gas constant (8,314 J K^{-1} mol^{-1}).

Further analysis of the sorption data can help to elucidate information related to the thermodynamic involved in sorption phenomenon. It can allow the interpretation of the experimental results in accordance with the statement of the theory [194] giving insight of the energies involved in the interactions between sorbant and sorbates. These functions include, with the heat of sorption, the variations of enthalpy and entropy. It has been suggested that enthapy change (ΔH) provides indications of the energy variation occurring on mixing water molecules with sorbents during sorption process [195]. Variations in entropy (ΔH) might be associated with spatial arrangements occurring in the water-matrix interface in a well defined state [196].

The enthalpy-entropy compensation theory can be applied to evaluate the physical and chemical processes such as sorption phenomenon [194]. This theory states that to reduce free energy during sorption a linear function occurs relating the variations of enthalpy to variations in entropy in the system [194]. The enthalpy-entropy compensation theory has been used to investigate the sorption phenomenon of various biological materials such as cottonseed [194], tropical woods [196] and starch based materials [195].

When complex polymeric structures are related chemically, changes in the enthalpy and entropy can occur. If there is an increase in the intermolecular forces, it can generate a reduction in the configurational freedom of the system [197]. The relation between the difference of enthalpy and the differences in entropy during the sorption process can be mathematically represented by the following equation [195]:

$$\ln Aw = -\Delta H/(RT) + (\Delta\Delta / R) \tag{3}$$

where

A_w = water activity in the food system, T= temperature (K) and R = universal gas constant (8,314 J mol^{-1} K^{-1}). The enthalpy-entropy compensation theory should follow a linear relationship as indicated by Eq. 4 that considers the variation in enthalpy, entropy and Gibbs free energy [193].

$$\Delta S = \frac{(\Delta\Delta + \Delta G)}{T_\beta} \tag{4}$$

where

β= isokinetic temperature (K), ΔH is the change of enthalpy, ΔS is change in entropy and ΔG is the variation in the Gibbs free energy. In order to validate the existence of a true compensation during the sorption process [199-200] proposed a comparison between T_β and the harmonic temperature (T_{hm}), defined in the next function:

$$T_{hm} = \frac{n}{\sum_{i=i}^{n}(1/T)} \qquad (5)$$

Where: n= number of isotherms and T = temperature of equilibrium (K). The parameter T_β can be estimated in terms of an interval at (1-α)100% confidence using the following equation:

$$\beta_{m-2,\alpha/2}\sqrt{Var(T_\beta\,)} \qquad (6)$$

where

$$-\overline{\Delta S}\,)^2 \qquad (7)$$

and

$$-\overline{\Delta S}\,)^2 \qquad (8)$$

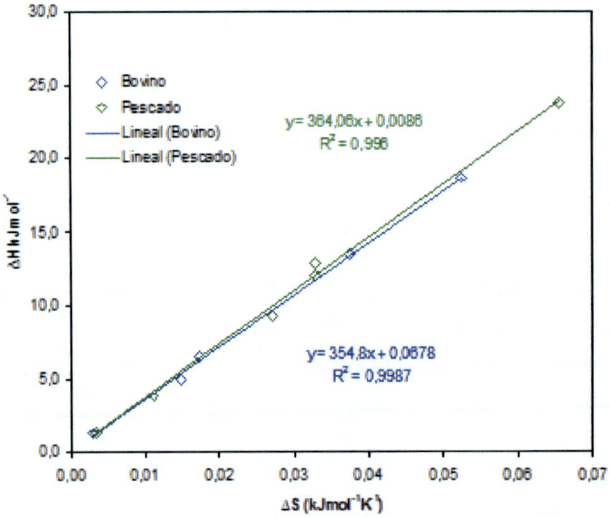

Figure 6. ΔH versus ΔS calculated for fish and bovine gelatin films obtained at 10, 20 y 30 °C. The equations represent the estimated linear regressions for each data set.

In Eq. 6, m is the number of ((ΔH), (ΔS)) data pairs, t is the t value at (m-2) degree of freedom, $\overline{\Delta H}$ is the average differential enthalpy (kJ mol^{-1}) and $\overline{\Delta S}$ is the average differential entropy (J mol^{-1} K^{-1}) [195]. According to [199-200] the compensation theory only exists if $T_\beta \neq T_{hm}$.

Figure 6 depicts the calculated values of ΔH and ΔS by equation 3 using the sorption data obtained from bovine and fish gelatin equilibrated under different relative humidities at 10, 20 and 30°C. This graph clearly shows a linear relationship between ΔH and ΔS ($R^2 > 0.99$) indicating the existence of compensation between both thermodynamic parameters.

The isokinetic temperature (Tβ) and ΔG were calculated using Eq. 4. The expression of Tβ in terms of an interval (95% confidences) was obtained using Eq. 6, 7 y 8 (Table 4).

Table 4. Parameters ΔG, T_{hm} y $T_\beta$obtained for from fish and bovine gelatin films

Source	ΔG ($kJmol^{-1}$)	T_{hm} (K)	T_β (K)
Fish	0,0086	292.9	364,1 ± 62.1
Bovine	0.0678	289.5	354,8 ± 19.8

(*) t value for n= 6 and α= 0.025)

The harmonic mean temperature (T_{hm}) was estimated using equation 5, giving values of T_{hm} ~ 292.9K and T_{hm} ~ 289.5K for fish and bovine gelatin films validating respectively. These values were significantly different to the obtained T_β as indicated in table 1, supporting the application of the compensation concept of moisture sorption by theses materials [199-200].

Leffler [201] proposed that if $T_\beta > T_{hm}$, the sorption process is enthalpy controlled and if $T_\beta < T_{hm}$ sorption would be entropy driven. In this work, T_β for both materials were significantly higher than T_{hm} (Table 4), indicating that the sorption mechanism can be considered enthalpy driven.

4. Usual Methods for Hydrocolloid Films and Coatings Formation

The conditions of the film and coating process have a significant effect on the physical properties of the resulting film. Principal technologies for producing edible films are similar to those for thermoplastic structures: solvent casting and extrusion. Obviously, the conditions are different, but the principles are allied. For the coating the main techniques are: spraying and dipping. Edible films and coating must be uniform and free from defects for optimize their functionalities. In Figure 7, three film defects are presented. The non uniformity, the insertion of bubble and the mechanical damage of the food.

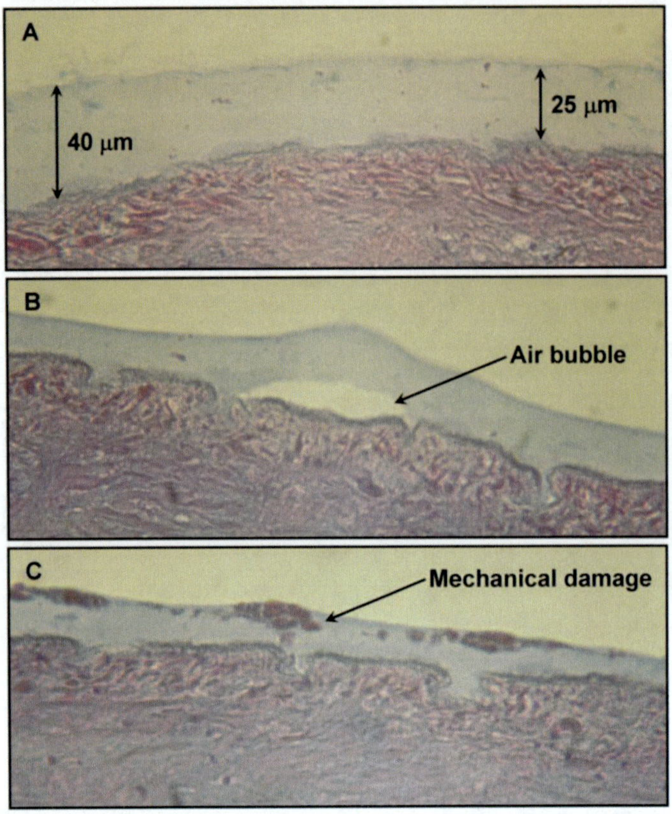

Figure 7. Histochemistry technique showing the cross-section of the alginate coating on clams and the defects: (a) non-uniformity, (b) insertion of air bubble and (c) mechanical damage of the food product. The coating was made by dipping.

4.1. Solvent Casting

Solvent casting is the most used techniques to form hydrocolloid edible films. Water or water–ethanol solutions or dispersions of the edible materials are spread on a suitable substrate (See paragraph 3.1) and later drying. During drying of the solution, solvent evaporation leads to a decrease in solubility of the polymer until polymer chains align themselves to form films. The choice of the substrate is important to obtain films, which can be easily peeled without any damage after that the solvent is evaporated. Generally, the film are air-dried during several hours in the ventilated oven [202]. Infrared drying chambers are advantageous in that they hasten the drying process [203]. Optimum moisture content (5–10% w/v) is desirable in the dried film. Film structures depend on the drying conditions (Temperature and relative humidity), wet casting thickness as well as the composition of the casting solution. Films prepared from pure polymers tend to be brittle and often crack upon drying therefore an addition of plasticizers is necessary [204]. In general, rapid drying of cast polymer solutions limits the development of intermolecular associations within the film structure since the mobility of the polymer chains are restricted when the solvent concentration decrease [160]. Thus, a rapid drying often causes defects such as cracks or curling in the films [205]. Alcantara *et al.* [206] reported that film formation from WPI with a rapid drying at 95°C were thinner, stronger and had a lower *E* than films dried more slowly at 21°C. Polysaccharide–lipid multilayer films with water vapour barrier have been also produced by casting [38-207]. The film forming solutions with a mixture of gelatine, while still hot, are cast by pipetting solutions onto acrylic or flat glass plates and dried at room temperature [208]. Microwave drying was also been investigated and compared with air-drying by Kaya and Kaya [209]. They reported that the film dried by microwave were less hazy and glossier than film dried at room condition.

4.2. Extrusion

In the plastics industry, the most synthetic polymer films are produced by extrusion. The application of extrusion technology to edible films, such as starch films, is reported by Finkenstadt and Willett [210] or Guan and Hanna [211]. The extrusion process is based on the thermoplastic properties of polymers when plasticized and heated above their glass transition temperature under low water content-conditions. The extrusion processing of polymer products often need the addition of plasticizers such as polyethylene glycol, sorbitol in an amount of 10% to 60% w/w (see paragraph 2.2). In comparison with the solvent casting method, the extrusion method is attractive for industrial processes since it does not require solvent addition and a time of evaporation. Pectin has also been combined with glycerol and extruded into films [115, 212]. Furthermore, the formation of extruded protein films produced several scientific patents. For example, Naga *et al.* [213] developed extruded films of soy protein with a plasticizers (polyhydric alcohols) and water whereas Koepff *et al.* [214] reported extruded gelatins films, smooth, uniform, transparent with slight opacity. Such films become increasingly flexible with increased hydration.

To form multilayer films the technique of co-extrusion may be used. However, differences in the physical and chemical properties of the constituent materials in such multilayer films can lead to formation of a number of defects. These defects may be detrimental in terms of the eventual mechanical, barrier, and optical performance of the films. Recently, a defect, the zigzag (or viscous fingers) at the interface of extruded films was evaluated based on steady-state calculations [215].

4.3. Dipping

Nowadays, dipping is a common method for applying coatings on fruits and vegetables [7]. The coating by dipping on the coating solution properties such as density, viscosity and surface tension, as well as food withdrawal speed from the coating solution [109]. Different theoretical approaches can be used to estimate film thickness from coating solution properties [158-159]. Generally the food was dipped into the film-forming solutions between 5 and 30 s [55,140].

Direct application of antimicrobial agents onto food surfaces by dipping has proven to be less effective as there is loss of activity because of leaching onto the food, enzymatic activity, and reaction with other food components [216]. However, the brushing method for the application of film solution to fresh beans and strawberries was found to be better than wrapping and dipping methods in terms of reducing the moisture loss [140].

The application of edible coatings to minimally processed fruits is not easy because it is difficult to obtain a good adhesion of the coating to the hydrophilic surface of the cut fruit. A solution to solve this problem could be a multilayer technique: the layer-by-layer electrodeposition where two or more layers of material with nanometer dimensions are physically or chemically bonded to each other [7, 217]. In Figure 7, the layer-by-layer technique is presented to coat highly hydrophilic food systems such as fresh-cut of apple fruit.

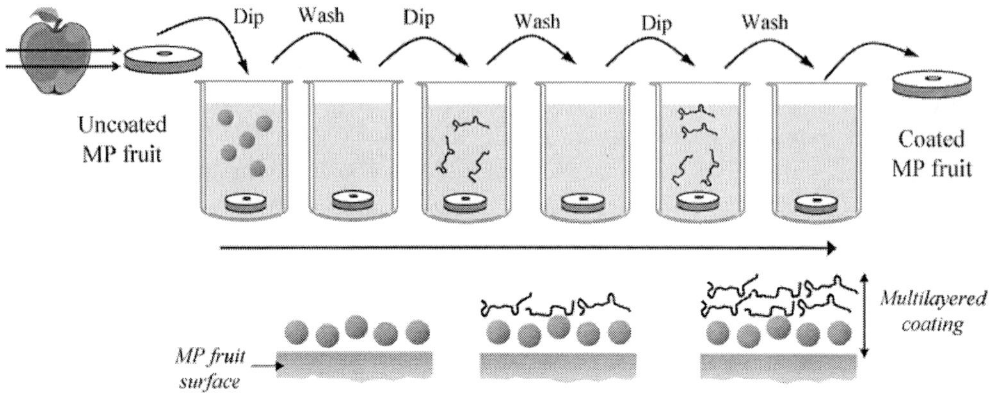

Figure 8. Schematic representation of coating a minimally processed (MP) fruit with a multilayered edible coating by using the layer-by-layer assembly with three dipping and washing steps (Reproduced with permission from [7]).

4.4. Spraying

In the food industry, coating by spraying is the conventional method generally used when the coating-forming solution is not very viscous. Indeed, highly viscous solution cannot be or very uneasily sprayed, and thus only dipping techniques can be apply, giving high thickness to the coating.

Initially, in the case of plan jet, the spray exists the nozzle as a two dimensional sheet of fluid. This sheet then breaks up into a series of cylindrical ribbons which eventually undergo the same type of breakup as laminar capillary jets [218]. The control of the final drop size and therefore of the quality of the coating depend on spray gun and nozzle, temperature, air and liquid flow rate, humidity of incoming air and of the polymer solution [219-220]. Nowadays, programmable spray systems are available for automation during such operations. Classic spraying system can produce a fine spray with relatively drop-size distribution up to 20 μm whereas electrospraying has shown to be capable of producing uniform particles of less than 100 nm from polymer and biopolymer solutions. Futhermore, other different factors are critical in the formation of polymeric films by spraying systems, such as drying time, drying temperature, drying method, etc.

There are some latent diseases in fruits, which usually occur in postharvest storage. Therefore, studies were focused on the effects of preharvest chitosan spray or/and postharvest chitosan coating to extended the protective effect against decay of fruit and to avoid the use of chemical products [221]. Recently, Meng at al. [222] are reported that the treatment with low chitosan concentration (1 g L^{-1}) before harvest and a coating of the fruit with a higher concentration (10 g L^{-1}) after harvest have a beneficial effect on fruit quality and resistance to fruit decay. A similar effect was reported by Bhaskara Reddy *et al.* [223] on strawberry fruit.

5. Future Needs in Edible Film and Coatings

5.1. Scale-Up of Edible Film and Coating in a Industrial Production

Despite the numerous published studies corroborating that hydrocollloid edible films and coatings are economical (as starch...), extend shelf life and enhance quality and microbial safety of food, commercial applications are still very limited. Generally the works report experimental set-up built at the laboratory scale and a coating process made by hand. Moreover some techniques are unsuitable for the industry. For example, the production of film by solvent casting is quite costly, time consuming (up to 48h) and the drying may generate, for the food product, surface dehydration and moisture loss. Spraying, due to the large droplet velocity, may cause damage to a delicate food product such as berries. Further research on the application of hydrocolloid coatings at the real scale is needed, in particular for resolving the problems of non-uniformity in the thickness and in the homogeneity of film.

5.2. Control of Microstructures for Edible Films

To keep the food product attractive for the consumers, the surface of edible films and coatings must appear uniform and without defects. However, the classic methods such as dipping, spraying, solvent casting not allow to control the film thickness. Moreover, most edible films and coatings are formed with little consideration of the internal structure created whereas when functional ingredients (antimicrobials, antioxidants, colour agents, emulsifiers, flavour, nutraceuticals) are incorporated, the film lost its homogeneity.

Microfluidics - the science of designing, manufacturing, and operating devices and processes that deal with small amounts of fluids (10^{-6} to 10^{-9} litres) - has the potential to significantly change the way of processing edible films. Microfluidic devices can be identified by the fact that they have channels with at least one dimension smaller than 1 mm. The devices themselves have dimensions ranging from millimeters down to micrometers. Recent investigations focused on generating and manipulating emulsions with microfluidic devices are motivated by the potential to use controlled flows and dispensing devices on the scale of the droplets or bubbles to tailor the properties of the emulsions and foams [224-225]. Examples of such a microfluidic device for creating a dispersion of water drops in oil or oil in water with starch granules are reported by Anna *et al.* [226] (shown in Figure 9) and Skurtys and Aguilera [227], respectively. The stream of the water is focused by the streams of oil into the flow-focusing orifice and breaks to release mono-dispersed or poly-dispersed water droplets into the outlet micro-channel. The oil forces the water to form a narrow thread that breaks off into mono-dispersed water droplets with diameters much smaller than the orifice diameter (Figure 9k and q). To prepare double emulsions, two T junctions can be used in a series as showed in Figure 10 [228].

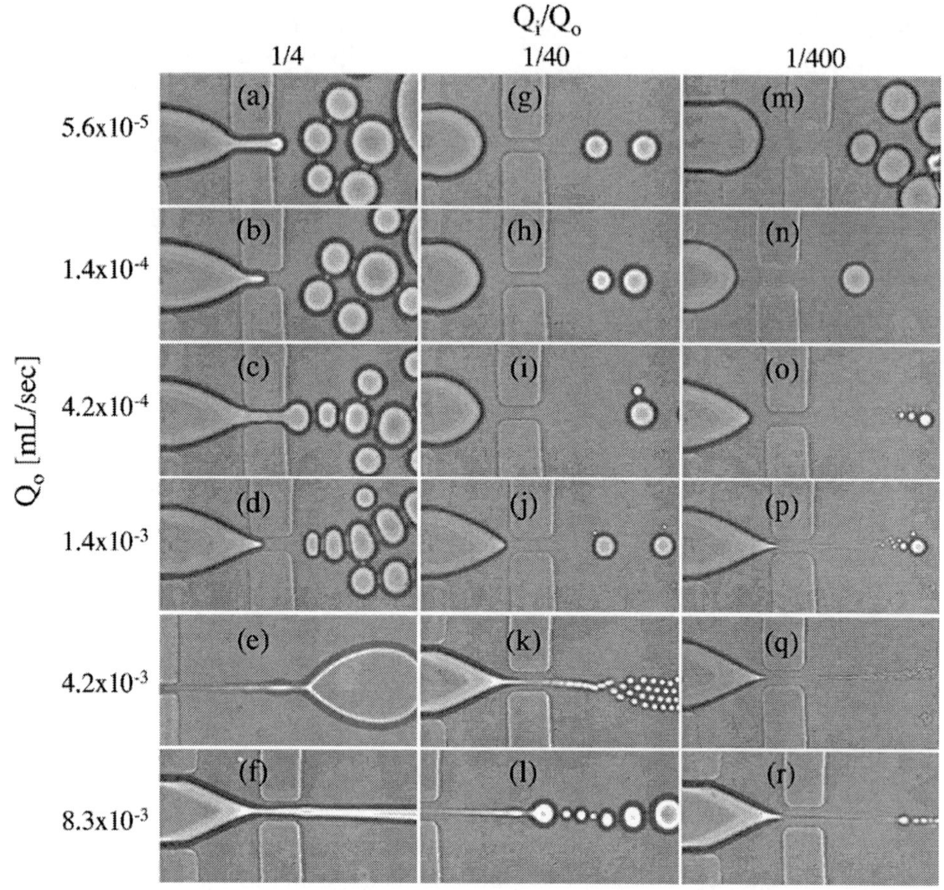

Figure 9. Micro device having a "flow-focusing orifice" before the outlet microchannel. Each image shows the drop size and pattern of drop stream at specific values of Q_o (rows) and Q_i / Q_o (columns), Q_i and Q_o being the flow rates of water and oil, respectively (Reproduced with permission from [226]).

Walther *et al.* [229] studied the deformation of droplets of a carrageenan solution immersed in sunflower oil (with different surfactant concentrations) after passing through a large flow-focusing orifice. Downstream the liquid carrageenan droplets were sheared by flow so that their shape was distorted.

Advantage of microfluidic techniques is that its possibility to produce a thin layer with a microstructures such as oil droplet, air bubbles (with or without aroma), encapsulated probiotics, etc. In the matrix, the distribution of the microstructures can be mono-dispersed in size and uniform dispersed therefore a homogeneous film is obtained. Moreover, the thickness of the layer can be also constant and controlled. Moreover, in comparison with the multilayer one process is necessary to produce the film.

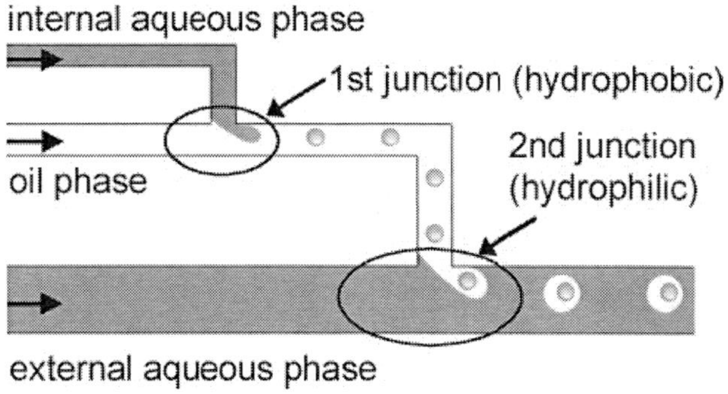

Figure 10. Scheme used to prepare double emulsions (W/O/W) using two T-shaped junctions in series (Reproduced with permission from [228]).

5.3. Control of Nanostructures for Edible Films

Solid-gel processing is widely used in the synthesis of inorganic and organic-inorganic hybrid materials and capable of producing nanoparticles, nanorods, thin films, and monolith. In food engineering and most particularly for the preparation of edible films, using atomic layer deposition can give to food scientists some advantages over conventional technologies. Atomic layer deposition (ALD) has demonstrated its capability of depositing 2 or more layers of materials and thus to create nanolaminates. Excellent reviews on ALD have been published by Ritala and Leskelä [230] for the materials science. As presented in the paragraph 4.3, the most interesting methods based on is based is the layer-by-layer (LbL) deposition technique which produce nanolaminates. The food surface is coated with interfacial films consisting of multiple nanolayers of different materials. The LbL deposition technology allows precise control over the thickness (0.2 to 500 nm), the composition, the structure and therefore over the properties of the interfacial films. Alloy macroscopic sample with a thickness up to 400 μm has been demonstrated.

5.4. Sensory Properties of Edible Coatings

Flavour is one of the most important characteristics of any food product. Its critical role in determining the way consumers assess food quality has made it a key area of research for the food industry [173]. However, the studies of sensory evaluation on the edible films and coatings are very scarce and more investigations need to be done. In particular, edible films can modified to flavour of the coated food and then affected the tolerability and the desirability by consumers. The taste, this is not the only sensory properties to investigate. There are also the colour and the texture of the coated food .

ACKNOWLEDGMENT

The authors (OS and CV) wish to thank CONICYT, sponsor of the Bicentenario Grant (PBCT PSD-62)

REFERENCES

[1] Debeaufort, F.; Quezada-Gallo, J. A.; Voilley, A. *Crit. Rev. Food Sci.* 1998, *38*, 299–313.

[2] Tharanathan, R. N. *Trends Food Sci. Tech.* 2003, *14*, 71-78.

[3] Cha, D. S.; Chinnan, M. S. *Crit. Rev. Food Sci.* 2004, *44*, 223–237.

[4] Siracusa, V.; Rocculi, P.; Romani, S.; Dalla Rosa, M. *Trends Food Sci. Tech.* 2008, *19*, 634-643.

[5] Bifani, V.; Ramírez, C.; Ihl, M.; Rubilar, M.; Garcia, A.; Zaritzky, N. *LWT-Food Sci. Technol.* 2007, *40*, 1473-1481.

[6] Lin, D.; Zhao, Y. *Comp. Rev. Food Sci. Food Saf.* 2007, *6*, 60-75.

[7] Vargas, M.; Pastor, C.; Chiralt, A.; McClements, D. J.; González-Martínez, C. *Crit. Rev. Food Sci.* 2008, *48*, 496-511.

[8] Donhowe G; Fennema O. In *Edible coatings and films to improve food quality*; Krochta, J.M.; Baldwin, E.A.; Nisperos-Carriedo, M.O.; Ed.; Lancaster: Technomic Publishing, 1994; pp 1-24.

[9] Williams, P. A.; Phillips, G. O. In *Handbook of hydrocolloids*; Phillips G.; Willians P.; Ed.; CRC Press, Cambridge, England, 2000; pp 1-19.

[10] Kester, J. J.; Fennema, O. R. *Food Technol.* 1986, *40*, 47-59.

[11] Baldwin, E. A.; Nisperos-Carriedo, M. O.; Baker, R. A. *Crit. Rev. Food Sci.* 1995, 35, 509–524.

[12] Armisén, R.; Galatas, F. In *Handbook of Hydrocolloids*; Phillips G.; Willians P.; Ed.; CRC Press, Cambridge, England, 2000; pp 21-40.

[13] Stanley, N. F. In *Food polysaccharides and their applications*; Stephen A. M.; Ed.; New York, Marcel Dekker, 1995, pp 187-199.

[14] Freile-Pelegrín, Y.; Madera-Santana, T.; Robledo, D.; Veleva, L.; Quintana, P.; Azamar, J. A. *Polym. Degrad. Stabil.* 2007, *92*, 244-252.

[15] Geraldine, R. M.; Soares, N. F. F.; Botrel, D. A.; Goncalves, L. A. *Carbohyd. Polym.* 2008, *72*, 403-409.

[16] Phan The, D.; Debeaufort, F.; Luu, D.; Voilley, A. *J. Agr. Food Chem.* 2005, *53*, 973-981.

[17] Phan The, D.; Debeaufort, F.; Voilley, A.; Luu, D., *Food hydrocolloid* 2009, *23*, 691-699.

[18] King, A. H. In Food hydrocolloids; Glicksman M.; Ed.; CRC Press Inc., 1982, pp 115–88.

[19] Rhim, J. W. *LWT-Food Sci. Technol.* 2004, *37*, 323–30.

[20] Guilbert, S. *Bull. Int. Dairy Fed.* 2000, *346*, 10–16.

[21] Borchard, W.; Kenning, A.; Kapp, A.; Mayer, C. *Int. J. Biol. Macromol.* 2005, *35*, 247–56.
[22] Zactiti, E.; Kieckbusch, T. *J. Food Eng.* 2006, *77*, 462–467.
[23] Tapia, M. S.; Rojas-Graü, M. A.; Rodrígez, F. J.; Ramírez, J.; Carmona, A.; Martin-Belloso, O. *J. Food Sci.* 2007, *72*, 190-196.
[24] Pranoto, Y.; Salokhe, V. M.; Rakshit, S. K. *Food Res. Int.* 2005, *38*, 267–272.
[25] Rojas-Graü, M. A.; Avena-bustillos, R. J.; Olsen, C.; Friedman, M.; Henika, P.; Martín-Belloso, O.; Pan, Z.; McHugh, T. H. *J. Food Eng.* 2007, *81*, 634–641.
[26] Wu, Y.; Weller, C. L.; Hamouz, F.; Cuppett, S.; Schnepf, M. *J. Food Sci.* 2001, *66*, 486–93.
[27] Volery, P.; Besson, R.; Schaffer-Lequart, C. *J. Agr. Food Chem.* 2004, *52*, 7457-7463.
[28] Shaw, C.; Secrist, J.; Tuomy, J. *U.S. Patent 4196219*, 1980.
[29] Lee, J. Y.; Park, H. J.; Lee, C. Y.; Choi, W. Y. *LWT-Food Sci. Technol.* 2003, *36*, 323–329.
[30] Ninomiya, H.; Suzuki, S.; Ishii, K. *U.S. Patent 5620757*, 1997.
[31] Karbowiak, T.; Debeaufort, F.; Champion, D.; Voilley, A. *J. Colloid Interf. Sci.* 2006, *294*, 400–410.
[32] Karbowiak, T.; Debeaufort, F.; Voilley, A. *Food Hydrocolloid* 2007, *21*, 879–888.
[33] Ribeiro, C.; Vicente, A. A.; Teixeira, J. A.; Miranda, C. *Postharvest Biol. Tec.* 2007, *44*, 63–70.
[34] Murray J. C. F. In *Handbook of hydrocolloids*; Phillips G.; Willians P.; Ed.; CRC Press, Cambridge, England, 2000; pp 219-230.
[35] Gennadios, A.; Hanna, M. A.; Kurth, B. *LWT-Food Sci. Technol.* 1997, *30*, 337–50.
[36] Turhan, K. N.; Sahbaz, F. *J. Food Eng.* 2004, *61*, 459–466.
[37] Morillon, V.; Debeaufort, F.; Bond, G.; Capelle, M.; Voilley, A. *Crit. Rev. Food Sci.* 2002, *42*, 67–89.
[38] Debeaufort, F.; Quezada-Gallo, J. A.; Delporte, B.; Voilley, A. *J. Membrane Sci.* 2000, *180*, 47–55.
[39] Maftoonazad, N.; Ramaswamy, H. S. *LWT-Food Sci. Technol.* 2005, *38*, 617–24.
[40] Shahidi, F.; Arachchi, J. K. V.; Jeon, Y. J. *Trends Food Sci. Tech.* 1999, *10*, 37–51.
[41] Tharanathan, R. N.; Kittur, F. S. *Crit. Rev. Food Sci.* 2003, *43*, 61–87.
[42] Jayakumar, R.; Nwe, N. T.; Tokura, S.; Tamura, H. *Int. J. Biol. Macromol.* 2007, *40*, 175–181.
[43] Rabea, E. I.; Badawy, M. E. T.; Stevens, C. V.; Smagghe, G.; Steurbaut, W. *Biomacromolecules* 2003, *4*, 1457–1465.
[44] Tang, R.; Du, Y.; Fan, L. *J. Polym. Sci.* 2003, *41*, 993–997.
[45] Park, S. Y.; Marsh, K. S.; Rhim, J. W. *J. Food Sci.* 2002, *67*, 104–197.
[46] Peniston, Q. P.; Johnson, E. *US Patent 41951755*, 1980.
[47] Sathivel, S.; Liu, Q.; Huang, J.; Prinyawiwatkul, W. *J. Food Eng.* 2007, *83*, 366–373.
[48] Coma, V. *Meat Sci.* 2008, *78*, 90–103.
[49] Fan, W.; Sun, J.; Chen, Y.; Qiu, J.; Zhang, Y.; Chi, Y. *Food Chem.* 2009, *115*, 66-70.
[50] Fernandez-Saiz, P.; Lagaron, J. M.; Ocio, M. J. *Food hydrocolloid* 2009, *23*, 913-921.
[51] Coma, V.; Martial-Gros, A.; Garreau, S.; Copinet, A.; Deschamps, A. *J. Food Sci.* 2002, 67, 1162–1169.

[52] Sebti, I.; Martial-Gros, A.; Carnet-Pantiez, A.; Grelier, S.; Coma, V. *J. Food Sci.* 2006, *70*, 100–104.
[53] Bordenave, N.; Grelier, S.; Coma, V. *J. Agr. Food Chem.* 2007, *55*, 9479–9488.
[54] Han, C.; Zhao, Y.; Leonard, S. W.; Traber, M. G. *Postharvest Biol. Tec.* 2004, *33*, 67–78.
[55] Vargas, M.; Albors, A.; Chiralt, A.; González-Martinez, C. *Postharvest Biol. Tec.* 2006, *41*, 164–71.
[56] Chien, P. J.; Sheu, F.; Yang, F. H. *J. Food Eng.* 2007, *78*, 225–229.
[57] Dutta, P. K.; Tripathi, S.; Mehrotra, G. K.; Dutta, J. *Food Chem.* 2009, 114, 1173-1182.
[58] Subhashchandra, S. C.; Bhaskara, N.; Bhandary, M. H.; Raghunath, B. S. *J. Sci. Food Agr.* 1996, *70*, 453-460.
[59] Mei, Y.; Zhao, Y.; Yang, J.; Furr, H. R. *J. Food Sci.* 2002, *67*, 1964-1968.
[60] Sahin, S.; Summu, G.; Altunakar, B. *J. Sci. Food Agr.* 2005, *85*, 2375-2379.
[61] Bosquez-Molina, E.; Guerrero-Lagarreta, I.; Vernon-Carter, E. J. *Food Res. Int.* 2003, *36*, 885–893.
[62] Díaz-Sobac, R.; García, H. S.; Beristain, C.I.; Vernon-Carter, E. J. *J. Food Process. Pres.* 2002, *26*, 129–141.
[63] Mazza, G.; Qi, H. *J. Agr. Food Chem.* 1991, *39*, 2163–2166.
[64] Hoagland, P.; Parris, N. *J. Agr. Food Chem.* 1996, *44*, 1915–1919.
[65] Marudova, M.; Lang, S.; Brownsey, G. J.; Ring, S. G. *Carbohydr. Res.* 2005, *340*, 2144–2149.
[66] Rojas-Graü, M. A.; Avena-Bustillos, R. J.; Friedman, M.; Henika, P. R.; Martín-Belloso, O.; McHugh, T. H. *IFT Annual Meeting* 2006, Abstract *389*.
[67] Lenart, A.; Dabrowska, R. *Dry. Technol.* 1999, *7–8*, 1359–1373.
[68] Lenart, A.; Piotrowski, D. *Dry. Technol.* 2001, *19*, 849–877.
[69] Baeva, M.; Panchev, I. *Food Chem.* 2005, *92*, 343-348.
[70] Walstra, P. In *Physical Chemistry of Foods*; Walstra, P.; Ed.; Marcel Dekker - New York. 2003.
[71] Claudia, A. R. B.; Bello-Pérez, L. A.; Gacia, M. A.; Martino, M. N.; Solorza-Feria, J.; Zaritzky, N. E. *Carbohyd. Polym.* 2005, *60*, 235-244.
[72] Copeland, L.; Blazek, J.; Salman, H.; Tang, M. C. *Food Hydrocolloid* 2009, *23*, 1527-1534.
[73] Xu, X. Y.; Kim, K. M.; Hanna, M. A.; Nag, D. *Ind. Crop. Prod.* 2005, *21*, 185-192.
[74] Myllärinen, P.; Buléon, A.; Lahtinen, R.; Forssell, P. *Carbohyd. Polym.* 2002, *48*, 41–8.
[75] Rindlav, A.; Hulleman, S. H. D.; Gatenholm, P. *Carbohyd. Polym.* 1997, *34*, 25–30.
[76] Krogars, K.; Heinämäki, J.; Karjalainen, M.; Rantanen, J.; Luukkonen, P.; Yliruusi, J. *Eur J. Pharm. Biopharm.* 2003, *56*, 215–221.
[77] Parra, D. F.; Tadini, C. C.; Ponce, P.; Lagão, A. B. *Carbohyd. Polym.* 2004, *58*, 475–481.
[78] Mali, S.; Grossmann, M. V. E.; García, M. A.; Martino, M. N. M.; Zaritzky, N. E. *Food Hydrocolloid* 2005, *19*, 157-164.
[79] Hu, G.; Chen, J.; Gao, J. *Carbohyd. Polym.* 2009, *76*, 291–298.
[80] Baldwin, E. A.; Baker, R.A. In *Protein based films and coatings* Gennadios A.; Ed.; Boca Raton, Fla., CRC Press. 2002; pp 501–515.
[81] Sothornvit, R.; Krochta, J. M. *J. Agr. Food Chem.* 2000; *48*, 6298–302.

[82] Sothornvit, R.; Krochta, J. M. *J. Food Eng.* 2001, *50*, 149–155.
[83] Sothornvit, R.; Reid, D. S.; Krochta, J. M. *Amer. Soc. Ag. Eng.* 2002, *45*, 1479–1484.
[84] Cho, S. Y.; Rhee, C. *LWT-Food Sci. Technol.* 2002, *35*, 151–157.
[85] Wang, V. C. H.; Kim, M. S.; Lee, S. Y. *J. Food Sci.* 2005, *70*, 387–391.
[86] Gómez-Guillén, M. C.; Turnay, J.; Fernández-Martín, F.; Ulmo, N.; Lizarbe, M. A.; Montero, P. *Food Hydrocolloid* 2002, *16*, 25-34.
[87] Mitchell, J. R. In *Functional properties of food macromolecules* Hill, S. E.; Ledward, D. A.; Mitchell, J. R.; Ed.; Aspen Publishers., 1998; pp. 50–73.
[88] Sobral, P. J. A.; Habitante, A. M. Q. B. *Food Hydrocolloid* 2001, *15*, 377–382.
[89] Ledward, D. A., In *Handbook of hydrocolloids*; Phillips G.; Willians P.; Ed.; CRC Press, Cambridge, England, 2000; pp 67-86.
[90] Johnston-Banks, F.A., In *Gelatin* Harris, P.; Ed.; Elsevier, London, 1990; pp. 233–289.
[91] Sobral, P. J. A.; Menegalli, F. C.; Hubinger, M. D.; Roques, M. A. *Food Hydrocolloid* 2001, *15*, 423–432.
[92] Arvanitoyannis, I.S. In *Protein-based Films and Coatings* Gennadios, A.; Ed.; CRC Press, Boca Ratón, Florida, 2002.
[93] Simon-Lukasik, K. V.; Ludescher, R. D. *Food Hydrocolloid* 2004, *18*, 621–630.
[94] Gómez-Estaca, J.; Montero, P.; Fernández-Martín, F.; Gómez-Guillén, M. C. *J. Food Eng.* 2009, *90*, 480–486.
[95] Gómez-Guillén, M. C.; Ihl, M.; Bifani, V.; Silva, A.; Montero, P. *Food Hydrocolloid* 2007, *21*, 1133–1143.
[96] Pérez–Gago, M. B.; Nadaud, P.; Krochta, J. M. *J. Food Sci.* 1999, *64*, 1034–1037.
[97] Ennis, M. P.; Mulvihill, D. M. In *Handbook of hydrocolloids*; Phillips G.; Willians P.; Ed.; CRC Press, Cambridge, England, 2000; pp 189-218.
[98] Pérez-Gago, M. B.; Krochta, J. M. In *Protein-Based Films and Coatings* Gennadios, A.; Ed.; CRC Press, Boca Raton, FL, 2002; pp. 159–180.
[99] McHugh, T. H.; Krochta, J. M. *Food Technol.* 1994, *48*, 97–103.
[100] Fairley, P.; Monahan, F. J.; German, J. B.; Krochta, J. M. *J. Agr. Food Chem.* 1996, 44, 438–443.
[101] Miller, K. S.; Krochta, J. M. *J. Appl. Polym. Sci.* 1997, *88*, 1095–1099.
[102] Hong, S.-I., Krochta, J. M. *J. Food Sci.* 2003, *68*, 224–228.
[103] Shaw, N. B.; Monahan, F. J.; O'Riordan, E. D.; O'Sullivan, M. *J. Food Sci.* 2002, *67*, 164–167.
[104] Perez-Gago, M. B.; Krochta, J. M. *J. Food Sci.* 1999, *64*, 695–698.
[105] Ozdemir, M.; Floros, J. D. *J. Food Eng.* 2008, *84*, 116–123.
[106] Osés, J.; Fabregat-Vázquez, M.; Pedroza-Islas, R.; Tomás, S. A.; Cruz-Orea, A.; Maté, J. I. *J. Food Eng.* 2009, *92*, 56–62.
[107] Hagenmaier, R. D.; Shaw, P. E. *J. Agr. Food Chem.* 1991, 38, 1799–1803.
[108] Park, H. J. *Trends Food Sci. Tech.* 1999, *10*, 254–260.
[109] Cisneros-Zevallos, L.; Krochta, J. M. *J. Food Sci.* 2003, *68*, 503-510.
[110] Ofori-Kwakye, K.; Fell, J. T. *Int J Pharm* 2003, *250*, 251–257.
[111] Dogan, N.; McHugh, T.M. *J. Food Sci.* 2007, *72*, 16–22.
[112] Pinotti, A.; García, M. A.; Martino, M. N.; Zaritzky, N E. *Food Hydrocolloid* 2007, *21*, 66–72.

[113] Srinivasa, P. C.; Ramesh, M. N.; Tharanathan, R. N., *Food Hydrocolloid* 2007, *21*, 1113-1122.

[114] Faber, J. N.; Harris, L. J.; Parish, M. E.; Beuchat, L. R.; Suslow, T. V.; Gorney, J. R.; Garrett, E. H.; Busta, F.F. *Crit. Rev. Food Sci.* 2003, *2*, 142–160.

[115] Liu, L.; Kerry, J. K.; Kerry, J. P. *Int. J. Food Sci. Tech.* 2006, *41*, 295–302.

[116] Veiga-Santos, P.; Oliveira, L. M.; Cereda, M. P.; Alves, A. J.; Scamparini, A. R. P. *Food Hydrocolloid* 2005, *19*, 341–349.

[117] Chen, C. H.; Lai, L. S., *Food Hydrocolloid* 2008, *22*, 1584-1595.

[118] Carvalho, R. A.; Sobral, P. J. A.; Thomazine, M.; Habitante, A. M. Q. B.; Giménez, B. *Food Hydrocolloid* 2008, *22*, 1117-1123.

[119] Jongjareonrak, A.; Benjakul, S.; Visessanguan, W.; Prodpran, T.; Tanaka, M. *Food Hydrocolloid* 2006, *20*, 492–501.

[120] Barreto, P. L. M.; Pires, A. T. N.; Soldi, V. *Polym. Degrad. Stabil.* 2003, *79*, 147–152.

[121] Myllärinen, P.; Partanen, R.; Sepalla, J.; Forsell, P. *Carbohyd. Polym.* 2002, *50*, 355–61.

[122] Jagannath, J. H.; Nanjappa, C.; Das Gupta, D. K.; Bawa, A. S. *Int. J. Food Sci. Tech.* 2006, *41*, 498-506.

[123] Cerqueira, M. A.; Lima, A. M.; Souza, B. W. S.; Teixeira, J. A.; Moreira, R. A.; Vicente, A. A. *J. Agr. Food Chem.* 2009, *57*, 1456-1462.

[124] Bourtoom, T.; Chinnan, M. S.; Jantawat, P.; Sanguandeekul, R. *Food Sci. Technol. Int.* 2006, *12*, 119-126.

[125] Veiga-Santos, P.; Oliveira, L. M.; Cereda, M. P.; Scamparini, A. R. P. *Food Chem.* 2007, *103*, 255-262.

[126] Chang, Y. P.; Cheah, P.B.; Seow, C. C. *J. Food Sci.* 2000, *65*, 1–7.

[127] Krochta, J. M. In *Protein-based Films and Coatings* Gennadios A.; Ed.; Boca Raton, CRC Press, 2002; pp 1-41.

[128] Mali, S.; Grossmann, M. V. E.; García, M. A.; Martino, M. N. M.; Zaritzky, N. E. *J. Food Eng.* 2006, *75*, 453–60.

[129] Cha, D. S.; Chinnan, M. S. *Crit. Rev. Food Sci.* 2004, *44*, 223–237.

[130] Appendini, P.; Hotchkiss, J. H. *Innovative Food Science and Emerging Technologies* 2002, *3*, 113–126.

[131] Sivarooban, T.; Hettiarachchy, N. S.; Johnson, M. G. *Food Res. Int.* 2008, *41*, 781-785.

[132] Scannell, A. G. M.; Hill, C.; Ross, R. P.; Marz, S.; Hartmeier, W.; Arendt, E. K. *Int. J. Food Microbiol.* 2000, *60*, 241–249.

[133] Santiago-Silva, P.; Soares, N. F. F.; Nóbrega, J. E.; Júnior, M. A. W.; Barbosa, K. B. F.; Volp, A. C. P.; Zerdas, E. R. M. A.; Würlitzer, N. J. *Food Control* 2009, *20*, 85–89.

[134] Suppakul, P.; Miltz, J.; Sonneveld, K.; Bigger, S.W. *J. Food Sci.* 2003, *68*, 408–420.

[135] Mitrakas, G.E.; Koutsoumanis, K.P.; Lazarides, H.N. *Innovative Food Science and Emerging Technologies* 2008, *9*, 550–555.

[136] Mecitoğlu, C., Yemenicioğlu, A. *Food Chem.* 2007, *104*, 726–733.

[137] Seifu, E.; Buys, E. M.; Donkin, E. F. *Trends Food Sci. Tech.* 2005, *16*, 137-154.

[138] Kussendrager, K. D.; van Hooijdonk, A. C. M. *Brit J. Nutr.* 2000, *84*, 19–25.

[139] Seydim, A.C.; Sarikus, G. *Food Res. Int.* 2006, *39*, 639–644.

[140] Ayranci, E., Tunc, S. *Z Lebensm Unters Forsch A* 1997, *205*, 470-473.

[141] Bico, S. L. S.; Raposo, R. M. S.; Morais, C.; Morais, A. M. M. B. *Food Control* 2009, *20*, 508–514.

[142] Burt, S. A. *Int. J. Food Microbiol.* 2004, *94*, 223–53.

[143] Zinoviadou, K. G.; Koutsoumanis, K. P.; Biliaderis, C. G. *Meat Sci.* 2009, *82*, 338-345.

[144] Maizura, M.; Fazilah, A.; Norziah, M. H.; Karim, A. A. *J. Food Sci.* 2007, *72*, 324–330.

[145] Ramesh, M. N.; Siddalingaiya, G. P. *Patent Wipo WO/2006/103698*; 2006.

[146] Park, S. I.; Zhao, Y. *J. Agr. Food Chem.* 2004, *52*, 1933–1939.

[147] Weaver, C. M.; Heaney, R. P. In *Modern Nutrition in Health and Disease* Shils M. E.; Olson J. A; Shike M.; Ross A. C.; Ed.; Philadelphia Lippincott Williams & Wilkins, 1998; pp 141-155.

[148] De Gennes, P. G.; Brochard-Wyart, F.; Quéré, D. In *Capillarity and Wetting Phenomena* De Gennes, P. G.; Brochard-Wyart, F.; Quéré, D.; Ed.; Springer-Verlag, 2004; pp 203.

[149] Miller, C. A.; Neogi, P. In *Interfacial phenomena: equilibrium and dynamic effects.* CRC Press Taylor & Francis Group, 2008; pp 61-107.

[150] Adamson, A. W.; Gast, A. P. In *Physical Chemistry of Surfaces*, Sixth Edition, Wiley-Interscience Publication, 1997; pp 347-430.

[151] Tghygesen, L. G.; Løkke, M. M.; Micklander, E.; Engelsen, S. B. *Trends Food Sci. Tech.* 2003, *14*, 50–57.

[152] McAnally, G. D.; Everall, N. J.; Chalmers, J. M.; Smith, W. E. *Appl. Spectrosc.* 2003, *57*, 44-49.

[153] Ishikawa, K.;Yamano, H.; Kagawa, K.; Asada, K.; Iwata, K.; Ueda, M. *Opt. Laser Eng.* 2004, *41*, 19–29.

[154] Hsu, B. L.; Weng, Y. M.; Liao, Y. H.; Chen, W. *J. Agr. Food Chem.* 2005, *53*, 5089–5095.

[155] Dai, H.; Shen, Y.; Zhou, H.; Cai, X. *Thin Solid Films* 2008, *516*, 1796–1802.

[156] Fukano, T.; Yamaguchi, I. *Opt. Lett.* 1966, *21*, 1942–1944.

[157] Kim, S. W.; Kim, G. H. *Appl. Optics* 1999, *38*, 5968–5973.

[158] Levich V. In *Physicochemical hydrodynamics*. Prentice-Hall, 1962.

[159] Snoeijer, J. H.; Ziegler, J.; Andreotti, B.; Fermigier, M.; Eggers, J. *Phys. Rev. Lett.* 2008, *100*, 244502.

[160] Sperling, L.H. In *Introduction to Physical Polymer Science*. 2nd ed., New York, NY, USA: John Wiley & Sons, Inc, 2006; Chp 11.

[161] Kristo, E.; Biliaderis, C. G.; Zampraka, A. *Food Chem.* 2007, *101*, 753–76.

[162] Butler, B. L.; Vergano, P. J.; Testin, R. F.; Bunn, J. M.; Wiles, J. L. *J. Food Sci.* 1996, *61*, 953-61.

[163] Park J. W.; Testin R. F.; Park H. J.; Vergano P. J.; Weller C. L. *J. Food Sci.* 1994, *59*, 916–919.

[164] Mark, J. E. In *Polymer Data Handbook*. Oxford University Press, Inc, 1999.

[165] Lazaridou, A.; Biliaderis, C. G. *Carbohydr. Polym.* 2002, *48*, 179-190.

[166] Wang, Y.; Padua, G. W. *Macromol. Mater. Eng.* 2003, *288*, 886–893.

[167] Pfeiffer, T.; Menner, M. *Fleischwirtschaft* 1999, *79*, 79-84.

[168] Crank, J. In *The Mathematics of Diffusion.* 2nd ed; Clarendon Press, Oxford, 1975.

[169] Lide, D. R. In *Handbook of Chemistry and Physics*. 87th Edition, CRC Taylor and Francis, 2007.

[170] Talasila, P. C.; Cameron, A. C. *J. Food Sci.* 1997, *62*, 926-930.

[171] Simpson, R.; Acevedo, C.; Almonacid, S. *J. Food Eng.* 2009, *92*, 233-239.

[172] Gontard, N.; Guilbert, S.; Cuq, J. L. *J. Food Sci.* 1993, *58*, 206–211.

[173] Voilley, A.; Souchon, I. In *Flavour in Food.* Voilley, A.; Etiévant, P.; Ed.; Woodhead Publishing Limited and CRC Press LLC, 2006; pp 117-132.

[174] Hirose, K.; Harte, B. R.; Giacin, J. R.; Miltz, J.; Stine, C. In *Food and Packaging Interactions*, Hotchkiss J. H.; Ed.; ACS Symposium Series 365, American Chemical Society, Washington, DC, 1998; pp 28-41.

[175] Miller, K. S.; Upadhyaya, S. K.; Krochta, J. M. *J. Food Sci.* 1998, *63*, 244-247.

[176] Williams, R.; Mittal, G. S. *LWT-Food Sci. Technol.* 1999, *32*, 440–445.

[177] Han, C.; Krochta, J. M. *J. Food Sci.* 2001, *66*, 294-299.

[178] García, M. A.; Ferrero, C.; Bértola, N.; Martino, M.; Zaritzky, N. *Innovative Food Science and Emerging Technologies* 2002, *3*, 391–397.

[179] García, M.; Bifani, V.; Campos, C.; Martino, M. N.; Sobral, P.; Flores, S.; Ferrero, C.; Bertola, N.; Zaritzky, N. E.; Gerschenson, L.; Ramírez, C.; Silva, A.; Ihl, M.; Menegalli, F. In *Food Engineering: Integrated Approches*, Gutiérrez-López, G. F.; Barbosa-Cánovas, G. V.; Welti-Chanes, J.; Parada-Arias, E.; Ed.; Springer, 2008; pp 225-241.

[180] Trezza, T. A.; Krochta, J. M. *J. Food Sci.* 2000, *65*, 658-662.

[181] Nikolova, K.; Panchev, I.; Sainov, S. *J. Optoelectron. Adv. M.* 2005, *7*, 1439-1444.

[182] Villalobos, R.; Chanona, J.; Hernández, P.; Gutiérrez, G.; Chiralt, A. *Food Hydrocolloid* 2005, *19*, 53–61.

[183] Fabra, M. J.; Talens, P.; Chiralt, A. *Food Hydrocolloid* 2009, *23*, 676–683.

[184] Hutchings, J. B. In *Food color and appearance,* Maryland: Aspen Publisher, 1999.

[185] Enrione, J. I. In PhD thesis *Mechanical Stability of Intermediate Moisture Starch-Glycerol Systems*. Division of Food Sciences. Nottingham, The University of Nottingham, 2005.

[186] Anderson, R. B. *J. Am. Chem. Soc.* 1946, *68*, 686-691.

[187] De Boar, J. H. In *The Dynamic Character of Adsorption*, Oxford, UK, Clarendon Press, 1953.

[188] Filho, J. F. L.; Romanelli, P. F.; Barboza, S. H. R.; Gabas, A. L.; Telis-Romero, J. *J. Food Eng.* 2002, *52*, 201-206.

[189] Vazquez, G.; Chenlo, F.; Moreira R. *J. Food Eng.* 2003, *60*, 449-452.

[190] Quirijns, E.; van Boxtel, A.; van Loon, W.; van Straten, G. *J. Sci. Food Agr.* 2005, *85*, 1805-1814.

[191] Enrione, J. I.; Hill, S. E.; Mitchell, J. R. *J. Agr. Food Chem.* 2007, *55*, 2956-2963.

[192] Yan, Z.; Soussa-Gallagher, J.; Oliveira Fernanda, A. R. *J. Food Eng.* 2008, *86*, 342-348.

[193] Gabas, A. L.; Menegalli, F. C.; Telis-Romero, J. *J. Food Sci.* 2000, *65*, 680-684.

[194] Tunc, S.; Duman, O. *J. Food Eng.* 2007, *81*, 133-143.

[195] McMinn, W. A. M.; Al-Muhtaseb, A. H.; Magee, T. R. A. *Food Res. Int.* 2005, *38*, 505-510.

[196] Nkolo, M. E. Y. N.; Noah, N. J.; Bardet, S. *Thermochim. Acta* 2008, *468*, 1-5.

[197] Liu, Y.; Feng. *J. Mater. Sci.* 2001, *36*, 1809-1815.

[198] Van den Berg, C.; Bruin, S. In *Water Activity and its Estimation in Food Systems:Theoretical Aspects,* Rockland, L. B.; Stewart, G. F.; Ed.; London, UK, Academic Press, 1981.

[199] Krug, R. R.; Hunter, W. G.; Greiger, R. A. *J. Phys. Chem.* 1976, *80*, 2335-2341.

[200] Krug, R. R.; Hunter, W. G.; Greiger, R. A. *J. Phys. Chem.* 1976, *80*, 2341-2352.

[201] Leffler, J. *J. Org. Chem.* 1955, *20*, 1202.

[202] Cha, D. S.; Cooksey, K.; Chinnan, M. S.; Park, H. J. *LWT-Food Sci. Technol.* 2003, *36*, 209–213.

[203] Tharanathan, R. N.; Srinivasa, P. C.; Ramesh, M. N. *Indian patent 0085/DEL/02*; 2002.

[204] Mastromatteo, M.; Chillo, S.; Buonocore, G. G.; Massaro, A.; Conte, A.; Bevilacqua, A.; Del Nobile, M. A. *J. Food Eng.* 2009, *92*, 467–473.

[205] Obara, S.; McGinity, J. W. *Int. J. Pharm.* 1995, *37*, 849-53.

[206] Alcantara, C. R.; Rumsey, T. R.; Krochta, J. M. *J. Food Process. Engr.* 1998, *21*, 387-405.

[207] Vargas, M.; Albors, A.; Chiralt, A.; González-Martinez, C. *Food Hydrocolloid* 2009, *23*, 536–547.

[208] Chambi, H.; Grosso, C. *Food Res. Int.* 2006, *39*, 458–466.

[209] Kaya, S.; Kaya, A. *J. Food Eng.* 2000, *43*, 91-96.

[210] Finkenstadt, V. L.; Willett, J. L. *Carbohydr. Polym.* 2004, *55*, 149–154.

[211] Guan, J.; Hanna, M. A. *Bioresource Technol.* 2006, *97*, 1716–1726.

[212] Fishman, M. L.; Coffin, D. R.; Onwulata, C. I.; Konstance, R. P. *Carbohydr. Polym.* 2004, *57*, 401–413.

[213] Naga, M.; Kirihara, S.; Tokugawa, Y.; Tsuda, F.; Hirotsuka., M. *US Patent 5569482*; 1996.

[214] Koepff, P.; Braumer, K.; Stahl, H., Dick, E. *US Patent 5316717*; 1994.

[215] Prakash, O. Comput. Mat. Sci. 2006, *37*, 7–11.

[216] Ouattara, B.; Simard, R. E.; Piette, G.; Bégin, A.; Holley, R. A. *Int. J. Food Microbiol.* 2000, *62*, 139-48.

[217] Weiss, J.; Takhistov, P.; McClements, D. J. *J. Food Sci.* 2006, *71*, 107–116.

[218] Stone, H. A. *Annu. Rev. Fluid Mech.* 1994, *26*, 65–102.

[219] Bergeron, V.; Bonn, D.; Martin, J. Y.; Vovelle, L. *Nature* 2000, *405*, 772.

[220] Bergeron, V. *C. R. Physique*, 2003, *4*, 211–219.

[221] Romanazzi, G.; Nigro, F.; Ippolito, A.; Divenere, D.; Salerno, M. *J. Food Sci.* 2002, *67*, 1862-1867.

[222] Meng, X.; Li, B.; Liu, J.; Tian, S. *Food Chem.* 2008, *106*, 501–508.

[223] Bhaskara Reddy, M. V.; Belkacemi, K.; Corcuff, R.; Castaigne, F.; Arul, J. *Postharvest Biol. Tec.* 2000, *20*, 39–51.

[224] Skurtys, O.; Aguilera, J. M. *J. Colloid Interf. Sci.* 2008. *318*, 380.

[225] Skurtys, O.; Aguilera, J. M. *Food Biophysics* 2008, *3*, 1.

[226] Anna, S. L.; Bontoux, N.; Stone, H. A. *Appl. Phys. Lett.* 2003, *82*, 364 – 366.

[227] Skurtys, O.; Aguilera, J. M. *Food hydroccoloid* 2009, In Press, doi:10.1016/j.foodhyd.2009.01.014

[228] Okushima, S.; Nisisako, T.; Torii T.; Higuchi, T. *Langmuir* 2004, *20*, 9905–9908.

[229] Walther, B. L.; Walkenström, P.; Hermansson, A. M. *J. Colloid Interf. Sci.* 2004, *270*, 195 – 204.
[230] Ritala, M.; Leskelä, M. In *Handbook of Thin Film Materials*, Nalwa, H. S.; Ed; Academic Press, San Diego, CA, Vol 1, 2002; pp.103.

In: Food Hydrocolloids: Characteristics, Properties... ISBN: 978-1-60876-222-4
Editor: C. S. Hollingworth, pp. 81-117

Chapter 3

Rheological Properties and Sensory Attributes of Mashed Potatoes Made with Mixtures of Sodium Caseinate and Xanthan Gum

Cristina Fernández, Wenceslao Canet and María Dolores Alvarez*

Department of Plant Foods Science and Technology,
Instituto del Frío-CSIC,
José Antonio de Novaís no. 10, 28040 Madrid, Spain

Abstract

Fresh (F) and frozen/thawed (F/T) mashed potatoes made with mixtures of sodium caseinate (SC) and xanthan gum (XG) were assessed for rheological properties as evaluated with two kinds of deformation tests—large deformation [instrumental texture profile analysis (ITPA) and cone penetration (CP)] and small deformation (oscillatory measurements)—colour, drip loss (DL), pH, and two different sensory analyses. The influence of SC and XG concentrations on rheological properties and quality characteristics was studied by means of response surface methodology (RSM) using a central composite rotatable experimental design with SC concentrations ranging between 2.5 and 7.5 g kg^{-1} and XG concentrations ranging between 0.5 and 2.5 g kg^{-1}, as independent variables. The effects were strongly dependent on the levels of SC and XG added. The effect of XG concentration on the oscillatory measurements was more significant, although hydrophobic interactions between SC and XG may well have influenced the viscoelastic behaviour of the products. F mashed potatoes with added biopolymer mixtures behaved like a weaker gel structure; this fact is attributed to repelling forces between negatively charged phosphate groups on potato starch granules and negative charges on XG molecules. Processing significantly reduced the effect of adding SC and XG mixtures on the gel strength and the viscoelastic behaviour of the mashed potatoes due to high local starch concentrations and retrogradation of both amylose and

* E-mail: mayoyes@if.csic.es

amylopectin. Addition of biopolymer mixtures modified the pH of the product, mainly due to the presence of SC. DL values revealed that addition of XG at concentrations of 1.5 g kg^{-1} or higher, either individually or mixed with SC, were effective in stabilizing mashed potatoes against syneresis. Scores for overall acceptability (OA) were higher in the F products than in their F/T counterparts, and when SC/XG mixtures (each biopolymer at 1.5 g kg^{-1}) were included in the formulation, the product exhibited very acceptable sensory quality thanks to their creamy mouthfeel conferred by this mixture.

Keywords Textural properties; Viscoelastic properties; Puréed cooked potatoes; Cryoprotectant mixtures; Freezing/thawing; Sensory analysis; Response surface.

INTRODUCTION

Food products are composed of numerous ingredients, in which physical properties such as texture and stability are mainly controlled by gels or foams built up with biopolymers including proteins and polysaccharides. It is therefore imperative to understand the underlying mechanisms that regulate the interactions between polysaccharides and proteins [1, 2] in order to be able to improve and modify the physical properties of food. Sodium caseinate (SC) and xanthan gum (XG) are used to improve food stability and texture [3].

SC, which is a mixture of four caseins (α_{s1}-, α_{s2}-, β- and κ-), is widely used as an emulsifier in food systems [4]. Because the individual caseins contain hydrophobic and hydrophilic regions, they adsorb rapidly at the oil-water interface and consequently provide emulsion stability through a combination of steric and electrostatic interactions [5-7]. However, the stability of emulsions made with SC is sensitive to the protein concentration. Above a certain critical protein concentration, caseinate-based emulsions can exhibit pronounced creaming or serum separation due to depletion flocculation induced by excess unadsorbed protein in the aqueous continuous phase [8, 9].

Polysaccharides are predominantly hydrophilic and therefore are not particularly surface-active [10]. Hence, they are usually added to the aqueous phase of emulsions as thickening agents in order to modify the rheological behaviour of the aqueous phase and thereby retard instability mechanisms. However, under specific conditions they can also cause phase separation and flocculation [7]. XG is one such example; thermodynamic incompatibility [5, 11] and isotropic/anisotropic phase separation [12], as well as depletion flocculation [5, 13], are all possible phenomena potentially causing phase separation in emulsions containing xanthan. Also, in the presence of XG, SC forms a gel upon acidification [14]. SC does not show a clear phase separation at neutral pH in the presence of low concentrations of XG (~0.1 %) [15].

Blends of starch with non-starchy hydrocolloids resulting in binary gels have been studied as structure providers in food production. XG is known specifically for its water-holding ability and is used in starch-based products to influence the gelatinization and rheological properties of starches. The addition of hydrocolloids (e.g., XG) increases the viscosity of starch pastes, influences the retrogradation rate and, most importantly, prevents the syneresis of starch [16-20]. Chaisawang and Suphantharika [16] investigated effects of XG (negatively charged gum) on the pasting and rheological properties of native tapioca starch (nonionic) and anionic tapioca starch (negatively charged). The addition of XG resulted in a significant reduction of peak, breakdown, setback and final viscosities of the anionic starch.

A rheological study of the effects of SC on potato starch was also conducted [21], as well as on F and F/T mashed potatoes [22-24]. The microstructural evolution of viscoelastic emulsions stabilized by SC and XG has been studied [7]. To investigate interactions between SC and XG, the effects of pH on structure of SC/XG complex were analyzed [3]. The influence of XG on the formation and stability of SC oil-in-water emulsions was studied [4]. The same authors also investigated the behaviour of SC/XG mixtures at neutral pH [5]. Aqueous solutions of SC and XG at pH 7 and containing 0.1 M NaCl, and mixtures of these, were investigated using dynamic light scattering [15]. Authors concluded that phase separation in XG-SC mixtures does not occur at XG concentrations below 0.5 wt.%.

The effect of monovalent (NaCl) and divalent ($CaCl_2$) salts on the rheology of SC solutions has been studied [25], and likewise the influence of ionic calcium on stability of SC emulsions. Also, Mohammed *et al.*, [26] investigated the promotion and inhibition of xanthan “weak-gel” rheology by calcium ions. Phase separation between a non-polyelectrolyte (starch polysaccharides) and a polyelectrolyte (caseinate) is encouraged by high salt concentrations, since this reduces the unfavourable entropy change produced by the difference in counter-ion concentration occurring upon phase separation [21]. The dependence of the ion type on the textural properties of heat-induced mixed protein-gum gels has also been studied [27].

However, mashed potatoes as prepared here can contain four biopolymers: denatured milk protein, native potato starch, SC and XG. They also contain salt (NaCl), which supplies sodium ions, and calcium ions and even fat, both from the milk, and hence their gel forming ability is influenced by very different complex interactions. To the best of our knowledge, this is the first time that the combined use of caseinate and xanthan in mashed potatoes has been reported in the literature.

The purpose of this chapter was to investigate how the addition of SC (negatively charged at neutral pH) and XG (negatively charged gum) mixtures affects in particular the texture, rheology and sensory characteristics of F and F/T mashed potatoes with native anionic potato starch (containing negatively charged phosphate groups). A second objective, based on the results, was to use response surface methodology (RSM) for interpretation of the experimental data.

Materials and Methods

Test Material

The potatoes used were fresh tubers (cv Kennebec) from Aguilar de Campoo (Burgos, Spain) grown in 2007, with weights (g) within the confidence interval ($95.69 \leq \mu \leq 111.81$) and specific weights (g cm^{-3}) within the interval ($1.0721 \leq \mu \leq 1.0787$); $P \leq 0.01$. Raw material was stored in a chamber at 4 ± 1 °C and 85 % relative humidity throughout the experiment [28, 29].

Preparation of Mashed Potatoes

Tubers were manually washed, peeled and diced. Mashed potatoes were prepared in 650-g batches from 607.7 g kg^{-1} of potatoes, 230.8 g kg^{-1} of semi-skimmed in-bottle sterilized milk (calcium content, 0.12 g kg^{-1}; fat content, 15.5 g kg^{-1}), 153.8 g kg^{-1} of water and 7.7 g kg^{-1} of salt (NaCl) using a Thermomix TM 21 food processor (Vorwerk España, M.S.L., S.C., Madrid, Spain). Biopolymers were added at this point; the appropriate amount of SC and XG was added to the rest of ingredients in the form of a dry powder and all the ingredients were then cooked for 20 min at 100 °C (blade speed: 100 rpm), and the amount of liquid evaporated was determined by weighing the ingredients before and after boiling. The evaporated liquid was then replaced by an equal weight of boiling water and the ingredients were again cooked at 100 °C for 5 min. The mash was immediately ground for 40 s (blade speed: 2000 rpm) and homogenized through a stainless steel sieve (diameter 1.5 mm). Following preparation, half of each F sample was immediately analysed and the other half was placed on flat freezing and microwave thawing trays and frozen and thawed according to the procedures indicated below.

Cryoprotectant Mixtures

Table 1. Notation system for F and F/T mashed potatoes and biopolymer contents

System notation	Description and biopolymer content (g kg^{-1})
F, F/T	F and F/T mashed potatoes without added biopolymers
F-SC5, F/T-SC5	F and F/T mashed potatoes with 5 g kg^{-1} added sodium caseinate (SC)
F-XG1.5, F/T-XG1.5	F and F/T mashed potatoes with 1.5 g kg^{-1} added xanthan gum (XG)
F-SC1.46/XG1.5, F/T-SC1.46/XG1.5	F and F/T mashed potatoes with 1.46 g kg^{-1} added SC and 1.5 g kg^{-1} added XG
F-SC2.5/XG0.5, F/T-SC2.5/XG0.5	F and F/T mashed potatoes with 2.5 g kg^{-1} added SC and 0.5 g kg^{-1} added XG
F-SC2.5/XG2.5, F/T-SC2.5/XG2.5	F and F/T mashed potatoes with 2.5 g kg^{-1} added SC and 2.5 g kg^{-1} added XG
F-SC5/XG0.09, F/T-SC5/XG0.09	F and F/T mashed potatoes with 5 g kg^{-1} added SC and 0.09 g kg^{-1} added XG
F-SC5/XG1.5, F/T-SC5/XG1.5	F and F/T mashed potatoes with 5 g kg^{-1} added SC and 1.5 g kg^{-1} added XG
F-SC5/XG2.91, F/T-SC5/XG2.91	F and F/T mashed potatoes with 5 g kg^{-1} added SC and 2.91 g kg^{-1} added XG
F-SC7.5/XG0.5, F/T-SC7.5/XG0.5	F and F/T mashed potatoes with 7.5 g kg^{-1} added SC and 0.5 g kg^{-1} added XG
F-SC7.5/XG2.5, F/T-SC7.5/XG2.5	F and F/T mashed potatoes with 7.5 g kg^{-1} added SC and 2.5 g kg^{-1} added XG
F-SC8.54/XG1.5, F/T-SC8.54/XG1.5	F and F/T mashed potatoes with 8.54 g kg^{-1} added SC and 1.5 g kg^{-1} added XG
F-SC2.5/XG1.5, F/T-SC2.5/XG1.5	F and F/T mashed potatoes with 2.5 g kg^{-1} added SC and 1.5 g kg^{-1} added XG

Sodium caseinate (SC) (EM-6) was supplied by Manuel Riesgo, S.A. (Madrid, Spain), and food-grade xanthan gum powder (XG) (Keltrol F [E]) was donated by Premium Ingredients, S.L. (Girona, Spain). Range-finding experiments were performed at the outset of this research in order to ascertain the maximum acceptable amount of SC and XG that could be added to the mashed potato on the basis of flavour, viscosity and colour [22-24]. From these preliminary results, the lower and upper levels of each cryoprotectant to be used in mixtures were set at 2.5 and 7.5 g kg^{-1} for SC and at 0.5 and 2.5 g kg^{-1} for XG. Table 1 shows the biopolymer concentrations in the mashed potatoes tested, together with the notations used to refer to each of the samples.

Freezing and Thawing Processes

Mashed potatoes were frozen by forced convection with liquid nitrogen vapour in an Instron programmable chamber (model 3119-05, −70/+250 °C) at −60 °C until their thermal centres reached −24 °C (freezing rate, 1 ± 0.10 °C min^{-1}). Air and product temperatures were monitored by T-type thermocouples (NiCr/NiAl; −200 to +1000 °C) using the MMS3000™ Multi Measurement System™ (Mod. T4, Commtest Instruments, Christchurch, New Zealand). After freezing, the samples were packed in 300 × 200 mm^2 rectangular polyethylene plastic bags, sealed under light vacuum (–0.05 MPa) on a Multivac packing machine (Sepp Haggenmüller KG, Wolfertschwenden, Germany), placed in a domestic freezer for storage at −24 °C and left there for at least 1 month before thawing. For microwave thawing, frozen mashed potato samples were unpacked and then thawed in a Samsung M1712N microwave oven (Samsung Electronics S.A., Madrid, Spain). Samples were irradiated for a total of 20 min at an output power rating of 600 W. Samples were first irradiated for 15 min, and then removed from the microwave and gently stirred by hand with a spoon for 1 min to achieve a uniform sample temperature. Next, samples were irradiated for an additional 5 min under the same conditions. After thawing, the temperature reached at the product thermal centre was measured in all cases (+50 ± 5 °C) [30].

Heating of Samples

All the samples were brought up to 55 °C by placing them in a Hetofrig CB60VS water-bath (Heto Lab Equipment A/S, Birker∅d, Denmark), where water and product temperatures were monitored by T-type thermocouples using a hardware and software system developed with the LabWindows/CVI package (National Instruments Spain S.L., Madrid, Spain) for automation of thermal process control [31]. Sample testing temperature was 55 °C, as previous studies showed that this is the preferred temperature for consumption of mashed potato [30].

Instrumental Objective Texture Measurements

ITPA and CP tests were carried out with a TA HDi Texture Analyser (Stable Micro Systems Ltd, Godalming, UK) using a 250 N load cell and the application program supplied with the apparatus (Texture Expert for Windows™, version 2. 61). During the tests, the mashed potatoes were kept at 55 °C by means of a temperature controlled Peltier cabinet (XT/PC) coupled to a separate heat exchanger and proportional-integral-derivative (PID) control unit. For ITPA tests, a flat 35 mm diameter aluminium plunger (SMS P/35) was used inside a 60 mm diameter stainless steel vessel containing 50 ± 1 g of sample. The following experimental conditions were selected for each ITPA test: deformation rate (180 mm min^{-1}), compression level (33.3 %), with a rest period of 5 s between cycles. Textural parameters such as consistency: CON (N), adhesiveness: ADH (N s), springiness: SPR (dimensionless), cohesiveness: COH (dimensionless) and gumminess: GUM (N) were automatically calculated by the software from the curve generated by these tests. Definitions of textural parameters are described elsewhere [30, 32]. A spreadability rig was used to perform the cone penetration tests, consisting of a 45° conical Perspex probe (P/45 degree) that penetrated a conical sample holder containing 7 ± 0.1 g of mashed potatoes [32]. Tests were done at a deformation rate of 180 mm min^{-1}. The maximum resistance to penetration, MRP (N), and the average penetration force, APF (N), were measured. Instrumental measurements were performed in quadruplicate and results averaged.

Oscillatory Rheological Measurements

Small amplitude oscillatory shear experiments (SAOS) were carried out with a controlled stress rheometer (Bohlin CVR 50, Bohlin Instruments, Gloucestershire, UK) at 55 °C with smooth plates (40 mm diameter and 2 mm gap) and a solvent trap to prevent water evaporation. Samples were allowed to relax for 5 min before conducting rheological measurements such as equilibration time after loading the sample on the sensor system. Temperature control was carried out with a Peltier Plate system (-40 to +180 °C; Bohlin Instruments, Gloucestershire, UK). Dynamic rheological tests were performed under the following conditions: (1) linear viscoelastic domain determined for each sample from stress sweeps at 1 rad s^{-1}, and (2) three frequency sweeps over the range 0.1-100 rad s^{-1} at very low strains, mostly below 10^{-3}. A new sample was used each time for rheological measurements, which are therefore average values of four determinations. Bohlin software was used to calculate the phase angle, δ (°), storage modulus, G' (Pa), loss modulus, G'' (Pa), and complex viscosity, η^* (Pa s) at 1 rad s^{-1}. A parameter was defined from the stress sweeps to characterize the fluid behaviour for the nonlinear viscoelastic range (fluid-like relative angle, α), namely the ratio of the phase angle (δ) measured at strain $\gamma = 2\ 10^{-1}$ to the phase angle corresponding to a pure fluid ($\delta = 90°$) [33]. In addition, a power-law type relationship was verified for the dynamic rheological data of mashed potatoes. Linear regressions of ln (G') and ln (G'') versus ln (ω) were carried out in each experimental condition and the magnitudes of slope and intercepts were computed as described previously [34, 35].

Other Quality Parameters

Measurement of Colour

Instrumental measurement of colour of the mashed potato samples in the pots was carried out with a CM-3500d (Konica Minolta) spectrophotometer, and the results were expressed according to the CIELAB system with reference to illuminant D65 and a visual angle of 10°. The parameters determined were L^* (L^* = 0 [black] and L^* = 100 [white]), a^* ($-a^*$ = greenness and $+a^*$ = redness), b^* ($-b^*$ = blueness and $+b^*$ = yellowness), recommended by the International Commission on Illumination [36]. Colour was also expressed as L^*/b^*, i.e. the white/yellow ratio [37].

Drip Loss

Drip loss (DL) was measured by centrifugal force. Tubes containing the sample (approximately 10 g of mashed potatoes) were centrifuged at 6000 rpm (15000×g) for 30 min in a Sorvall®, RC-5B apparatus. DL was expressed as the percentage of liquid separated per total weight of sample in the centrifuge tube [38].

Determination of pH

The pH of the mashed potatoes was measured with a Schott CG pH meter (Model 842; Schott-Geräte GmbH, Mainz, Germany). Measurements of all the quality parameters were performed in quadruplicate and the results averaged.

Sensory Analyses

Samples were subjected to a descriptive quantitative method adapted for each product [39]. Sensory evaluation was conducted by a four-member panel with 6 years of specific training in potato purees [30, 32, 40, 41]. Each sample was tested twice and average scores calculated, so that each sample was tested eight times in all. Panellists were asked to evaluate samples for texture (consistency, adhesiveness, creaminess and fibrousness), appearance (authentic colour, off-colours, shine and uniformity), and taste (sweetness, authentic taste and off-taste). Scores were awarded on a scale of 1–5, in which 1 indicated total absence of the sensory attribute and 5 a very definite attribute. F and F/T mashed potatoes were also subjected to an overall acceptability (OA) test based on all sensory attributes (texture, colour, taste) on a nine-point hedonic scale (9=like extremely, 1=dislike extremely). In this case, sensory assessment was conducted by a fourteen-member untrained panel.

Experimental Design and Data Analysis

The effect of SC and XG concentration on the instrumental and sensory parameters used to evaluate the quality of mashed potatoes was studied using RSM. Statgraphics® software version 5.0 (STSC Inc., Rockville, MD, USA) was used to provide the experimental designs, calculate equations, do statistical evaluation and print out data. A central composite rotatable experimental design was used with SC concentration ranging between 2.5 and 7.5 g kg^{-1} and

XG concentration ranging between 0.5 and 2.5 g kg^{-1} as independent variables. In the design, coded variables are related to uncoded ones by equations: $X_1 = (\xi_1\text{-}a)/b$ and $X_2 = (\xi_2\text{-}c)/d$, where X_1, X_2, ξ_1 and ξ_2 are SC and XG concentrations in coded and uncoded mode respectively; a, c are the central points of the uncoded SC and XG concentrations ranges studied (5 and 1.5 g kg^{-1}, respectively); b, d are calculated as the difference between mentioned central points and the upper and lower conditions of the uncoded SC and XG concentrations ranges studied (2.5 and 1 g kg^{-1}, respectively).

The design required 13 variable combinations that were performed in random order, including five replicates of the centre region to generate contours of equal response [42]. Each combination of concentrations was repeated twice in the mashed potato samples and all the results averaged. The following dependent variables (responses) were analysed by RSM: objective texture parameters from ITPA and CP tests (CON, ADH, SPR, COH, GUM, MRP, and APF), oscillatory measurements (δ, G', G'', η^*, α, n', n'', $\ln(K')$, $\ln(K'')$), colour parameters (L^*, a^*, b^*, L^*/b^*), DL, pH, and the OA of the samples.

Data were analysed for coefficient of determination (R^2) value, probability, lack-of-fit test and regression coefficients (b_0, b_1, b_2, b_{11}, b_{22} and b_{12}). The relationship between the dependent and independent variables was expressed in terms of a second order polynomial equation having the form:

$$Y = b_0 + b_1X_1 + b_2X_2 + b_{11}X_1^2 + b_{22}X_2^2 + b_{12}X_1X_2$$

where

Y = response, X_1 = SC concentration (g kg^{-1}), X_2 = XG concentration (g kg^{-1}) and b_0, b_1, b_2, b_{11}, b_{22}, b_{12} = coefficients (constants) which measure linear, quadratic and interaction effects. Thus, the optimum ranges for the two factors tested (SC and XG concentrations) were interpreted from contour plots, and stationary points of X_1 and X_2 respectively, which give the optimum response, were calculated, and likewise the value of the estimated response at one random point [42, 43].

The effect of each added ingredient on the specific instrumental parameters was also studied by means of correlation analysis (Table 2). One-way analysis of variance was used to determine the differences in the instrumental measurements and in the OA at the different SC and XG concentrations in the central composite design, and in the controls for F and F/T mashed potatoes prior to optimization The least significant difference (LSD) tests were used with confidence intervals of 99 % for comparison of instrumental parameters and 95 % for OA.

Sensory attributes measured by the descriptive quantitative method (discrete values) were subjected to Friedman two-factor ranked analysis of variance ($k = 16$ and $N = 8$), and where differences existed between samples, a regular Wilcoxon signed rank test was used to pinpoint them [44]. Non-parametric tests were carried out using SPSS 14.0 software (SPSS, Inc., Chicago, IL). F and F/T mashed potatoes without and with individually added biopolymers at the centre points of the range studied for each biopolymer (i.e. SC at 5 g kg^{-1} and XG at 1.5 g kg^{-1}) were used as controls to study the effects of SC and XG mixtures on the quality of the mashed potatoes (samples F, F-SC5, F-XG1.5, F/T, F/T-SC5, F/T-XG1.5, Table 1).

Table 2. Correlation matrix of biopolymer and response data for SC and XG mixtures

F mashed potatoes

	SC	XG	SC*SC	XG*XG	SC*XG	CON	ADH	SPR	COH	GUM	MRP	APF	δ	G'	G''	η^*	α	n'	Ln(K')	n''	Ln(K'')
SC	1	-	-	-	-	0.27	0.26	0.03	0.25	0.41	0.28	0.44	0.30	-0.19	-0.12	-0.20	0.42	0.27	-0.13	_0.58_	-0.09
XG	-	1	-	-	-	0.07	0.07	-0.12	0.11	0.09	0.54	_0.56_	_0.83_	_-0.86_	_-0.74_	_-0.85_	_0.61_	_0.84_	_-0.86_	0.52	_-0.78_
SC*SC	-	-	1	-	-	0.19	0.18	0.02	0.37	0.39	0.45	_0.56_	0.34	-0.23	-0.14	-0.24	0.40	0.32	-0.20	_0.65_	-0.16
XG*XG	-	-	-	1	-	0.28	0.25	-0.23	-0.06	0.23	0.51	_0.60_	_0.69_	_-0.70_	_-0.63_	_-0.70_	0.40	_0.69_	_-0.73_	0.53	_-0.69_
SC*XG	-	-	-	-	1	0.15	0.14	0.01	0.26	0.25	_0.56_	_0.74_	_0.80_	_-0.74_	_-0.59_	_-0.73_	_0.70_	_0.81_	_-0.70_	_0.70_	_-0.59_

	SC	XG	SC*SC	XG*XG	SC*XG	L*	a*	b*	L*/b*	pH	OA
SC	1	-	-	-	-	0.21	0.27	_0.64_	_-0.77_	_0.76_	-0.33
XG	-	1	-	-	-	0.44	-0.10	0.24	-0.20	0.03	_0.76_
SC*SC	-	-	1	-	-	0.06	0.43	0.51	_-0.66_	_0.61_	-0.38
XG*XG	-	-	-	1	-	0.42	-0.04	0.20	-0.19	0.01	_0.62_
SC*XG	-	-	-	-	1	0.39	0.09	0.51	-0.54	0.46	0.32

F/T mashed potatoes

	SC	XG	SC*SC	XG*XG	SC*XG	CON	ADH	SPR	COH	GUM	MRP	APF	δ	G'	G''	η^*	α	n'	Ln(K')	n''	Ln(K'')
SC	1	-	-	-	-	-0.34	-0.41	0.13	0.22	-0.17	0.01	-0.11	0.25	-0.21	0.09	-0.18	0.29	0.39	-0.28	_0.68_	-0.09
XG	-	1	-	-	-	0.04	-0.14	0.21	_0.59_	0.33	0.07	0.13	_0.74_	-0.38	0.05	-0.47	0.03	_0.66_	-0.43	0.23	0.12
SC*SC	-	-	1	-	-	-0.49	-0.54	0.07	0.25	-0.29	0.05	-0.14	0.31	-0.26	0.10	-0.21	0.32	0.43	-0.30	_0.73_	-0.06
XG*XG	-	-	-	1	-	0.05	-0.18	0.00	0.39	0.24	0.04	0.14	_0.60_	-0.14	0.18	-0.25	0.01	0.49	-0.21	0.08	0.29
SC*XG	-	-	-	-	1	-0.21	-0.36	0.28	_0.55_	0.10	-0.06	-0.03	_0.76_	-0.49	-0.02	_-0.56_	0.40	_0.78_	-0.57	_0.58_	-0.03

Table 2. (Continued)

F/T mashed potatoes

	SC	XG	SC*SC	XG*XG	SC*XG	*L**	*a**	*b**	*L**/*b**	pH	OA
SC	1	-	-	-	-	0.08	-0.46	*0.65*	*-0.75*	*0.80*	-0.30
XG	-	1	-	-	-	0.53	-0.27	0.18	-0.05	0.23	*0.68*
SC*SC	-	-	1	-	-	-0.06	-0.35	0.50	*-0.61*	*0.86*	-0.22
XG*XG	-	-	-	1	-	*0.60*	-0.34	0.31	-0.17	0.29	0.48
SC*XG	-	-	-	-	1	0.42	-0.52	0.55	-0.51	*0.62*	0.33

Figures in italics are significant, $P < 0.05$.

Results and Discussion

Data Analysis of Rheological Properties and Quality Parameters

The effect of each supplementary cryoprotectant on the specific textural and viscoelastic properties and quality parameters of F and F/T mashed potatoes with added SC and XG mixtures can be seen in Table 2. Neither linear, quadratic nor interaction terms showed a significant correlation with the ITPA parameters in the F products, while increasing XG levels correlated significantly with augmented COH, and interaction also correlated significantly with this textural parameter in the F/T products. In F mashed potatoes, only the interaction term for SC and XG correlated significantly with the MRP parameter from CP tests, while XG concentration correlated significantly with APF and the quadratic terms for SC and XG concentrations and the interaction also correlated significantly with APF. Neither linear, quadratic nor interaction terms showed a significant correlation with CP parameters in F/T products.

Addition of XG to F mashed potatoes increased phase shift (δ) and rheological characteristic n' (slope of storage modulus), while increasing XG levels correlated significantly with reduced G', G'', η^* and rheological characteristics lnK' and lnK'' (ln of intercepts of ln(ω) versus ln(G') and ln(G'') regressions [34, 35]). Tis indicates that F mashed potatoes with added XG/SC mixtures behaved like a weaker gel structure, mainly due to the presence of XG in the systems, confirming previous findings [22, 23]. Also, according to Hemar *et al.*, [5], for XG concentrations up to 1 wt% the viscosity of commercial milk protein/XG mixtures is dominated by the viscosity of XG. The cause could be repelling forces between the potato starch granules and XG molecules at the surface due to incompatible charges in the chain structure, negatively charged gum and anionic potato starch [16]. In any event, in the case of δ, G', G'', η^*, n', lnK' and lnK'', quadratic terms for XG and an interaction term for SC and XG concentrations were significant. Unexpectedly, SC concentration correlated significantly with rheological characteristic n'' (slope of loss modulus); the quadratic term for SC and interaction also correlated significantly with n'' in the F products. Only α had a significant positive linear correlation with XG content, and again the interaction term correlated significantly with the fluid-like relative angle. According to Kobori *et al.* [3], at pH 6.6, negatively-charged SC is loosely associated with XG fibres by means of hydrophobic interactions. The significance of the interaction term for SC and XG concentrations for some rheological properties indicates that hydrophobic interactions between SC and XG probably also influenced the viscoelastic behaviour of the F mashed potatoes. Hydrophobic interactions between SC and XG would be weakly attractive due to electrostatic repulsion [3].

In contrast, the correlation matrix calculated for the F/T dataset (Table 2) showed that only δ had a significant positive linear and quadratic correlation with XG content, while the interaction term for SC and XG also correlated significantly with the phase angle. In F/T mashed potatoes only the interaction term correlated significantly with η^*. In the case of n' and n'', the linear term for XG and the quadratic term for SC correlated significantly with each rheological characteristic respectively; and again the interaction term correlated significantly with n' and n'' in F/T products. This means that processing significantly reduced the effect of SC/XG mixtures on the gel strength and the viscoelastic behaviour of the mashed potatoes as compared to their F counterparts. Freezing of mashed potatoes probably produced high local starch concentrations and allowed chain crystallization of both amylose and amylopectin to occur [45, 46]; in addition, the loose associations between SC particles and XG probably do not survive on thawing. Ice crystal growth leads to contraction and disruption of starch granules, while cell wall disruption leads to a redistribution of amylose and amylopectin, producing local differences in concentration. The locally concentrated amylopectin is prone to molecular aggregation leading to shrinkage and water expulsion [38]. Obviously, interactions between the different biopolymers are also affected by freezing and by the swelling capacity of potato starch granules on thawing.

Addition of SC to both F and F/T mashed potatoes increased b^*, while increasing SC levels correlated significantly with reduced L^*/b^* ratios (Table 2). Quadratic term for SC also correlated significantly with b^* in both F and F/T products. Thus, the addition of SC and XG mixtures produces a darker-coloured mashed potato, mainly associated with incorporation at high concentrations of SC. In both F and F/T mashed potatoes, there was a significant linear and quadratic correlation between pH and SC content, while in the F/T products the interaction term for SC and XG concentrations also correlated significantly with the pH. This suggests that κ-casein distribution in SC, which is responsible for the negative charge on the surface, and the hydrophobic region on the surface area in SC, would influence the cologarithm of the activity of dissolved hydrogen ions of the samples more. Finally, both linear and quadratic terms for XG significantly correlated with the overall acceptability (OA) of F mashed potatoes, while only linear term for the gum correlated significantly with the OA of the processed samples.

Data were compiled for SC and XG mixtures; mean values of all the objective texture parameters from ITPA and CP tests obtained for F and F/T mashed potatoes are shown in Table 3. Textural analysis provided information on the structure of the product and the changes that occurred when force was applied [47]. The concentrations of each added SC and XG mixture significantly affected all the objective texture measurements ($P < 0.01$) in mashed potatoes. In both F and F/T products, the highest CON values were recorded in the samples corresponding to design point 10 (F-SC5/XG2.91 and F/T-SC5/XG2.91 respectively), indicating that mashed potatoes with added SC/XG mixtures at the highest XG concentration (2.91 g kg^{-1}) had stronger structures than the controls. In F samples, the highest ADH value was also recorded in F-SC5/XG2.91. CON and ADH measure cell adhesion in the mashed potatoes [30], which was superior at the highest XG concentration, even after freezing. In caseinate-stabilized emulsions containing XG [7], increasing the polysaccharide concentration causes immobilisation of the microstructure due to an increase in the local viscoelasticity: that is, the emulsion structure cannot easily rearrange to expel xanthan-enriched aqueous serum phase because a weak gel-like network is generated. Probably, the presence of monovalent (NaCl) and divalent cations (Ca^{2+}) in the mashed potatoes also influenced the consistency of both samples F-SC5/XG2.91 and F/T-SC5/XG2.91. The effect of electrolytes on XG solutions is dependent upon gum concentration, and the addition of NaCl increase the structure hardness at higher XG concentrations. XG forms weak gels based on the end-to-end association of helical conformation, forming a double-helix structure, and the addition of salt increases viscosity when the concentration of XG is sufficiently high [27]. The presence of an ionic material stabilizes the three-dimensional XG network providing stable viscosity and reducing electrostatic repulsion between carboxylate anions [48, 49]. The gelation of XG dispersion can be enhanced by the addition of divalent, trivalent or borate ions

[48], although it has been reported that the addition of sodium to XG slightly increases gel rigidity [50].

In any event, in the processed samples no difference was found between CON values recorded at the replicated centre points (samples F/T-SC5/XG1.5) and at design point 10. The effect of freezing, which raised CON values as compared to their F counterparts, could result from a slow, retrogradation-like mutual orientation of the components of the mashed potatoes, as has been proposed for dessert sauces thickened by starch-XG combinations [47]. On the other hand, in the F samples the lowest CON value was recorded in F-SC1.46/XG1.5, the sample with the lowest aggregate concentration of the two added biopolymers. It has been stated that moderately low levels of XG addition lead to depletion flocculation and gravity-induced phase separation in caseinate-stabilized emulsions containing XG [7]. On the other hand, caseinate levels as low as 0.01% dramatically reduce the swelling volume of potato starch and hence the viscosity of the system [21].

Similarly, in the F/T samples the lowest CON values were found in F/T-SC5, F/T-SC7.5/XG2.5, F/T-SC8.54/XG1.5 and F/T-SC1.46/XG1.5, in that order. Moreover, in both F and F/T mashed potatoes, addition of SC alone reduced CON values as compared to controls without added biopolymers (F and F/T respectively). Therefore, there was softening of mashed potatoes associated with the addition of SC at high concentrations, probably due to phase separation, producing a continuous casein phase in the intergranular space that prevents the formation of a continuous amylose network. It is further suggested that phase separation between starch and SC is encouraged at high salt concentration [21].

Comparison of the data without and with processing showed that the freezing and thawing of the mashed potatoes caused (except in mashed potatoes with 5 g kg^{-1} added SC) an increase in CON and ADH values as compared to their F counterparts. In commercial mashed potato, penetration and oscillatory parameters indicated that freezing and thawing resulted in a firmer product than the fresh control; this was attributed to retrogradation of gelatinized starch [30]. Also, Giannouli and Morris [51] found that much stronger and more cohesive networks are formed when solutions of XG are frozen and thawed.

With respect to dimensionless parameters—those that could be related to the internal composition or forces that maintain the integrity of the mashed potatoes, i.e., SPR and COH—in both F and F/T products, the lowest SPR values were recorded for F and F/T-SC5 controls respectively. Therefore, in the case of F samples, the addition of both biopolymers, either alone or mixed, increased retarded recoverable elasticity. In both F and F/T mashed potatoes, the lowest COH values were again recorded in control samples F and F-SC5 and F/T and F/T-SC5 respectively, indicating that the mashed potatoes with added XG alone and with added SC/XG binary mixtures were also more cohesive. Certainly, a spongy structure was observed in thawed samples which had been frozen without cryoprotectants. In starch pastes submitted to freezing and frozen storage, hydrocolloids prevent the formation of a sponge-like structure and the production of exudates due to amylose retrogradation [52, 53]. Specifically, it has been stated that in starch pastes with added XG, amylose-XG interaction competes with amylose-amylopectin aggregation, reducing or retarding the probability of amylose retrogradation. Consequently, the swelling volume of the potato starch and hence the cohesiveness of the system is increased on thawing.

Table 3. Coded and uncoded variables of the response surface design and instrumental objective texture measurements for F and F/T mashed potatoes with different combinations of concentrations added, and controls

System notation (design point)	Coded		Uncoded		CON	ADH	SPR	COH	GUM	MRP	APF
	X_1	X_2	ξ_1: SC (g kg^{-1})	ξ_2: XG (g kg^{-1})	(N)	(N s)			(N)	(N)	(N)
F					1.94 f	-3.83 b	0.86 a	0.63 a	1.22 c-f	14.15 b	1.56 a, b
F-SC5					1.66 d	-3.24 c, d	0.90 b	0.64 a	1.06 a	15.38 a	1.51 b
F-XG1.5					1.66 d	-4.00 b	0.96 c, d	0.82 f-h	1.36 g, h	12.45 c	1.51 b
F-SC1.46/XG1.5 (1)	-1.41	0	1.46	1.5	1.37 a	-3.17 d	0.95 c, d	0.83 g, h	1.13 a, b	12.19 c, d	1.32 e, f
F-SC2.5/XG0.5 (2)	-1	-1	2.5	0.5	1.66 d	-3.90 b	0.97 d	0.80 d-h	1.33 g, h	9.35 f	1.18 e
F-SC2.5/XG2.5 (3)	-1	1	2.5	2.5	1.57 b, c	-3.65 b, c	0.95 c, d	0.77 c-f	1.21 b-e	13.53 b	1.41 c
F-SC5/XG0.09 (4)	0	-1.41	5	0.09	1.79 e	-4.02 b	0.94 c	0.66 a, b	1.19 b-d	6.00 i	0.93 h
F-SC5/XG1.5 (5)	0	0	5	1.5	1.62 c, d	-3.87 b	0.97 c, d	0.81 e-h	1.31 f, g	8.81 f, g	1.14 e, f
F-SC5/XG1.5 (6)	0	0	5	1.5	1.65 c, d	-3.97 b	0.95 c, d	0.75 c, d	1.23 d-f	7.95 h	1.08 f, g
F-SC5/XG1.5 (7)	0	0	5	1.5	1.50 b	-3.62 b, c	0.96 c, d	0.78 d-g	1.17 b-d	8.64 f-h	1.03 g
F-SC5/XG1.5 (8)	0	0	5	1.5	1.60 c, d	-3.84 b	0.96 c, d	0.80 d-h	1.24 d-f	8.48 g, h	1.10 e-g
F-SC5/XG1.5 (9)	0	0	5	1.5	1.63 c, d	-3.90 b	0.97 c, d	0.81 e-h	1.28 e-g	8.65 f-h	1.12 f, g
F-SC5/XG2.91(10)	0	1.41	5	2.91	2.36 h	-5.78 a	0.95 c, d	0.72 b, c	1.69 h	11.59 d, e	1.52 b
F-SC7.5/XG0.5 (11)	1	-1	7.5	0.5	2.16 g	-5.64 a	0.97 c, d	0.85 h	1.82 i	11.00 e	1.42 c
F-SC7.5/XG2.5 (12)	1	1	7.5	2.5	1.65 c, d	-3.96 b	0.97 c, d	0.86 h	1.40 g	14.27 b	1.55 a, b
F-SC8.54/XG1.5 (13)	1.41	0	8.54	1.5	1.49 b	-3.16 d	0.95 c, d	0.83 g, h	1.24 d-f	16.12 a	1.62 a
F-SC2.5/XG1.5 (14)*	-1	0	2.5	1.5	1.50 b	-3.31 c, d	0.94 c, d	0.76 c-e	1.14 a-c	14.08 b	1.43 c
LSD (99%)					*0.0872*	*0.4174*	*0.0332*	*0.0576*	*0.0889*	*0.8374*	*0.0750*
F/T					2.08 d-f	-4.03 c, d	0.94 b	0.78 a, b	1.63 b	13.99 e-g	1.61 c-e
F/T-SC5					1.52 a	-2.85 e	0.92 a	0.84 a-c	1.27 a	11.83 b, c	1.42 b
F/T-XG1.5					2.08 d-f	-4.35 b-d	0.96 b-d	0.96	1.99 e-g	13.24 d, e	1.61 c-e
F/T-SC1.46/XG1.5 (1)	-1.41	0	1.46	1.5	1.89 b-d	-4.48 a-c	0.96 b-d	0.96	1.82 c, d	12.67 c, d	1.41 b
F/T-SC2.5/XG0.5 (2)	-1	-1	2.5	0.5	2.17 e-g	-5.19 a, b	0.96 b-d	0.86 b-d	1.86 d, e	14.46 f, g	1.76 f, g
F/T-SC2.5/XG2.5 (3)	-1	1	2.5	2.5	2.06 d-f	-4.37 b-d	0.95 b-d	0.91 c-e	1.87 d, e	14.27 e-g	1.69 e, f

F/T-SC5/XG0.09 (4)	0	-1.41	5	0.09	2.01 c-e	-4.37 b-d	0.95 b, c	0.77 a	1.55 b	11.13 a, b	1.39 a, b
F/T-SC5/XG1.5 (5)	0	0	5	1.5	2.27 f, g	-5.27 a	0.97 d	0.96 e	2.18 h, i	13.17 d, e	1.68 e, f
F/T-SC5/XG1.5 (6)	0	0	5	1.5	2.17 e-g	-4.73 a-c	0.96 b-d	0.95 e	2.07 g, h	13.32 d, e	1.60 c-e
F/T-SC5/XG1.5 (7)	0	0	5	1.5	2.18 e-g	-5.03 a, b	0.96 b-d	0.94 d, e	2.05 f-h	13.57 d-f	1.57 c, d
F/T-SC5/XG1.5 (8)	0	0	5	1.5	2.21 e-g	-5.01 a, b	0.96 b-d	0.95 e	2.10 g, h	13.36 d, e	1.62 c-e
F/T-SC5/XG1.5 (9)	0	0	5	1.5	2.22 e-g	-5.11 a, b	0.97 c, d	0.95 e	2.11 g, h	13.27 d, e	1.60 c-e
F/T-SC5/XG2.91(10)	0	1.41	5	2.91	2.40 g	-5.14 a, b	0.96 b-d	0.97 e	2.30 i	14.95 g	1.82 g
F/T-SC7.5/XG0.5 (11)	1	-1	7.5	0.5	2.02 d, e	-4.61 a-c	0.96 b-d	0.95 e	1.92 d-f	14.91 g	1.65 d, e
F/T-SC7.5/XG2.5 (12)	1	1	7.5	2.5	1.67 a, b	-3.61 d, e	0.96 b-d	0.94 e	1.58 b	10.65 a	1.31 a
F/T-SC8.54/XG1.5 (13)	1.41	0	8.54	1.5	1.79 b, c	-3.95 c, d	0.96 b-d	0.96 e	1.69 b, c	14.99 g	1.66 d-f
F/T-SC2.5/XG1.5 (14)*	-1	0	2.5	1.5	2.12 e, f	-4.96 a, b	0.97 c, d	0.98 e	2.07 g, h	12.64 c, d	1.54 c
LSD (99%)				1.5	*0.2292*	*0.8556*	*0.0209*	*0.0823*	*0.1404*	*1.0981*	*0.1044*

Different letters in the same column indicate significant differences $P < 0.01$. LSD, least significant difference. * Design point number 14 was carried out to test the accuracy of the models.

In both F and F/T mashed potatoes, the lowest GUM values, calculated as CON×COH, were recorded in control samples F-SC5 and F/T-SC5, again indicating that starch granule swelling is discouraged by SC at intermediate concentration. Note that COH and GUM values were also higher in the F/T mashed potatoes than in their F counterparts. During freezing, xanthan chains are forced to align and associate by conversion of water to ice crystals, producing a cryogel network [51].

The lowest MRP and APF values in F and F/T mashed potatoes were recorded at design points 4 and 12 respectively (samples F-SC5/XG0.09 and F/T-SC7.5/XG2.5), and in these samples the CP parameters were significantly lower than in the three controls. In the case of F mashed potato, MRP and APF values recorded at the replicated centre points (samples F-SC5/XG1.5) and at the design point 2, where the XG concentration was low (0.5 g kg^{-1}), were also lower than in the controls. This result demonstrates that gel strength in the systems with added SC and XG mixtures was enhanced only at the highest SC concentration (8.54 g kg^{-1}).In the case of emulsions exhibiting gravity creaming, a low xanthan concentration generally induces visible phase separation, attributable mainly to reversible depletion flocculation of protein-coated emulsion droplets by the non-adsorbing polysaccharide [7]. However, according to Kobori *et al.*, [3], SC mixed with XG does not show a clear phase separation at neutral pH. Also, CP parameters were higher in the F/T samples than in the F equivalents.

Mean values of all the dynamic properties and rheological characteristics obtained for the F and F/T mashed potatoes are shown in Table 4. Also, the different concentrations of each added SC and XG mixture significantly affected all the oscillatory measurements ($P < 0.01$). In both F and F/T mashed potatoes, the lowest δ values were recorded in control samples without and with added SC alone (samples F, F/T, F-SC5 and F/T-SC5), as well as in the products corresponding to design point 4 (samples F-SC5/XG0.09 and F/T-SC5/XG0.09). Most of the F mashed potatoes had lower phase angles than their F/T counterparts. In the F samples, the highest G', G'' and η^* values were again recorded in F samples, F-SC5, F-SC5/XG0.09 and F-SC2.5/XG0.5. Similarly, in the F/T samples the highest G' and η^* values were found in F/T, F/T-SC5, F/T-SC7.5/XG0.5 and F/T-SC5/XG0.09, in that order. In the processed samples, the highest G'' value was recorded for design point 11 (F/T-SC7.5/XG0.5), although the differences with respect to the G'' values recorded for design point 10 (F/T-SC5/XG2.91) and controls (F/T, F/T-SC5 and F/T-XG1.5) were not significant.

The oscillatory properties, then, indicate that mashed potatoes without added XG were more solid-like. Lai *et al.*, [54] stated that there is a thermodynamic incompatibility between starch polysaccharides and XG, as exclusion effects of swollen granules increase the concentration of other polysaccharides in the continuous phase. On the other hand, the addition of salts to XG is known to reduce viscosity because of the sharp change from random coil to rigid ordered conformation [55], whereas the addition of salt to starch increases the cold paste viscosity. It appears that in the presence of salt the synergistic interaction of starch and XG is less effective in increasing the cold paste viscosity, probably because of the conformational change in XG [56]. In binary potato starch gels, XG has been reported to inhibit formation of the polymeric network [57]. Moreover, Arltoft *et al.* [58] recently suggested that since XG is a non-gelling agent, its role in mixed gels could be confined to act as a filler rather than as an active component. The results reported in this chapter also show

Table 4. Oscillatory dynamic measurements for F and F/T mashed potatoes with different combinations of concentrations added, and controls

System notation (design point)	δ (°)	G' (Pa)	G'' (Pa)	η^* (Pa s)	α	n'	ln (K')	n''	ln (K'')
F	8.70 a	8957.75 a	1463.25 a	8606.25 a	0.1033 a	0.11 a	9.04 a	0.06 a, b	7.40 i
F-SC5	10.85 c	7144.50 b	1321.00 c, d	6899.00 b	0.1022 a	0.12 a	8.83 b	0.05 a	7.28 g-i
F-XG1.5	12.62 d, e	5393.00 d	1293.75 c, d	5248.25 c	0.1444 e	0.14 b	8.52 d, e	0.06 b	7.22 f, g
F-SC1.46/XG1.5 (1)	13.82 f-h	4175.00 f, g	1026.10 f, g	4031.25 e, f	0.1489 f	0.16 d, e	8.23 g, h	0.09 d-f	6.97 c, d
F-SC2.5/XG0.5 (2)	11.87 d	7061.25 b	1456.75 a, b	6817.00 b	0.1222 c	0.14 b	8.75 b, c	0.08 c	7.36 h, i
F-SC2.5/XG2.5 (3)	15.90 j	2956.25 i	836.55 h	2881.50 g	0.1678 g	0.18 g, h	7.89 h	0.10 f, g	6.76 a
F-SC5/XG0.09 (4)	9.80 b	7052.50 b	1187.50 e	6707.75 b	0.1178 b	0.12 a	8.80 b	0.08 c-e	7.22 f, g
F-SC5/XG1.5 (5)	14.87 i	4137.50 f, g	1075.15 f	3999.25 e, f	0.1822 j	0.17 e, f	8.28 f-h	0.09 d-f	7.06 d, e
F-SC5/XG1.5 (6)	14.40 h, i	4043.00 g	1008.50 f, g	4113.75 e	0.1900 l	0.17 e, f	8.28 f-h	0.07 c	7.03 c-e
F-SC5/XG1.5 (7)	13.50 f, g	4314.00 f	1065.50 f	4272.50 e	0.1722 h	0.16 d, e	8.33 f, g	0.10 f, g	7.05 d, e
F-SC5/XG1.5 (8)	14.24 g-i	4160.25 f, g	1050.10 f	4130.50 e	0.1820 j	0.17 d-f	8.30 f-h	0.09 d-f	7.10 e, f
F-SC5/XG1.5 (9)	14.54 h, i	4180.50 f, g	1070.75 f	4156.25 e	0.1878 k	0.17 e, f	8.32 f, g	0.09 e, f	7.09 d-f
F-SC5/XG2.91(10)	14.87 i	3718.75 h	1018.20 f, g	3722.75 f	0.1722 h	0.17 f, g	8.18 g	0.12 h	6.97 c, d
F-SC7.5/XG0.5 (11)	13.37 e, f	5820.00 c	1369.25 b, c	5533.25 c	0.1411 d	0.15 b, c	8.63 c, d	0.10 f, g	7.26 g, h
F-SC7.5/XG2.5 (12)	16.97 k	3127.00 i	954.65 g	3098.00 g	0.1778 i	0.19 h	8.02 h	0.11 g, h	6.92 b, c
F-SC8.54/XG1.5 (13)	15.75 j	3097.75 i	861.58 h	3010.25 g	0.2067 m	0.18 h	7.97 h	0.13 i	6.82 a, b
F-SC2.5/XG1.5 (14)*	14.27 g-i	5025.50 e	1266.00 d, e	4816.50 d	0.1667 g	0.16 c, d	8.41 e, f	0.08	7.20 f, g
LSD (99%)	*0.7980*	*252.5000*	*88.9615*	*367.0460*	*0.0022*	*0.0107*	*0.1376*	*0.0111*	*0.1263*
F/T	11.10 a	5490.25 g	1077.75 e	5244.00 f	0.1344 a, b	0.12 a	8.54 g	0.06 a	7.12 g
F/T-SC5	11.27 a	4494.25 f	955.57 b-e	4392.50 e	0.1500 a-d	0.12 a	8.40 f	0.07 a, b	6.94 d-f
F/T-XG1.5	18.17 f	3087.75 b	982.15 c-e	3038.50 a, b	0.1756 c-e	0.17 h, i	7.93 a, b	0.07 a, b	6.96 e, f
F/T-SC1.46/XG1.5 (1)	13.90 b-d	3389.50 c, d	842.15 a-c	3291.00 b, c	0.1678 b-e	0.15 c-e	8.09 d	0.08 a-d	6.85 b-e
F/T-SC2.5/XG0.5 (2)	12.42 a, b	3922.75 e	862.65 a-d	3761.25 c, d	0.1644 a-e	0.14 b	8.23 e	0.09 b-d	6.87 c-e
F/T-SC2.5/XG2.5 (3)	13.92 b-d	3483.00 d	860.82 a-d	3363.25 b, c	0.1722 c-e	0.15 c-f	8.09 d	0.08 a-d	6.83 b-e
F/T-SC5/XG0.09 (4)	11.10 a	4282.50 f	764.87 a, b	4085.25 d, e	0.1300 a	0.11 a	8.30 e, f	0.06 a	6.80 a-c

Table 4. (Continued)

System notation (design point)	δ (°)	G' (Pa)	G'' (Pa)	η^* (Pa s)	α	n'	ln (K')	n''	ln (K'')
F/T-SC5/XG1.5 (5)	13.62 b-d	3396.75 c, d	863.80 a-d	3331.00 b, c	0.1656 a-e	0.16 e-g	8.08 d	0.10 c, d	6.84 b-e
F/T-SC5/XG1.5 (6)	14.65 d, e	3160.25 b, c	826.57 a-c	3058.00 a, b	0.1644 a-e	0.16 f-h	8.00 b-d	0.11 c, d	6.76 a-c
F/T-SC5/XG1.5 (7)	12.52 a, b	3177.00 b, c	697.97 a	3031.00 a, b	0.1433 a-c	0.14 b, c	7.97 a-d	0.09 b-d	6.68 a
F/T-SC5/XG1.5 (8)	13.60 b-d	3244.67 b-d	796.12 a-c	3140.00 a, b	0.1578 a-e	0.15 c-f	8.02 b-d	0.10 c, d	6.73 a, b
F/T-SC5/XG1.5 (9)	13.80 b-d	3285.33 b-d	813.21 a-c	3220.00 b	0.1596 a-e	0.16 d-g	8.03 b-d	0.10 c, d	6.78 a-c
F/T-SC5/XG2.91(10)	13.92 b-d	4290.75 f	1058.45 d, e	3920.75 d, e	0.1567 a-e	0.15 c-f	8.27 e	0.10 b-d	7.03 f, g
F/T-SC7.5/XG0.5 (11)	12.90 b, c	4346.25 f	1135.90 e	4250.50 d, e	0.1322 a, b	0.15 b-d	8.28 e, f	0.11 d, e	6.96 e, f
F/T-SC7.5/XG2.5 (12)	15.60 e	2787.75 a	785.80 a-c	2662.75 a	0.1933 e	0.18 i	7.87 a	0.11 c, d	6.77 a-c
F/T-SC8.54/XG1.5 (13)	14.35 c-e	2800.50 a	773.07 a, b	2853.25 a, b	0.1811 d, e	0.17 g-i	7.95 a-c	0.14 e	6.77 a-c
F/T-SC2.5/XG1.5 (14)*	14.35 c-e	3516.25 d	869.32 a-d	3282.25 b, c	0.1578 a-e	0.15 c-f	8.05 c, d	0.08 a-c	6.81 b-d
LSD (99%)	*1.5093*	*285.9820*	*204.1860*	*535.4260*	*0.0372*	*0.0109*	*0.1236*	*0.0309*	*0.1316*

Different letters in the same column indicate significant differences $P < 0.01$. LSD, least significant difference. * Design point number 14 was carried out to test the accuracy of the models.

that the viscoelastic gel-like structure of the fresh products was weakened by freezing and thawing, confirming previous findings [24]. Oscillatory parameters were used to measure the intercellular friction, which was less after freezing [30].

Table 4 also shows the effect of SC and XG concentrations on the α parameter characterizing the nonlinear viscoelastic domain, and likewise on the slopes (n' and n") and ln of intercepts of ln(ω) versus ln(G') and ln(G") regressions. The lowest α values were found in the F samples without added biopolymers, and with added SC at 5 g kg^{-1}, as one single component or mixed with XG at the lowest concentration (samples F-SC5 and F-SC5/XG0.09). The F mashed potatoes that displayed a strong structure over the linear viscoelastic range, then, were also more solid-like, with high viscosity after breakdown ($\alpha \rightarrow 0$). Of the processed samples, F/T, F/T-SC5 and F/T-SC5/XG0.09 also displayed lower α values, although the differences between α values after processing were less significant. Since foodstuffs are subject to large deformations, it is definitely important to analyse both linear and nonlinear viscoelastic ranges in order to determine a product's performance under actual processing conditions [33].

In both F and F/T samples, n' values were higher than n" values, indicating that G' was more frequency-dependent than G" [35]. For example, n' values were higher at centre points than in F, F/T, F-SC5 and F/T-SC5 controls, again indicating that the replicated mashed potatoes (samples F-SC5/XG1.5 and F/T-SC5/XG1.5) had a weaker gel structure. At the replicated centre points 5-9, n' values were lower and n" values were higher in the F/T samples, indicating that G' was more independent of frequency, whereas G" was more dependent in the processed samples. Freezing, then, affected the viscous component more than the elastic component of the mashed potatoes.

Table 5 shows the effect of SC and XG concentrations on the other quality parameters and the OA of the mashed potatoes. Again, concentrations of each mixture significantly affected all the quality parameters considered and the OA of the samples ($P < 0.01$, and $P < 0.05$ for OA). As compared to F control without added biopolymers, L^* decreased when SC and XG were added to the mashed potatoes, either individually or combined in binary mixtures. Addition of SC alone at 5 g kg^{-1} to F and F/T controls significantly reduced the L^*, a^*, and b^* values and increased the L^*/b^* ratio, i.e. produced lighter-coloured mashed potatoes. Conversely, as compared to F-SC5 control, a^* values increased significantly when 1.5 g kg^{-1} XG was added to F mashed potatoes, although no differences were detected after freezing and thawing. These results mean that addition of SC alone increased the greenness of F mashed potatoes, whereas addition of XG unaccompanied caused loss of greenness. In both F and F/T products, addition of both biopolymers at the lowest total concentrations (design points 1-3) produced a lighter-coloured mashed potato (higher L^*/b^* ratio) than either of the controls without added biopolymers. Note however that in comparison to F-SC5 and F/T-SC5 controls, when higher concentrations of SC and XG were added (design points 4-13), the L^*/b^* ratio decreased, causing darkening, and its value was similar to that of the F and F/T controls. Therefore, it would seem that addition of XG and SC together mitigated the effect on colour produced by the addition of SC alone. Despite the significant changes found, it is worth noting that differences in the sample colour do not seem to be of great practical importance.

Table 5. Colour parameters, other quality measurements and overall acceptability for F and F/T mashed potatoes with different combinations of concentrations added, and controls

System notation (design point)	*L**	*a**	*b**	*L**/*b**	DL (%)	pH	OA
F	68.38 a	-3.23 a-c	8.33 a	8.21 a	9.11 a	6.06 d	6.57 e
F-SC5	66.48 e-g	-3.35 a	6.75 h, i	9.84 f, g	8.40 b	6.09 g	4.75 d
F-XG1.5	66.22 g, h	-2.71 g-i	7.58 d-f	8.74 b, c	0.00 e	6.04 b	8.62 j
F-SC1.46/XG1.5 (1)	65.48 j	-3.06 b-d	6.83 h	9.60 e, f	0.00 e	6.03 a	8.02 i
F-SC2.5/XG0.5 (2)	64.42 k	-3.03 c-e	6.41 i	10.06 g	1.49 c	6.05 c	5.12 d
F-SC2.5/XG2.5 (3)	65.93 h, i	-2.96 d, e	7.07 g, h	9.32 d, e	0.00 e	6.06 d	8.92 j
F-SC5/XG0.09 (4)	65.49 j	-2.97 d, e	7.46 e-g	8.78 b, c	1.96 d	6.08 f	4.35 c
F-SC5/XG1.5 (5)	67.62 b	-2.62 i	8.21 a, b	8.25 a	0.00 e	6.09 g	7.42 g
F-SC5/XG1.5 (6)	66.51 e-g	-2.71 g-i	7.79 c-f	8.55 a, b	0.00 e	6.08 f	7.87 h, i
F-SC5/XG1.5 (7)	66.40 f, g	-3.03 c-e	7.30 f, g	9.11 c, d	0.00 e	6.08 f	7.60 g, h
F-SC5/XG1.5 (8)	66.80 c-e	-2.81 f-i	7.80 b-f	8.72 b, c	0.00 e	6.09 g	7.45 g
F-SC5/XG1.5 (9)	67.10 c	-2.94 e-g	8.16 a-c	8.75 b, c	0.00 e	6.08 g	7.67 g-i
F-SC5/XG2.91(10)	67.69 b	-2.92 e-h	7.89 b-d	8.59 a, b	0.00 e	6.07 e	6.97 f
F-SC7.5/XG0.5 (11)	66.93 c, d	-3.09 b-d	8.00 a-c	8.36 a, b	0.00 e	6.08 f	4.87 d
F-SC7.5/XG2.5 (12)	66.72 d-f	-3.06 b-d	8.00 a-c	8.34 a, b	0.00 e	6.09 g	3.92 b
F-SC8.54/XG1.5 (13)	64.68 k	-3.27 a, b	7.48 d-g	8.65 a-c	0.00 e	6.08 f	3.35 a
F-SC2.5/XG1.5 (14)*	65.63 i, j	-3.40 a	7.48 d-g	8.77 b, c	0.00 e	6.05 c	6.42 e
LSD (99%)	*0.3535*	*0.2325*	*0.4155*	*0.4529*	*0.2907*	*0.0099*	*0.3662*
F/T	66.37 d	-3.23 e, f	7.86 e, f	8.45 a-c	21.12 a	6.15 a, b	4.70 b
F/T-SC5	65.45 b, c	-3.36 a, b	6.66 a	9.83 f	5.37 b	6.14 a	2.32 a
F/T-XG1.5	65.83 c	-3.38 a	7.41 b-d	8.88 d, e	0.00 c	6.14 a	5.80 d
F/T-SC1.46/XG1.5 (1)	65.39 b	-3.33 a-c	6.76 a	9.67 f	3.66 d	6.14 a	7.65 h
F/T-SC2.5/XG0.5 (2)	65.40 b	-3.25 d-f	7.20 b	9.08 e	14.55 e	6.15 a, b	5.12 c

F/T-SC2.5/XG2.5 (3)	67.41 g, h	-3.29 b-e	7.38 b, c	9.13 e	0.00 c	6.17 a-c	7.37 g
F/T-SC5/XG0.09 (4)	66.50 d, e	-3.20 f, g	8.19 f, g	8.13 a	18.75 f	6.16 a-c	2.20 a
F/T-SC5/XG1.5 (5)	67.36 g, h	-3.02 k	7.81 e	8.62 c, d	0.00 c	6.16 a-c	6.25 f
F/T-SC5/XG1.5 (6)	66.59 d, e	-3.10 i, j	7.75 d, e	8.60 c, d	0.00 c	6.15 a, b	6.12 e, f
F/T-SC5/XG1.5 (7)	66.49 d, e	-3.32 a-d	7.82 e	8.51 b, c	0.00 c	6.16 a-c	6.15 e, f
F/T-SC5/XG1.5 (8)	66.81 e, f	-3.15 g-i	7.80 e	8.58 c, d	0.00 c	6.15 a, b	6.17 e, f
F/T-SC5/XG1.5 (9)	67.14 f, g	-3.19 f-h	7.81 e	8.59 c, d	0.00 c	6.15 a, b	6.19 f
F/T-SC5/XG2.91(10)	67.69 h	-2.99 k	8.26 g	8.20 a, b	0.00 c	6.17 a-c	5.95 d, e
F/T-SC7.5/XG0.5 (11)	66.65 d, e	-3.11 h, i	7.84 e, f	8.50 b, c	0.00 c	6.19 b, c	4.97 c
F/T-SC7.5/XG2.5 (12)	67.68 h	-3.03 j, k	8.32 g	8.14 a	0.00 c	6.20 c	6.25 f
F/T-SC8.54/XG1.5 (13)	64.85 a	-3.25 c-f	7.60 c-e	8.53 b-d	0.00 c	6.20 c	5.85 d
F/T-SC2.5/XG1.5 (14)*	66.67 d, e	-3.34 a, b	7.83 e	8.52 b, c	0.00 c	6.15 a, b	6.32
LSD (99%)	*0.4125*	*0.0774*	*0.3483*	*0.3570*	*1.3395*	*0.0410*	*0.2289*

Different letters in the same column indicate significant differences $P < 0.01$. LSD, least significant difference. * Design point number 14 was carried out to test the accuracy of the models.

In the case of F mashed potatoes, DL percentages were significantly higher in F and F-SC5 controls than in the rest of the samples (Table 5). Addition of at 5 g kg^{-1} SC alone only reduced DL from 9.11 to 8.40 %. Water loss resulting from the addition of SC alone may be caused by its displacement from interchain spaces in the potato starch matrix as a result of the occupation of these spaces by the protein [59]. On the other hand, there was no drip loss in the F-XG1.5 control, which is consistent with the well-established ability of XG to immobilize water and thus reduce DL as reported elsewhere [23, 60]. Relaxation times of water molecules in binary potato starch gels with XG have been measured [57]. According to the authors, the amount of free water and its mobility was limited in the presence of XG. At a concentration of 0.5/100 g suspension, XG showed reduced exudate production more effectively than guar gum during refrigerated storage of yam starch pastes [53]. Our results, then, also corroborate previous findings. Addition of SC and XG mixtures reduced drip loss (DL) to 0.00 %, except when the lowest XG concentrations were added to F mashed potatoes (design points 2 and 4). In F/T control, F/T-SC2.5/XG0.5 and F/T-SC5/XG0.09, DL percentages were higher than in their F counterparts. Moreover, freezing and thawing increased DL from 0 to 3.66 % with respect to the F counterparts in mashed potatoes with 1.46 g kg^{-1} added SC and 1.5 g kg^{-1} added XG. The syneresis and related physical property changes induced by freezing and thawing could be reduced by adding a small amount of hydrocolloids, but the cryoptotecting mechanism of the hydrocolloids is not fully understood [52]. Unexpectedly, however, the water adsorption capacity of sample F/T-SC5 was higher than in its F counterpart.

In fact in this study DL values revealed that addition of XG at 1.5 g kg^{-1} or higher, either alone or mixed with SC, effectively stabilized mashed potatoes against syneresis. On the other hand, SC alone exhibited no water-holding capability as a result of pronounced serum separation due to depletion flocculation induced by excess unadsorbed protein in the aqueous continuous phase, as reported elsewhere [7-9].

Of the fresh control samples, the lowest pH value was recorded in F-XG1.5 control; when SC was added individually to F mashed potatoes at 5 g kg^{-1}, the pH value was higher than in the F control without added biopolymers. Of the F products containing SC and XG mixtures, the pH value was significantly lower in F-SC1.46/XG1.5 (corresponding to the lowest SC concentration) than in the rest of the samples, although pH values in F-SC2.5/XG0.5 and F-SC2.5/XG1.5 were also quite low. Addition of SC and XG mixtures to F mashed potatoes did not increase the pH of the samples as compared to F-SC5 control. This means that contribution of SC to the pH of the F mashed potatoes was more significant than that of XG. On the other hand, addition of SC at 5 g kg^{-1} or at a higher concentration, either individually or mixed with XG, raised the pH of F mashed potatoes. In spite of the significant differences found, the pH was similar to the usual pH (6) of a well-washed native starch sample according to Rasper [61]. Freezing and thawing masked the effects of both SC and XG concentrations on the pH values, and only at the design points 11-13, corresponding to the samples with the highest SC concentrations, were the pH values significantly higher than in the rest of the F/T products. The pH values were notably higher in all the F/T samples than in their F counterparts. The pH usually drops after freezing and then undergoes inflection, the intensity and direction of which varies according to the temperature and product [62].

Finally, in F mashed potatoes the panellists scored the F-XG1.5 control significantly higher for OA than F and F-SC5 controls (Table 5). Of the samples with added SC/XG mixtures, panellists awarded the highest scores for OA to F-SC1.46/XG1.5 and F-

SC2.5/XG2.5, to which 1.5 and 2.5 g kg^{-1} of each single component was added (design points 1 and 3); on the other hand they awarded lower OA scores to F-SC7.5/XG2.5 and F-SC8.54/XG1.5 than to the three controls. For the F product, high OA scores were also recorded at the replicated centre points; these were significantly higher than the scores for the F and F-SC5 controls. In the case of F/T mashed potato, panellists once more scored F/T-SC1.46/XG1.5 and F/T-SC2.5/XG2.5 highest for OA; and again, these scores were significantly higher that those of the three F/T controls. Scores for OA were higher in the F mashed potatoes than in their F/T counterparts, highlighting the tendency of freezing and thawing processes to reduce the quality of the product. Note that the panellists awarded the same OA scores to the F/T-SC5/XG1.5 samples, corresponding to the replicated centre points, as to the F control without added biopolymers. In other words, according to sensory analysis mashed potatoes with low and equal SC and XG concentrations were preferred in terms of OA.

Optimization of Cryoprotectant Concentrations

Based on the results of the first statistical analysis, RSM was used to identify any optimum SC and XG concentrations for instrumental measurements and OA in the F and F/T mashed potatoes. Regression coefficients of each response model fitted were tested for significance. These are presented in Table 6, together with their coefficients of determination, R^2, *F*-Ratios and *P*-values, and lack-of-fit tests. Response models of instrumental parameters which gave R^2-values < 0.75 have been omitted, as these models showed very low percentages of explained variability, indicating a significant lack-of-fit [63]. In addition, models fitted for the response variables were used to generate contour plots to identify the main effects of SC and XG concentrations (Figures 1, 2). In F mashed potatoes, independent variables had approximately the same effect on CP parameters, with significant linear and quadratic terms for SC (b_1 and b_{11} respectively) and significant linear terms for XG concentrations (b_2); note however that in the case of the APF model the quadratic term for XG concentration (b_{22}) was also significant. Both CP models had very high R^2-values (~ 0.95), with non-significant lack-of-fit tests. In F mashed potatoes, minimum APF values were recorded in the contour plot at approximately 4 and 0.5 g kg^{-1} of SC and XG respectively (Figure 1a). According to Moschakis *et al.* [7], moderately low levels of xanthan addition lead to depletion flocculation and gravity-induced phase separation in emulsions stabilized by SC and XG. Increasing the polysaccharide concentration causes immobilization of the microstructure due to an increase in the local viscoelasticity: in other words, the structure cannot easily rearrange to expel the xanthan-enriched aqueous serum phase because a weak gel-like network is generated. This would account for the increase of APF with increasing XG concentration.

Table 6. Regression coefficients, coefficient of determination (R^2) and analysis of variance of the quadratic models for ITPA, CP, oscillatory parameters, colour, pH and OA of fresh and frozen/thawed mashed potatoes at the design response surface

F mashed potatoes

Coefficient	MRP (N)	APF (N)	δ (°)	G' (Pa)	η^* (Pa s)	α	n'	ln (K')	n''	ln (K'')	b^*	L^*/b^*	pH	OA
b_0	8.50	1.10	14.31	4167.05	4134.45	0.18	0.17	8.30	0.09	7.07	7.85	8.68	6.08	7.61
b_1	0.99**	0.10**	0.66*	-324.24**	-313.89**	0.01**	6.80E-3*	-0.05**	0.01*	-0.02	0.43*	-0.50*	0.02**	-0.55**
b_2	1.92**	0.14**	1.85**	-1439.08**	-1324.02**	0.02**	0.02**	-0.29**	9.91E-3*	-0.16**	0.16	-0.13	7.32E-4	1.28**
b_{11}	2.97**	0.19**	0.48	-207.81**	-253.24**	-4.81E-3	4.16E-3	-0.09**	8.03E-3	-0.06**	-0.36	0.26	-0.01	-0.60**
b_{22}	0.29	0.08**	-0.75*	666.82**	594.01**	-0.02**	-9.64E-3**	0.10**	3.121E-3	0.03*	-0.10*	0.03	-3.25E-3**	-0.71**
b_{12}	-0.23	-0.03	-0.10	353.00**	375.06**	-2.22E-3	8.33E-5	0.06**	-3.91E-3	0.07**	-0.16	0.18	0.00	-0.63**
R^2	0.98	0.94	0.93	0.97	0.96	0.92	0.96	0.93	0.80	0.80	0.76	0.80	0.91	0.99
F-*Ratio*	58.78	22.30	17.65	42.49	35.90	15.51	30.17	18.55	5.56	5.43	4.41	5.65	14.33	95.46
P-value	0.000	0.000	0.001	0.000	0.000	0.001	0.000	0.000	0.022	0.023	0.039	0.021	0.001	0.000
Lack of fit	0.060	0.070	0.204	0.005**	0.005**	0.121	0.184	0.000**	0.333	0.003**	0.551	0.490	0.215	0.271

F/T mashed potatoes

Coefficient	COH	δ (°)	G' (Pa)	η^* (Pa s)	n'	ln (K')	n''	L^*	b^*	L^*/b^*	pH	OA
b_0	0.95	13.64	3252.8	3156.00	0.16	8.02	0.10	66.88	7.80	8.58	6.15	6.18
b_1	0.01**	0.35	-138.09*	-103.79	7.22E-3*	-0.05*	0.02**	0.09	0.35**	-0.40**	0.02**	-0.48**
b_2	0.04**	1.02*	-248.32**	-277.30**	0.01**	-0.07**	5.40E-3*	0.59*	0.10**	-0.03	5.52E-3*	1.10**
b_{11}	4.61E-3	0.34	-92.87	-48.97	3.48E-3	-8.91E-3	5.67E-3*	-0.71**	-0.31**	0.28**	0.01**	0.42**
b_{22}	-0.04**	-0.47	502.94**	416.47**	-8.95E-3*	0.12**	-0.01**	0.28	0.21**	-0.19**	8E-3*	-0.92**
b_{12}	-0.01*	0.30	-279.69**	-297.44**	4.55E-3	-0.07*	3.50E-4	-0.24	0.08**	-0.10**	-2.5E-3	-0.24**

R^2	0.76	0.81	0.83	0.83	0.87	0.82	0.81	0.77	0.97	0.98	0.92	0.95
F-*Ratio*	4.47	5.99	6.91	7.01	9.69	6.57	5.82	4.58	45.52	75.66	16.05	24.33
P-value	0.038	0.018	0.012	0.012	0.005	0.014	0.019	0.036	0.000	0.000	0.001	0.000
Lack of fit	0.001**	0.792	0.007**	0.030*	0.432	0.037*	0.019*	0.092	0.005**	0.052	0.167	0.000**

Subscripts: 1 = SC concentration; 2 = XG concentration. Significant level: *0.05; **0.01.

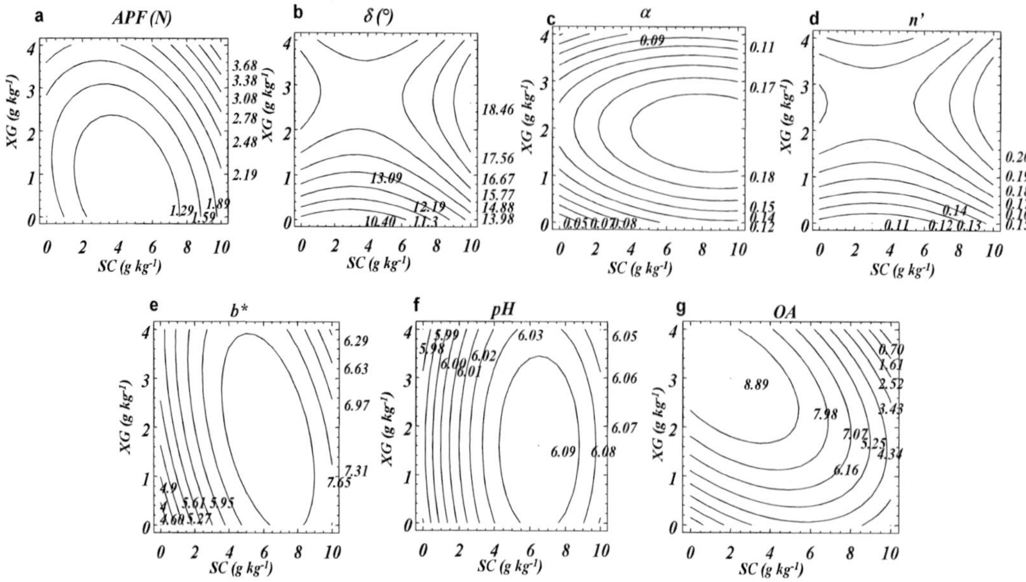

Figure 1. Quality parameter and OA contour plots as functions of SC and XG concentrations in F mashed potatoes: (a) average penetration force (APF); (b) phase angle (δ); (c) fluid-like relative angle (α); (d) slope of storage modulus (n'); (e) b^* colour parameter; (f) pH; (g) overall acceptability (OA).

For oscillatory measurements, independent variables also had the same effect on G', η^* and ln (K'), with significant interaction (b_{12}), linear and quadratic terms for SC (b_1 and b_{11} respectively) and XG concentrations (b_2 and b_{22} respectively); relatively very high R^2-values (> 0.85) were also recorded, but with significant lack-of-fit tests. Models with high R^2-values and significant lack-of-fit tests indicate that higher-order models are probably required to describe the effects of the independent variables and would be used only for trend analysis [63, 64]. δ, α and n' models had significant linear terms for SC and significant linear and quadratic terms for XG concentrations, with similar very high R^2-values (> 0.90) but non-significant lack-of-fit tests. In F samples, the n'' model only presented significant linear terms for SC and XG concentrations. The R^2-value for n'' was quite low (< 0.85), but the lack-of-fit test was non-significant. Finally, model ln (K'') had significant interaction and quadratic terms for SC, and linear and quadratic terms for XG concentrations (b_2 and b_{22} respectively), but the model presented a relatively low R^2-value (< 0.85) and a significant lack-of-fit test.

In F mashed potatoes, b^* and pH models had significant linear terms for SC and quadratic terms for XG concentrations, whereas for the L^*/b^* ratio model only the linear term for SC concentration was significant. For both colour parameters, R^2-values were low (< 0.85), with non-significant lack-of-fit tests. The R^2-value was very good for the pH (> 0.85) with non-significant lack-of-fit test. Finally, in F samples, the OA model had significant interaction, linear and quadratic terms for SC and XG concentrations, a very high R^2-value (> 0.85), and a non-significant lack-of-fit test. For δ and n' from oscillatory measurements (Fugres 1b, 1d), the contour plots were saddle-shaped. These topologies are one of the surface types characteristic of models involving quadratic terms [65], identifying a set of biopolymer concentrations that will produce maximum oscillatory parameters. The fluid-index α (Figure 1c), b^* (Figure 1e), the pH (Figure 1f) and the OA (Figure 1g) all increased with SC and XG concentrations up to certain levels beyond which the independent variables decreased, thus defining an obvious optimum region corresponding to maximum responses. It can be seen that α and OA were more sensitive to changes in XG concentration than in SC, while b^* and pH were more sensitive to changes in SC concentrations.

In F/T mashed potatoes, the COH model presented significant interaction and linear terms for SC and linear and quadratic terms for XG concentrations. However, the R^2-value was very low (~ 0.75), and the lack-of-fit test was significant. In the processed samples, for their part, the δ model only presented a significant linear term for XG concentration, with quite a low R^2-value (almost 0.80), but the lack-of-fit test was non-significant. Independent variables also had similar effects on G', η^* and ln (K') in the F/T products, with significant interaction and linear and quadratic terms for XG concentration; in the case of G' and ln (K') models, the linear term for SC concentration was also significant. The three models had relatively good R^2-values (> 0.80), but with significant lack-of-fit tests. The models for the slopes (n' and n'') had significant linear and quadratic terms for XG and linear terms for SC concentrations, and in the case of the n'' model the quadratic term for SC content was also significant. However, the n' model presented a very high R^2-value (> 0.85) with a non-significant lack-of-fit test, whereas the R^2-value corresponding to n'' model was only good (~ 0.80), and the lack-of-fit test was significant. In the case of the L^* model the linear term for XG and the quadratic term for SC concentrations were the only significant ones; the R^2-value was relatively low (< 0.80), but with a non-significant lack-of-fit test. Independent variables had similar effects on the b^* and L^*/b^* models, with significant interaction, linear and quadratic terms for SC and significant quadratic terms for XG concentrations; in the case of the b^* model, the linear term for XG concentration was also significant. Both colour models had very high R^2-values (> 0.95), although with a significant lack-of-fit test for b^* model, and a non-significant lack-of-fit test for L^*/b^* model.

In F/T samples, the pH model had significant linear and quadratic terms for SC and XG concentrations; the R^2-value for pH was very high (> 0.90), and the lack-of-fit test was non-significant. Finally, the OA model of the processed samples had significant interaction, linear and quadratic terms for SC and XG concentrations, and also a relatively very high R^2-value (0.85), but with a significant lack-of-fit test. In the case of F/T product (Figure 2), the phase angle (Figure 2a), the slope of the storage modulus (Figure 2b) and the L^*/b^* ratio (Figure 2c) also presented a saddle effect caused by the incorporation of SC. Unexpectedly, the contour plot for pH (Figure 2d) showed minimum values towards the bottom left-hand corner of the plot, then clearly increasing with the increasing of both SC and XG concentrations.

Some of the fitted models that presented high R^2-values (> 0.85) also had significant lack-of-fit tests. Among the common causes of poor modelling are a skewed response distribution, curvature in the relationship between a response variable and a design factor, large between-replicate variation and outlying experiments [65]. Only models with high percentages of explained variability (R^2 > 0.85) and non-significant lack-of-fit tests would be considered sufficiently accurate to make predictions [63]. Of the models fitted in this study, only the APF, δ, α, n', b^*, pH and OA models in the F product and the δ, n', L^*/b^* ratio and pH models in the F/T product definitely fulfil both conditions. Therefore, only the full models fitted for these response variables were used for further analysis to obtain stationary points [42], which were calculated for these responses (Table 7). In F fresh mashed potato, all the

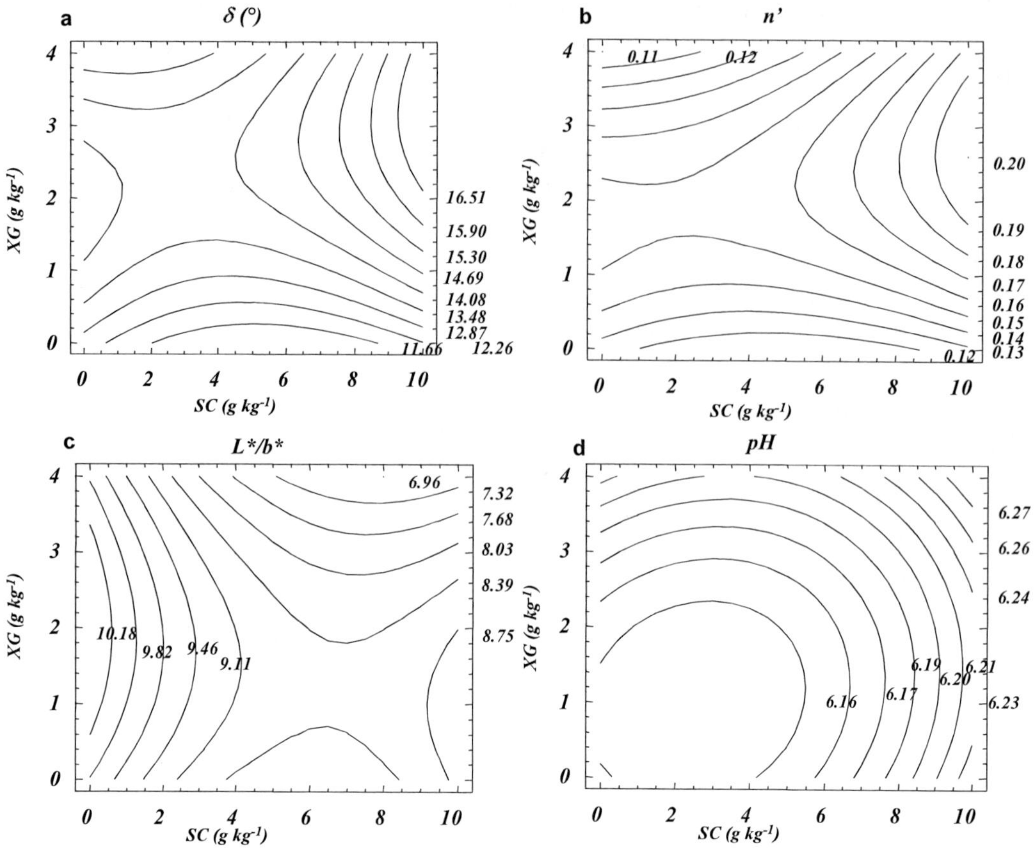

Figure 2. Quality parameter contour plots as functions of SC and XG concentrations in F/T mashed potatoes: (a) phase angle (δ); (b) slope of storage modulus (n'); (c) L*/b* ratio from colour measurement; (d) pH.

critical values lay in the region delimited by the ranges studied: APF showing minimum firmness; δ showing maximum structural weakening in the linear viscoelastic range; α showing maximum fluid-like characteristics after structural breakdown in the nonlinear viscoelastic range; *n*' showing maximum *G*' dependence of frequency; *b** evidencing maximum yellowness; pH showing maximum vicinity to neutrality; and OA showing maximum preference for the product. In F/T mashed potatoes, the critical values of stationary points for δ, *n*', *L*/b** ratio showing maximum darkness and pH showing minimum trend to neutrality, also lay in the region delimited by the ranges studied. In F and F/T products, in the cases of δ and *n*', and also the *L*/b** ratio in the processed samples, the critical values are saddle points, that is a maximum point when one seeks a point of minimum response, and therefore these stationary points are of no practical significance or the stationary point is remote from the design region [43].

Table 7. Stationary points, experimental data and values predicted by the models in F and F/T mashed potatoes with 2.5 g kg-1 added SC and 1.5 g kg-1 added XG

Quality parameter	SC concentration ($g\ kg^{-1}$)	XG concentration ($g\ kg^{-1}$)	Estimated response at stationary point	Experimental value at design point number 14	Predicted value at design point number 14	Residual percentage
F mashed potatoes						
APF (N)	4.196	0.539	1.010	1.430	1.190	+20.168
δ(°)	3.628	2.779	15.165	14.270	14.130	+0.991
α	8.361	1.899	0.196	0.167	0.164	+1.829
n'	2.928	0.414	0.129	0.160	0.163	-1.840
b^*	6.281	1.871	7.992	7.480	7.062	+5.919
pH	6.541	1.613	6.089	6.050	6.054	-0.066
OA	1.935	2.946	8.883	6.420	7.572	-17.944
F/T mashed potatoes						
δ(°)	2.821	2.320	13.908	14.350	13.633	+5.259
n'	1.810	1.864	0.153	0.150	0.151	-0.662
L^*/b^*	3.340	1.248	8.451	8.520	9.258	-8.971
pH	2.547	1.002	6.143	6.150	6.145	+0.081

The accuracy of the predictive models was further tested by adding 2.5 $g\ kg^{-1}$ of SC and 1.5 $g\ kg^{-1}$ of XG to mashed potatoes (design point 14, Table 3). Predicted values for the models were very close to the experimental values determined under these conditions for either the oscillatory properties and characteristics δ, α and n' or the pH for F samples, and for either the slope n' or for the pH in the processed products. The greatest prediction deviations (higher residual percentages) were produced by the APF and OA models. The explanation for the latter could be due, at least partially, that the random point chosen to test the accuracy of the models (design point 14) was not centred in the ranges studied.

Data Analysis of Sensory Attributes

The Friedman test showed that F samples with different cryoprotectant mixtures added presented significant differences in terms of consistency, adhesiveness, creaminess and fibrousness (χ^2 = 97.47, χ^2 = 99.90, χ^2 = 117.34 and χ^2 = 105.06 respectively; p = 0.000), authentic colour, shine and uniformity (χ^2 = 85.24, χ^2 = 82.85 and χ^2 = 91.71 respectively; p = 0.000) and sweetness, authentic taste and off-taste (χ^2 = 96.28, χ^2 = 97.40 and χ^2 = 114.12 respectively; p = 0.000). Panellists detected no effect of the cryoprotectant combinations on off-colours in F products. F/T samples presented significant differences in terms of consistency, adhesiveness, creaminess and fibrousness (χ^2 = 94.30, χ^2 = 108.53, 124.72 and 107.27 respectively; p = 0.000), authentic colour and shine (χ^2 = 74.93 and χ^2 = 87.27 respectively; p = 0.000) and authentic taste and off-taste (χ^2 = 114.52 and χ^2 = 99.25 respectively; p = 0.000). Panellists detected no effect of the cryoprotectant combinations on off-colours, uniformity or sweetness in F/T products.

Figure 3 show radar plots constructed to represent perceived sensory attributes. As regards sensory texture attributes of F mashed potatoes (Figure 3a), panellists distinguished significantly between F-SC2.5/XG1.5 and the rest of the samples (except for F, F-SC5, F-SC1.46/XG1.5 and F-SC2.5/XG0.5) with regard to consistency. F-SC2.5/XG1.5 scored and F-SC7.5/XG0.5 scored highest for consistency and adhesiveness. In the case of adhesiveness, panellists distinguished significantly between F-SC7.5/XG0.5 and the rest of the samples (except for F-SC5/XG2.91). Sensory analysis distinguished the F mashed potatoes containing 7.5 g kg^{-1} SC and 0.5 g kg^{-1} XG as better for thickness. For creaminess, panellists again distinguished significantly between F-XG1.5 control and the rest of the samples. F-XG1.5 scored highest for creaminess and presented a softer mouthfeel. Therefore, addition to F mashed potatoes of SC at any concentration used, either individually or mixed with XG, reduced the creaminess of the samples as compared to the product with 1.5 g kg^{-1} added XG alone, confirming previous findings [22-24]. In these experiments, five different hydrocolloids and two dairy proteins were added individually to mashed potatoes; in both F and F/T mashed potatoes with added XG, panellists scored samples with 0.5 and 1.5 g kg^{-1} added xanthan highest for creaminess. In the experiments referred to in this chapter, panellists also scored samples F-SC7.5/XG0.5 higher for fibrousness, although the differences between F-SC7.5/XG0.5, F-SC7.5/XG2.5 and F-SC8.54/XG1.5 were non-significant. Similarly, in F samples fibrousness increased with increasing SC concentration [24]. The superior fibrousness perceived at the highest SC concentrations could be ascribed to depletion flocculation induced by excess unadsorbed protein in the aqueous continuous phase, occurring above a certain critical protein concentration [7-9]. Certainly, a high SC concentration induced visible phase separation.

In the F/T mashed potatoes (Figure 3b), panellists scored F/T-SC5/XG2.91 higher than the rest of the samples for consistency and adhesiveness, although differences between F/T-SC5/XG2.91 and three of the replicated centre points (samples F/T-SC5/XG1.5) were non-significant for adhesiveness. It seems probable that an aggregated particle network is formed at reasonably high XG concentrations. F/T-SC1.46/XG1.5 scored highest for creaminess, although differences between the latter and F/T-XG1.5 were non-significant. Panellists perceived greater fibrousness than in the F/T control only in F/T-SC5 control and in sample F/T-SC5/XG0.09, both containing no XG. Therefore, it was clear that increased fibrousness was related to the addition of SC, whereas improved creaminess was associated with XG addition. Note however that as in the processed products, mashed potatoes with 1.46 g kg^{-1} added SC and 1.5 g kg^{-1} added XG were considered very acceptable either for their low fibrousness or their high creaminess. Creaminess is a pleasant sensation perceived on eating. It is associated with indicators of richness and high quality of food products, especially those containing fat [66]. The latter authors showed that apparent viscosity is the most significant factor in determining creaminess perception, and a high correlation between this rheological property and XG concentration has been reported [24]. Probably, XG enhances creamy/soft sensations thanks to the coating properties of the gum reducing starch retrogradation.

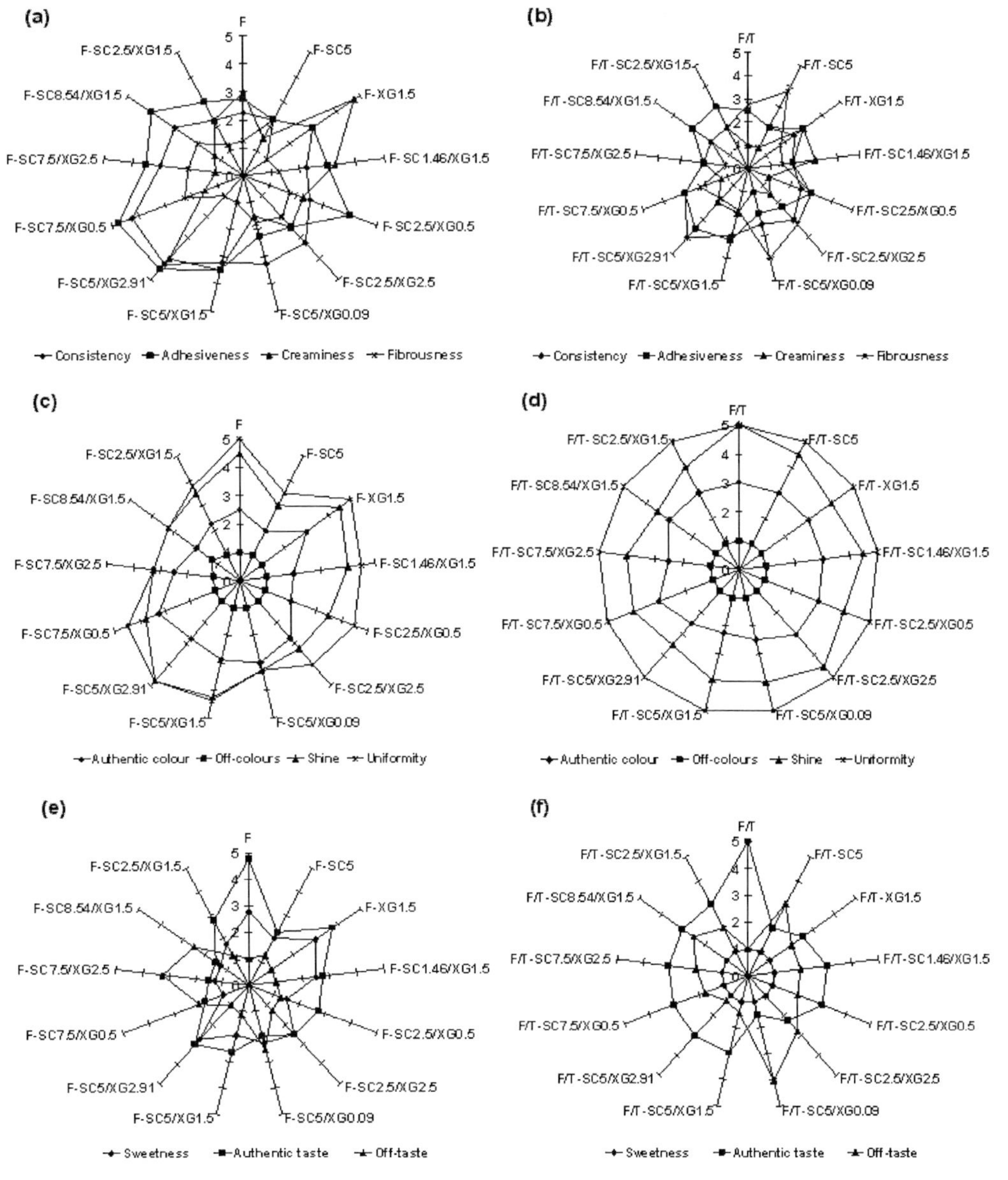

Figure 3. Sensory attributes as functions of SC and XG concentrations in F and F/T mashed potatoes: (a) texture sensory attributes of F mashed potatoes; (b) texture sensory attributes of F/T mashed potatoes; (c) appearance sensory attributes of F mashed potatoes; (d) appearance sensory attributes of F/T mashed potatoes; (e) taste sensory attributes of F mashed potatoes; (f) taste sensory attributes of F/T mashed potatoes. Identification of system notation in Table 1.

As regards sensory appearance attributes (Figure 3c), sample F-SC7.5/XG0.5 scored highest for authentic colour, although differences between scores for this sample, F-SC5 and F-XG5 controls, F-SC5/XG0.09 and replicated centre points (F-SC5/XG1.5) were non-significant. Panellists scored sample F-SC5/XG2.91, which had the highest XG concentration, highest for shine, although the difference with F control was not significant. Panellists also distinguished significantly between F and F-XG5 controls and the rest of the F samples in terms of uniformity. Both controls scored highest for this colour attribute. In the F/T products (Figure 3d), samples at the replicated centre points (F/T-SC5/XG1.5) scored lowest for authentic colour, whereas F/T control without added biopolymers scored highest for shine. This could be due to the water holding capacity exhibited by the samples with added cryoprotectant mixtures. Finally, turning to sensory taste attributes, in F samples panellists scored the ones with the highest SC concentrations (F-SC7.5/XG0.5, F-SC7.5/XG2.5 and F-SC8.54/XG1.5) significantly lower than the rest for sweetness (Figure 3e). For authentic taste, panellists distinguished significantly between the F control, which scored highest, and the rest of the samples. Conversely, samples F-SC7.5/XG2.5 scored highest for off-taste. Fernández *et al.* [24] also observed that individual addition of SC at 7.5 and 10 g kg^{-1} conferred an unfamiliar taste and odour on the mashed potatoes. Panellists also scored the F/T control higher than the rest for authentic taste. However, they considered that the samples without added biopolymers produced an unpleasant taste in the mouth due to an excessively strong potato flavour. Certainly, samples with intermediate scores for authentic taste were preferred by the trained panel. Finally, the samples with added SC alone, i.e. F/T-SC5 and F/T-SC5/XG0.09, which contained hardly any XG, scored highest for off-taste. Binding of flavour compounds and reduction of taste perception due to casein and whey proteins have been reported [67].

CONCLUSION

In general terms, freezing and thawing of the mashed potatoes enhanced textural properties as compared to their F counterparts. The cell adhesion of the F mashed potatoes with added XG/SC mixtures was superior at the highest XG concentration (2.91 g kg^{-1}), even after freezing. On the other hand, F mashed potatoes with added XG/SC mixtures presented a weaker gel structure than the mashed potatoes made without biopolymers, mainly due to the presence of XG in the systems. The cause could be incompatibility of charges on chain structure, negatively charged gum and anionic potato starch, although weakly attractive hydrophobic interactions between SC and XG appeared to play a role in the viscoelastic behaviour of the mashed potatoes. Processing significantly reduced the effect of SC/XG mixtures on the gel strength and the viscoelastic behaviour of the mashed potatoes as compared to their F counterparts. Ice crystal growth led to contraction and disruption of starch granules, while cell wall disruption led to a redistribution of amylose and amylopectin, producing local differences in concentration. Interactions between the different biopolymers were also affected by freezing and by the swelling capability of the potato starch granules on thawing. Addition of SC/XG mixtures mitigated the effect of the addition of SC alone on the colour of the mashed potatoes. Addition of XG at a concentration of 1.5 g kg^{-1} or higher, either alone or mixed with SC, was effective in stabilizing mashed potatoes against syneresis, while samples with added SC alone did not exhibit water-holding capability. SC contributed more significantly to the pH of the mashed potatoes than XG, and the pH of samples was higher in all the processed products than in their fresh counterparts. RSM analysis evidenced that the oscillatory properties and the pH of the mashed potatoes are all useful predictive tools for designing F and F/T mashed potatoes with specified quality parameters. The addition of low and equal SC and XG concentrations (1.5 g kg^{-1}) has obvious potential for the production of mashed potatoes with improved sensory quality and freeze-thaw stability. Further research is needed to gain a better understanding of the roles of SC and XG in such systems, particularly studies of mashed potato microstructure and DSC (differential scanning calorimetry).

ACKNOWLEDGEMENTS

The authors wish to thank the Spanish Ministry of Science and Innovation for financial support (AGL2007-62851) and Premium Ingredients, S.L. for donating ingredients.

REFERENCES

[1] Dickinson, E. (1995). Emulsion stabilization by polysaccharide and protein–polysaccharide complexes. In A. M. Stephen (Ed.), *Food polysaccharides and their applications* (pp. 501-515), New York: Marcel Dekker.
[2] Tolstoguzov, V. B. (1997). Protein-polysaccharide interactions. In S. Damodaran, & A. Paraf (Eds.), *Food proteins and their applications* (pp. 171-198). New York: Marcel Dekker.
[3] Kobori, T., Matsumoto, A., & Sugiyama, S. (2008). pH-dependent interaction between sodium caseinate and xanthan gum. *Carbohydrate Polymers*, doi: 10.1016/j.carbpol.2008.10.008.
[4] Hemar, Y., Tamehana, M., Munro, P. A., & Singh, H. (2001a). Influence of xanthan gum on the formation and stability of sodium caseinate oil-in-water emulsions. *Food Hydrocolloids*, *15* (4-6), 513-519.
[5] Hemar, Y., Tamehana, M., Munro, P. A., & Singh, H. (2001b). Viscosity, microstructure and phase behavior of aqueous mixtures of commercial milk protein products and xanthan gum. *Food Hydrocolloids*, *15* (4-6), 565-574.
[6] Dickinson, E. (1999). Caseins in emulsions: interfacial properties and interactions. *International Dairy Journal*, *9*, 305-312.
[7] Moschakis, T., Murray, B. S., & Dickinson, E. (2005). Microstructural evolution of viscoelastic emulsions stabilised by sodium caseinate and xanthan gum. *Colloid and Interface Science*, *284*, 714-728.
[8] Dickinson, E., & Golding, M. (1997a). Depletion flocculation of emulsions containing unadsorbed sodium caseinates. *Food Hydrocolloids*, *11*, 13-18.
[9] Dickinson, E., & Golding, M. (1997b). Rheology of sosium caseinate stabilized oil-in-water emulsion. *Journal of Colloid and Interface Science*, *191* (1), 166-176.
[10] Dickinson, E. (2003). Hydrocolloids at interfaces and the influence on the properties of dispersed systems. *Food Hydrocolloids*, *17* (1), 25-39.
[11] Syrbe, A., Bauer, W. J., & Klostermeyer, H. (1998). Polymer science concepts in dairy systems-An overview of milk protein and food hydrocolloid interaction. *International Dairy Journal*, *8*, 179–193.
[12] Koczo, K., Wasan, D. T., Borwankar, R. P., & Gonsalves, A. (1998). Flocculation of food dispersions by gums: isotropic/anisotropic dispersion separation by xanthan gum. *Food Hydrocolloids*, *12* (1), 43-53.

[13] Dickinson, E., Ma, J. G., & Povey, M. J. W. (1994). Creaming of concentrated oil-in-water emulsions containing xanthan. *Food Hydrocolloids*, *8*, 481-494.
[14] Braga, A. L. M., & Cunha, R. L. (2004). The effects of xanthan conformation and sucrose concentration on the rheological properties of acidified sodium caseinate-xanthan gels. *Food Hydrocollids*, *18*, 977-986.
[15] Nash, W., Pinder, D. N., Hemar, Y., & Singh, H. (2002). Dynamic light scattering investigation of sodium caseinate and xanthan mixtures. *International Journal of Biological Macromolecules*, *30*, 269-271.
[16] Chaisawang, M., & Suphantharika, M. (2006). Pasting and rheological properties of native and anionic tapioca starches as modified by guar gum and xanthan gum. *Food Hydrocolloids*, *20*, 641–649.
[17] Dolz, M., Hernández, M. J., & Delegido, J. (2006). Oscillatory measurements for salad dressings stabilized with modified starch, xanthan gum, and locust bean gum. *Journal of Applied Polymer Science*, *102*, 897–903.
[18] Mandala, I. G., & Bayas, E. (2004). Xanthan effect on swelling, solubility and viscosity of wheat starch dispersions. *Food Hydrocolloids*, *18*, 191–201.
[19] Mandala, I., Michon, C., & Launay, B. (2004). Phase and rheological behaviors of xanthan/amylose and xanthan/starch mixed systems. *Carbohydrate Polymers*, *58*, 285–292.
[20] Shi, X. H., & BeMiller, J. M. (2002). Effects of food gums on viscosities of starch suspensions during pasting. *Carbohydrate Polymers*, *50*, 7–18.
[21] Kelly, R. J., van Wagenbergb, M., Lathamc, J., & Mitchella, J. R. (1995). A rheological comparison between the effects of sodium caseinate on potato and maize starch. *Carbohydrate Polymers*, *28*, 347–350.
[22] Alvarez, M. D., Canet, W., & Fernández, C. (2008a). Effect of addition of cryoprotectants on the mechanical properties, colour and sensory attributes of fresh and frozen/thawed mashed potatoes. *European Food Research and Technology*, *226*, 1525-1544.
[23] Alvarez, M. D., Fernández, C., & Canet, W. (2008b). Oscillatory rheological properties of fresh and frozen/thawed mashed potatoes as modified by different cryoprotectants. *Food and Bioprocess Technology: An International Journal*, doi: 10.1007/s11947-007-0051-9.
[24] Fernández, C., Alvarez, M. D., & Canet, W. (2008). Steady shear and yield stress of fresh and frozen/thawed mashed potatoes: effect of biopolymers addition. *Food Hydrocolloids*, *22*, 1381-1395.
[25] Carr, A. J., Munro, P. A., & Campanella, O. H. (2002). Effect of added monovalent or divalent cations on the rheology of sodium caseinate solutions. *International Dairy Journal*, *12*, 487-492.
[26] Mohammed, Z. H., Haque, A., Richardson, R. K., & Morris, E. R. (2007). Promotion and inhibition of xanthan 'weak-gel' rheology by calcium ions. *Carbohydrate Polymers*, *70*, 38-45.
[27] Totosaus, A., Guerrero, I., & Montejano, J. G. (2005). Effect of added salt on textural properties of heat-induced gels made from gum-protein mixtures. *Journal of Texture Studies*, *36*, 78-92.
[28] Smith, O. (1987). Transport and storage of potatoes. In W. F. Talburt & O. Smith (Eds.), *Potato Processing* (pp. 203–285), New York: Van Nostrand Reinhold.

[29] Nourian, F., Ramaswamy, H. S., & Kushalappa, A.C. (2003). Kinetics of quality change associated with potatoes stored at different temperatures. *Lebensmittel-Wissenschaft und-Technologie*, *36*, 49-65.

[30] Alvarez, M. D., Canet, W., & Fernández, C. (2005). Effect of freezing/thawing conditions and long-term frozen storage on the quality of mashed potatoes. *Journal of the Science of Food and Agriculture*, *85*, 2327–2340.

[31] Rico, R., Alvarez, M.D. & Canet, W. (1995). Sistema de adquisición y análisis de datos. *Eurofach Electrónica*, *18*, 60–65.

[32] Canet, W., Alvarez, M. D., Fernández, C., & Tortosa, M. E. (2005a). The effect of sample temperature on instrumental and sensory properties of mashed potato products. *International Journal of Food Science and Technology*, *40*, 481–493.

[33] Navarro, A. S., Martino, M. N., & Zaritzky, N. E. (1997). Correlation between transient rotational viscometry and a dynamic oscillatory test for viscoelastic starch based systems. *Journal of Texture Studies*, *28*, 365-385.

[34] Alvarez, M. D., Fernández, C., & Canet, W. (2004). Rheological behaviour of fresh and frozen potato puree in steady and dynamic shear at different temperatures. *European Food Research and Technology*, *218*, 544–553.

[35] Alvarez, M. D., Canet, W., & Fernández, C. (2007a). The effect of temperature, geometry, gap and surface friction on oscillatory rheological properties of mashed potatoes. *Journal of Food Process Engineering*, *30*, 267-292.

[36] CIE. (1978). *Recommendations on Uniform Colour Spaces-Colour Difference Equations, Psychometric Colour Terms*. Supplement No. 2 to CIE Publication No. 15(E-1-3.1) 1971/(TC-1-3). Paris: CIE.

[37] O'Leary, E., Gormley, T. R., Butler, F., & Shilton, N. (2000). The effect of freeze-chilling on the quality of ready-meal components. *Lebensmittel-Wissenschaft und-Technologie*, *33*, 217–224.

[38] Eliasson, A. C., & Kim, H. R. (1992). Changes in rheological properties of hydroxypropyl potato starch pastes during freeze-thaw treatments I. A rheological approach for evaluation of freeze-thaw stability. *Journal of Texture Studies*, *23*, 279-295.

[39] Stanley, R. (1988). The suitability of soft fruit varieties for processing. 1987 trials. Technical memorandum 494. Chipping, Campden: The Campden Food Preservation Research Association.

[40] Alvarez, M. D., Canet, W., & Fernández, C. (2007b). Effect of modified starch concentration and freezing and thawing rates on properties of mashed potatoes (cv. Kennebec). *Journal of the Science of Food and Agriculture*, *87*, 1108–1122.

[41] Fernández, C., Alvarez, M. D., & Canet, W. (2006). The effect of low-temperature blanching on the quality of fresh and frozen/thawed mashed potatoes. *International Journal of Food Science and Technology*, *41*, 577–595.

[42] Myers, R. H., & Montgomery, D. C. (1995). *Response Surface Methodology: Process and Product in Optimization Using Designed Experiments* (1st edition). New York: John Wiley & Sons, Inc.

[43] Watts, D. G. (1995). Understanding canonical analysis. *Journal of Quality Technology*, 27 (1), 40-44.

[44] O'Mahony, M. (1986). Additional nonparametric tests. In S. R. Tannenbaum, & P. Walstra (Eds.), *Statistical methods and procedures* (pp. 303–365). New York: Marcel Dekker Inc.
[45] Ferrero, C., Martino, M. N., & Zaritzky, N. E. (1993). Effect of freezing rate and xanthan gum on properties of corn starch and wheat flour pastes. *International Journal of Food Science and Technology*, *28*, 481–498.
[46] Ferrero, C., & Zaritzky, N. E. (2000). Effect of freezing rate and frozen storage on starch-sucrose-hydrocolloid systems. *Journal of the Science of Food and Agriculture*, *80*, 2149–2158.
[47] Sikora, M., Kowalski, S., Tomasik, P., & Sady, M. (2007). Rheological and sensory properties of dessert sauces thickened by starch-xanthan gum combinations. *Journal of Food Engineering*, *79*, 1144-1151.
[48] Morris, V. J. (1990). Science, structure and application of microbial polysaccharides. In G. O. Phillips, P. A. Williams , & D. J. Wedlock (Eds.), *Gums and Stabilizers for the Food Industry – 5* (pp. 315-327). Oxford, U. K.: Oxford University Press.
[49] Kang, K. S., & Petitt, D. J. (1993). Xanthan, gellan, wellan and rhamsan. In R. L. Whistler, & J. N. Bemiller (Eds.), *Industrial Gums*: *Polysaccharides and Their Derivates* (3rd, pp. 341-397). San Diego, CA: Academic Press.
[50] Ma, L., & Barbosa-Cánovas, G. V. (1997). Viscoelastic properties of xanthan gels interacting with cations. *Journal of Food Science*, *62*, 1124-1128.
[51] Giannouli, P., & Morris, E. R. (2003). Cryogelation of xanthan. *Food Hydrocolloids 17*, 495-501.
[52] Lee, M. H., Baek, M. H., Cha, D. S., Park, H. J., & Lim, S.T. (2002). Freeze–thaw stabilization of sweet potato starch gel by polysaccharide gums. *Food Hydrocolloids*, *16*, 345–352.
[53] Mali, S., Ferrero, C., Redigonda, V., Beleia, A. P., Grossmann, M. V. E., & Zaritzky, N. E. (2003). Influence of pH and hydrocolloids addition on yam (Dioscorea alata) starch pastes stability. *Lebensmittel-Wissenschaft und-Technologie*, *36*, 475–481.
[54] Lai, V. M. F., Huang, A. L., & Lii, C. Y. (1999). Rheological properties and phase transition of red algal polysaccharides–starch composites. *Food Hydrocolloids*, *13*, 409–418.
[55] Norton, I. T., Goodall, D. M., Frangou, S. A., & Morris, E. R. (1984). Mechanism and dynamics of conformational ordering in xanthan polysaccharide. *Journal of Molecular Biology*, *175*, 371–394.
[56] Sudhakar, V., Singhal, R. S., & Kulkarmi, P. R. (1995). Studies on starch–hydrocolloid interactions: effect of salts. *Food Chemistry*, *53*, 405–408.
[57] Baranowska, H. M., Sikora, M., Kowalski, S., & Tomasik, P. (2008). Interactions of potato starch with selected polysaccharide hydrocolloids as measured by low-fields NMR. *Food Hydrocolloids*, *22*, 336-345.
[58] Arltoft, D., Madsen, F., & Pisen, R. (2008). Relating the microstructure of pectin and carrageenan in dairy desserts to rheological and sensory characteristics. *Food Hydrocolloids*, *22*, 660-673.
[59] Liehr, M., & Kuliche, W. M. (1996). Rheological examination of the influence of hydrocolloids on the freeze-thaw-stability of starch gels. *Starke (Weinheim)*, *48*, 52–57.

[60] Downey, G. (2002). Quality changes in frozen and thawed, cooked puréed vegetables containing hydrocolloids, gums and dairy powders. *International Journal of Food Science & Technology*, *37*, 869-877.

[61] Rasper, V. (1980). Theoretical aspects of amylographology. In W. C. Shuey, & K. H. (Eds.), *The amylograph handbook* (pp. 1–6), St. Paul: The American Association of Cereal Chemists.

[62] Canet, W., & Alvarez, M. D. (2006). Quality and safety of frozen vegetables. In Da-Wen (Ed.), *Handbook of Frozen Food Processing and Packaging* (pp. 377-415), London, New York: CRC Taylor & Francis Group, Boca Raton.

[63] Henika, R. G. (1982). Use of response surface methodology in sensory evaluation. *Food Technology*, *36*, 96-101.

[64] Alvarez, M. D., Morillo, M. J., & Canet, W. (1999). Optimisation of freezing process with pressure steaming of potato tissues (cv. Monalisa). *Journal of the Science of Food and Agriculture*, *79*, 1237–1248.

[65] Downey, G. (2003). Effects of cryoprotectant mixtures on physical properties of frozen and thawed puréed cooked potatoes: some introductory studies. *International Journal of Food Science and Technology*, *38*, 857–868.

[66] Akhtar, M., Murray, B. S., & Dickinson, E. (2006). Perception of creaminess of model oil-in-water dairy emulsions: Influence of the shear-thinning nature of a viscosity-controlling hydrocolloid. *Food Hydrocolloid*, *20*, 839-847.

[67] Koksoy, A., & Kilic, M. (2004). Use of hydrocolloids in textural stabilization of a yoghurt drink, ayran. *Food Hydrocolloids*, *18*, 593–600.

In: Food Hydrocolloids: Characteristics, Properties... ISBN: 978-1-60876-222-4
Editor: C. S. Hollingworth, pp. 119-144

Chapter 4

PROPERTIES OF EDIBLE SODIUM CASEINATE FILMS CONTAINING LIPIDS

Maria José Fabra, Pau Talens* and Amparo Chiralt
Department of Food Technology-Institute of Food Engineering for Development.
Universidad Politécnica de Valencia. Camino de Vera, s/n 46022. Valencia. Spain

ABSTRACT

Sodium caseinate has been used to obtain edible films with interesting properties due to their high nutritional quality, excellent sensory properties and potential to adequately protect food products from their surrounding environment. Caseinates provide good mechanical properties and are effective barriers to aroma compounds and gases such as oxygen and carbon dioxide. However, the highly hydrophilic nature of sodium caseinate films limits their ability to provide a significant moisture barrier. In this sense, the incorporation of lipids to the film matrix can improve the caseinate functional properties.

In this chapter, the effect of the incorporation of lipid mixtures (oleic acid: OA and beeswax: BW) on different properties of the caseinate based films has been discussed. The kind and amount of plasticizer have been analyzed in order to obtain films with adequate flexibility. The protein:lipid ratio and the OA:BW ratio in the lipid mixture were varied to find the best film formulation for a determined use. The effect of moisture content and temperature on the film properties was also studied. Likewise the influence of calcium on some film properties was analyzed.

Barrier properties to water and aroma compounds which are the main justification for the film application were discussed. Tensile properties were commented on as a function of the above mentioned variables. Optical properties (gloss and transparency) which are decisive to define the final appearance of the coated products were also described. All these aspects were related to the film's microstructure which was analysed through Scanning Electron Microscopy (SEM) and Atomic Force Microscopy (AFM).

---

* Corresponding author
Pau Talens.
Tel.: +34 963879836
Fax: +34 963877369.
E-mail address: pautalens@tal.upv.es

The protein:glycerol ratio which gives the best film properties was 1:0.30. The lipid ratios in the protein matrix which made the film preparation possible ranged between 0.15 and 0.50. Oleic acid had a plasticizing effect on the films, increasing their flexibility and stretchability, and reducing water vapor permeability with respect to sodium caseinate films. Aroma barrier efficiency is dependent on oleic acid:beeswax ratio and physico-chemical properties of aroma compounds. A close correlation was established between optical and microstructural parameters of the films. The addition of calcium caseinate to sodium caseinate-oleic acid-beeswax films modified the mechanical, barrier and optical properties of the films. Calcium ions promote inter- and intramolecular cross-linking, increasing brittleness and limiting elongation and biopolymer segmental mobility.

1. INTRODUCTION

The use of milk proteins (casein, whey protein) to form edible films has been extensively studied due to their high nutritional quality, excellent sensory properties and their numerous functional properties, such as their solubility in water and ability to act as emulsifiers (McHugh & Krochta, 1994a, Chen, 1995, Chen, 2002, Khwaldia, Perez, Banon, Desobry & Hardy 2004a, Longares, Monahan, O´Riordan & O´Sullivan, 2005, Shaw, Monahan, O' Riordan & O' Sullivan 2002). Sodium caseinate forms films from aqueous solutions without further treatment due to its random coil nature and ability to form extensive intermolecular hydrogen, electrostatic, and hydrophobic bonds, resulting in an increase of the interchain cohesion (Avena-Bustillo & Krochta, 1993; Brault, D'Aprano & Lacroix, 1997, Chen 2002). However, if films are formed with purely polymeric ingredients they tend to be brittle, especially under low moisture conditions. Thus, plasticizers are added to the polymeric matrix to facilitate the formation of flexible films. In this chapter, the effects of glycerol and sorbitol on caseinate films are discussed, since they have been described as good plasticizers (Siew, Heilmann, Easteal & Cooney, 1999, Audic & Chaufer, 2005, Cheuk-Hang, Kim & Lee, 2005). Their effects are based on their ability to reduce intermolecular hydrogen bonding, while increasing intermolecular spacing and reducing mechanical strength of the films.

Of the caseinates, sodium (NaCas) and calcium (CaCas) are the most common, but magnesium (MgCas) and potassium caseinates (KCas) are also commercially available. Some studies reflect that calcium caseinates improve barrier properties but provide films with more rigid structure (Chen, 2002). The increase in protein hydrophobicity could be attributed to divalent calcium cations which promote cross-linking between negatively charged carboxyl groups of the polypeptide chain, thus preventing those charge sites from interacting with water, rendering films with a more rigid structure. This effect suggests that films obtained from mixtures of sodium and calcium caseinates could provide films with more adequate flexibility and water vapour barrier properties than those obtained with pure sodium or calcium caseinates.

Even with calcium, caseinate films are highly hydrophilic and the incorporation of lipid materials (fatty acids or wax) to the film matrix can improve the caseinate functional properties. Each hydrophobic substance has particular properties and interactions in the matrix which leads to different types of behaviour with respect to moisture transfer. Waxes are the most efficient substances to reduce moisture permeability because of their high hydrophobicity (high content in long chain fatty alcohols and alkanes) (Morillon *et al.,* 2002).

Unsaturated fatty acids control moisture migration less efficiently because they are more polar than saturated lipids. The mechanical properties of the film are also greatly affected by the kind of lipid. Studies into soy protein-fatty acid composite films showed that oleic acid promoted greater flexibility in the films than saturated fatty acids (Rhim, Wu, Weller & Schnepf, 1999).

Physical and barrier properties of sodium caseinate-lipid films are also significantly influenced by the moisture content in the film (Gennadios *et al.,* 1993, Gontard *et al.,* 1993). In this sense, due to their hydrophilic nature, sodium caseinate based films tend to adsorb large quantities of water at elevated relative humidity conditions, resulting in plasticized film matrices that have weakened barrier and mechanical properties due to the water plasticizing effect. In this sense, the values of water vapor permeability of the films at relatively small relative humidity gradients, such as 33/58, 58/75 and 58/100%, are more useful because these conditions are far more common to food than 0/100 or 0/90% relative humidity ranges which are often reported in several works (Avena-Bustillos & Krochta, 1993, Chick & Ustunol, 1998).

In this chapter, the physical and barrier properties of sodium caseinate based films are discussed, by analyzing the effect of the kind and amount of plasticizer, the influence of the type of lipid (fatty acids and beeswax and their mixtures), as well as the protein:lipid ratio and the oleic acid-beeswax ratio in the lipid mixture. Likewise, both the effect of moisture content and temperature on the film properties and the partial substitution of sodium for calcium caseinate in the film were also analysed.

2. Barrier Properties

2.1. Water Vapor Permeability (WVP)

Like most protein-based films, those obtained with sodium caseinate (NaCas) have limited moisture barrier ability compared to commonly used synthetic plastic films, due to their hydrophilic nature. The effect of several factors such as the addition of plasticizers, lipids or other hydrocolloids and relative humidity gradient and temperature on water vapour permeability of caseinate films are discussed below.

Table 1 compares water vapor permeability of NaCas based films containing glycerol and sorbitol, as a function of the plasticizer: protein ratios. As has been described by other authors (Chick & Ustunol, 1998, Siew *et al.*, 1999), sorbitol is less hygroscopic than glycerol, giving rise to films with lower water affinity at the same level of relative humidity. At the same plasticizer- protein ratio, sorbitol-NaCas films were more effective moisture barriers than glycerol-NaCas films. The increase in the glycerol-protein ratio (0.3 to 0.9) provoked an exponential increase in the vapor transfer rate, whereas the increase in sorbitol:protein ratio (0.3 to 0.6) did not change significantly the WVP values and only caused a slight increase when the ratio was 0.9 This points out that glycerol, being more hydroscopic, absorbs more moisture, which acts as additional plasticizer in the film matrix.

Table 1. Tensile parameters and water vapor permeability values of films (70μm) prepared with different sodium caseinate:plasticizer ratios, measured at 25°C

NaCas:plasticizer ratio	*Glycerol*			*Sorbitol*			WVP ($g\text{-}mm\text{-}kPa^{-1}\text{-}h^{-1}\text{-}m^{-2}$)	
	EM (MPa)	TS (MPa)	E(%)	EM (MPa)	TS (MPa)	E(%)	*Glycerol*	*Sorbitol*
1:0.3	850 $(10)^a$	14.4 $(0.6)^a$	4.0 $(0.6)^a$	973 $(34)^a$	17 $(3)^a$	2.1 $(0.3)^a$	4.3 $(0.3)^a$	2.1 $(0.3)^e$
1:0.4	257 $(22)^b$	5.8 $(0.8)^b$	16 $(3)^b$	725 $(21)^b$	15 $(2)^{ab}$	6 $(1)^b$	4.4 $(0.4)^a$	2.1 $(0.4)^e$
1:0.5	121 $(21)^c$	3.2 $(0.3)^c$	21 $(3)^{bc}$	542 $(26)^c$	12 $(1)^b$	12 $(2)^c$	5.5 $(0.5)^b$	2.4 $(0.3)^e$
1:0.6	70.0 $(5)^d$	2.4 $(0.2)^d$	24 $(3)^c$	380 $(30)^d$	9.6 $(0.8)^c$	15 $(2)^c$	7.3 $(0.4)^c$	2.3 $(0.4)^e$
1:0.9	16.2 $(2.5)^e$	1.6 $(0.3)^e$	32 $(5)^d$	146 $(15)^e$	6.2 $(0.3)^d$	26 $(4)^d$	8.9 $(0.3)^d$	3.1 $(0.2)^f$

Mean values (standard deviation)

a-e: Different superscripts within a column indicate significant differences among formulations (p<underline 0.05)

As has already been pointed out in several works (Mc Hugh *et al.*, 1994, Chen, 2002, Fabra *et al.*, 2008a) films made from sodium caseinate have high water vapor permeability due to "clustering" of water molecules in their diffusion through microcavities, which constitutes one of the major drawbacks for their use in commercial applications. One way of improving moisture barrier properties of sodium caseinate films is to include lipids. Among these, beeswax is one of the most effective materials employed to decrease WVP of edible films due to its high hydrophobicity and solid state at room temperature (Yang & Paulson, 2000, Morillon *et al.*, 2002, Fabra *et al.*, 2008 a y b). Oleic acid, despite its relatively high polar and surfactant character, has been found to reduce WVP of protein films (Rhim *et al.*, 1999, Fabra *et al.*, 2008a y b, Monedero *et al.*, 2009), while having a plasticizing effect on protein matrices which improves the mechanical properties of the film (Fabra *et al.*, 2008a y b, Monedero *et al.*, 2009). As shown in table 2, both lipids reduced the WVP values due to the increase of the film hydrophobicity (Chick & Hernández, 2002, Debeaufort, Quezada-Gallo, Delporte, & Voilley, 2000, Morillon *et al.*, 2002, Park, Testin, Vergano & Weller, 1994, Pérez-Gagó & Krochta, 2001, Shellhammer & Krochta, 1997). The higher the lipid ratio in the film, the greater the WVP reduction, especially at high RH gradients, where the lowest lipid ratio hardly affected the permeability values. Beeswax was more efficient than oleic acid, especially at low relative humidity. This effect can be attributed to the greater polarity of the unsaturated fatty acid. The strong interactions between the polar groups of oleic acid and water allow the OA molecules to penetrate into the NaCas matrix film and establish attractive forces between plasticizer molecules and chain segments. These attractive forces reduce the cohesive forces between the polymer chains and increase the segmental mobility thus favoring water molecule diffusivity (Morillon *et al.,* 2002). Indeed, water vapor transfer depends on both type and physical state of the lipid, as has been shown in previous works on NaCas films (Fabra *et al.*, 2008a).

Table 2 Tensile parameters and water vapor permeability values of films (100 μm) prepared with 1:0.15 and 1:0.5 sodium caseinate:lipid ratio measured at 25°C

NaCas:lipid ratio	Type of lipid	*33% RH*			*58% RH*			WVP ($g*mm*kPa^{-1}*h^{-1}*m^{-2}$)	
		EM (MPa)	TS (MPa)	E (%)	EM (MPa)	TS (MPa)	E (%)	*33-58% RH*	*58-100% RH*
control		579 (9)^{a}	7.3 (0.2)^{a}	19.4 (1.4)^{a}	132 (14)^{a}	3.2 (0.4)^{a}	19.4 (3.8)^{a}	0.58 (0.03)^{a}	11.9 (0.4)^{e}
1:0.15	OA	517 (9)^{b}	10.0 (0.2)^{b}	3.6 (0.2)^{b}	31 (6)^{b}	2.3 (0.3)^{bc}	34 (3)^{b}	0.48 (0.02)^{b}	11.4 (0.6)^{e}
1:0.5	OA	45.6 (0.8)^{c}	2.4 (0.2)^{c}	38 (1)^{c}	27 (3)^{b}	1.8 (0.2)^{b}	40 (2)^{c}	0.45 (0.06)^{bc}	6.1 (0.2)^{f}
1:0.15	BW	606 (20)^{ad}	14.1 (0.7)^{d}	3.4 (0.2)^{b}	162 (13)^{c}	3.4 (0.3)^{a}	3.7 (1.4)^{d}	0.40 (0.04)^{c}	11.0 (0.8)^{e}
1:0.5	BW	611 (13)^{d}	16.0 (0.3)^{e}	4.2 (1.4)^{b}	181 (13)^{c}	2.36 (0.04)^{c}	6.0 (0.7)^{e}	0.29 (0.06)^{d}	5.7 (0.20)^{f}

Mean values (standard deviation)

a-e: Different superscripts within a column indicate significant differences among formulations (p<underline 0.05)

The effect of saturated fatty acids on sodium caseinate based films has been described in a previous work (Fabra *et al.*, 2009a). Some authors claim that the longer the tail of the saturated fatty acids, the more effective the WVP reduction, owing to the differences in the mobility of the chain (McHugh & Krochta, 1994b, Rhim *et al.,* 1999). Nevertheless, neither Tanaka *et al.* (2001) nor the results obtained for sodium caseinate based films (Fabra *et al.*, 2009a) showed a direct correlation between the length of the aliphatic chain of saturated fatty acids and the barrier properties of protein films. A lower water vapour barrier efficiency of OA was observed as compared with saturated fatty acids, which was attributed to its high molecular volume associated with the unsaturation and the subsequent low melting temperature, which implies chain mobility (Grenner & Fennema, 1992). On the contrary, saturated fatty acid molecules are self-associated as bilayers of different sizes in the film forming dispersions and these laminar structures grow and remain in the dried film. Micrographs from these films showed that the length of lipid layers in the matrix is greater for lauric acid and decreases as the chain length increases. This indicates that the growth of molecule aggregates during the film drying is more effective when the fatty acid chain is shorter, probably due to the greater molecular mobility. These successive lipid layers in the film matrix greatly contribute to reduce water permeability, thus improving the water barrier properties (Fabra *et al.*, 2009a).

Saturated fatty acids and beeswax give rise to better WVP properties but worse tensile properties than oleic acid and, to optimize film formulation, mixtures of lipids could be used.

In this sense, oleic acid and beeswax were combined in different OA:BW ratios (100:0, 70:30, 50:50, 30:70, 0:100) in films containing 0.3 glycerol-protein ratio and 0.5 lipid ratio and the effect of the relative humidity gradient and temperature on WVP was analyzed. Table 3 shows these values for three relative humidity ranges (33-58, 58-75 and 58-100%) and two temperatures (5 and 25°C).

Table 3 Water vapour permeability ($g*mm*kPa^{-1}*h^{-1}*m^{-2}$) values of films (100 μm) prepared with 1:0.5 NaCas:lipid ratio. Lipid phase was composed of mixtures of OA and BW

OA:BW ratio	33-58% HR		58-75% HR		58-100% HR	
	5°C	25°C	5°C	25°C	5°C	25°C
Control	0.34 (0.03)^{ax}	0.58 (0.03)^{ax}	4.2 (0.2)^{bx}	10.0 (0.9)^{cx}	5.9 (0.8)^{bx}	11.9 (0.3)^{dw}
100:0	0.07 (0.02)^{ay}	0.45 (0.06)^{by}	0.6 (0.1)^{by}	3.4 (0.4)^{cy}	4.4 (0.3)^{cxy}	6.1 (0.2)^{dx}
70:30	0.027 (0.003)^{az}	0.20 (0.04)^{bz}	0.4 (0.2)^{by}	0.5 (0.1)^{bz}	2.9 (0.4)^{cy}	2.6 (0.2)^{cy}
50:50	0.051(0.001)^{ay}	0.14 (0.02)^{bz}	0.6 (0.1)^{cy}	0.6 (0.1)^{cz}	2.7 (0.1)^{dy}	1.6 (0.1)^{ez}
30:70	0.10 (0.03)^{ay}	0.20 (0.06)^{az}	1.2 (0.4)^{bz}	1.05 (0.06)^{bz}	4.3 (0.2)^{cxy}	3.05 (0.03)^{dy}
0:100	0.14 (0.01)^{ay}	0.29 (0.06)^{bz}	2.2 (0.7)^{cz}	3.6 (0.3)^{dy}	3.7 (0.1)^{dy}	5.7 (0.2)^{ex}

Mean values (standard deviation)

$^{a-e:}$ Different superscripts within a file indicate significant differences among temperature and relative humidities ($p < 0.05$).

$^{x-z:}$ Different superscripts within a column indicate significant differences among formulations ($p < 0.05$)

For all relative humidity gradients tested, water vapor permeability decreased when the OA:BW ratio increased, in agreement with previous studies working with different protein:lipid ratios (Fabra *et al.*, 2008a) The high efficiency of beeswax was attributed to the high content of esters of long chain fatty alcohols and acids, as well as long chain alkanes, which limits diffusion of water (Bravin *et al.*, 2004, Shellhammer & Krochta, 1997, Debeaufort *et al.*, 2000, Morillon *et al.*, 2002, Kristo *et al.*, 2007, Fabra *et al.*, 2008a). Nevertheless, when the beeswax content added comes to more than 30% of the total lipid phase (70:30 OA:BW ratio) no further reduction in the WVP of sodium caseinate films was observed. This effect was also observed by other authors in beeswax-sodium caseinate emulsified films (Avena-Bustillos & Krochta, 1993), in beeswax-whey protein emulsified films (Pérez-Gagó & Krochta, 2001) and in oleic acid-beeswax-soy protein emulsified films (Monedero *et al.*, 2009).

As expected, the water vapour barrier properties of the films were affected by the RH gradient and temperature. As reported in several works (Lim, Mine & Tung, 1999, Fabra *et al.*, 2009a), the increase of water content in the film structure implies an increase in the molecular mobility due to the plasticization effect and therefore, in all the properties which depend on molecular diffusion, such as transport properties (Roos, 1995). This effect was also observed in fatty acid-sodium caseinate films (Fabra *et al.*, 2009a). The temperature increase also supposed an increase in molecular diffusion. However, the effect of RH or temperature was much less marked when lipids were present in the films, especially when these were mixtures of OA and beeswax. This seems to indicate that, with a protein-lipid ratio of 1:0.5, the hydrophobic character of the dispersed phase greatly contributes to control the water transport, despite an increased molecular mobility of the matrix. In this sense, the greater efficiency of OA-BW mixtures as compared with pure components is remarkable. It is likely that the greater hydrophobic nature of BW together with the surfactant character of OA, which promotes interfacial interactions with protein matrix, made the mixture more efficient as a water barrier in the matrix. The influence of temperature in films containing lipid mixtures is not significant and there was even an unexpected increase in WVP observed at the highest RH when the temperature decreased. In this sense, three different effects of temperature and RH must be considered: the influence on molecular mobility in the continuous matrix, the effect on the mobility in the lipid phase (with low impact on WVP) and the influence on molecular interactions which could also affect the transport properties. Nevertheless, from a practical point of view, the much lower impact of both RH and temperature on WVP of films containing lipids is remarkable, which improves their ability to act as efficient water vapour barriers in different ambient conditions. This is particularly clear when OA-BW mixtures are used.

The effect of the BW:OA ratio, temperature and relative humidity has been modelled through a step-wise multiple regression. Eq. (1) gives the obtained equation (r^2=0.87) which allows us to obtain the Ln WVP values with an error of estimation of 0.6. Figure 1 shows the predicted vs.: the experimental values where the accuracy of predictions can be observed.

$$\text{Ln (WVP)} = -3.398 - 2.7789*BW - 2.62458*OA + 0.0764384*HR - 15.4919*T^{-1} + 0.186962*HR*T^{-1} \quad (1)$$

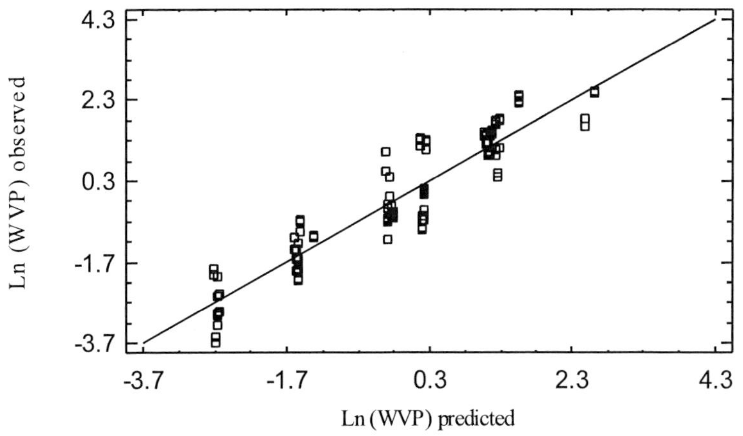

Figure 1. Comparison between observed values of Ln(WVP) and those predicted by Eq.(1) for the different films with 1:0.5 NaCas:lipid ratio.

By considering films obtained from protein, polysaccharide mixtures, some works indicate that the incorporation of polysaccharides into globular protein matrices may extend the functional properties of these ingredients (Zaleska *et al.*, 2000, Turgeon & Beaulieu, 2001, Arvanitoyannis *et al.*, 1996). These combinations could give rise to synergistic effects between the components leading to reduced water vapour permeability, optimum flexibility, transparency and gloss. Little information exists about the film-forming ability of such combinations of protein and polysaccharide components. In a recently published work (Fabra *et al.*, 2008b), the effect of λ-carrageenan and alginate on water vapour permeability properties of sodium caseinate based films have been discussed together with the effect of lipid incorporation (OA:BW mixtures) in the matrices. In lipid-free films, both polysaccharides increased the water vapour permeability values of the matrix; this effect being more notable for λ-carrageenan, depending on the polysaccharide concentration. Similar results were observed by other authors mixing different proteins and polysaccharides (Parris *et al.,* 1995, Sabato *et al.,* 2001, Yu *et al.*, 2004, Coughlan *et al.*, 2004). When these matrices contained lipids, the presence of λ-carrageenan implied a slight modification of the film properties with respect to the protein films, but the presence of alginate produced more permeable films.

In other studies, sodium caseinate was added to corn starch matrices in an effort to restrict the WVP. Avanitoyannis *et al.*, 1996 observed that WVP decreased when the sodium caseinate content increased, which was attributed to the less hydrophilic nature of NaCas compared to corn starch.

Another factor that can affect the water barrier properties of sodium caseinate films is the presence of calcium caseinate. Several authors report that calcium improves barrier properties but provides films with a more rigid structure (Ho, 1992, Avena-Bustillos & Krochta, 1993, Chen, 2002, Khwaldia *et al.*, 2004a). The increase in protein hydrophobicity could be attributed to divalent calcium cations which promote cross-linking between negatively charged carboxyl groups of the polypeptide chain, thus preventing those charge sites from interacting with water. Nevertheless, even with calcium, caseinate films are highly hydrophilic and the incorporation of lipid materials to promote water barrier properties is an interesting alternative.

As shown in table 4, a significant decrease in the WVP was observed after the substitution of sodium caseinate for calcium caseinate, mainly when lipids are present. The promotion of intermolecular cross-linking by calcium tightens the protein matrix and limits biopolymer segmental mobility, thus reducing WVP (Chen, 2002). Chen, 2002, reported a correlation between the WVP values of simple caseinate films and the size of cations. The smaller ions occupy less space between protein strands, thus resulting in tighter structures with a greater resistance to mass transfer. In this sense, WVP of magnesium caseinate films is lower than that of sodium caseinate films. Similarly, the WVP of sodium caseinate films is lower than that of potassium caseinate films (Ho, 1992). The incorporation of lipids (OA:BW mixtures) into the different caseinate matrices had a similar effect to that described in a previous work (Fabra *et al.* 2008a). The addition of pure OA reduces WVP, but the incorporation of BW into the lipid blend (OA:BW ratio of 70:30,) supposes a more effective increase in water barrier properties. Small additions of BW contribute to increase the hydrophobic character of the film giving rise to an increase in water vapour barrier properties. The slight effect of greater amounts of BW has been attributed to the structural effects in the matrix (Fabra *et al.* 2008a, b). Likewise, it is remarkable that the film containing the sodium-calcium caseinate ratio of 1:0.5 and 70 % of OA in the lipid phase showed WVP values very close to those obtained for the calcium caseinate film with the same lipid composition, although the tensile properties of the first are better due to its greater flexibility.

Table 4. Mean values (and standard deviation) of water vapor permeability (g-mm-kPa^{-1}-h^{-1}-m^{-2}) for films prepared with different NaCas:CaCas ratios, with and without OA:BW lipid mixtures

NaCas:CaCas ratio	OA:BW ratio		
	control	100:0	70:30
1:0	5.9 (0.8)^{ay}	4.4 (0.3)^{by}	2.9 (0.4)^{cy}
1:0.5	6.41 (0.04)^{ay}	3.5 (0.4)^{bz}	2.0 (0.5)^{cy}
0:1	3.9 (0.5)^{az}	3.1 (0.5)^{az}	1.51 (0.03)^{bz}

Mean values (standard deviation)

$^{a-c:}$ Different superscripts within a file indicate significant differences among formulations ($p < 0.05$).

$^{y-z:}$ Different superscripts within a column indicate significant differences among sodium caseinate:calcium caseinate ratio ($p < 0.05$)

From the results discussed, water vapor transport in sodium caseinate based films is a complex phenomenon due to the strong interaction of adsorbed water molecules on the NaCas matrix. Specific interactions among the film components underlined the difficulties in modelling the water vapour permeability of films as a function of composition. Furthermore, film thickness plays an important role in the functional properties of edible films. Although in ideal polymer structures, gas and water permeabilities are independent of film thickness (Swartzberg, 1986), there is experimental evidence (McHugh *et al.*, 1993, Park & Chinnan, 1995) of non-ideal behaviour. Figure 2 compares water vapour permeability values of sodium caseinate edible films with different thicknesses at 5°C and 58-100 % RH gradient where the increase of WVP with film thickness can be observed. It may be noted that no data could be obtained for films composed of 1:0.3:0.5 NaCas:Gly:BW ratio and 70 μm because this film was too brittle to take off the casting plate. This positive deviation from the ideal behaviour indicates that sodium caseinate films have an affinity for moisture that is not taken into account in Fick's and Henry's laws. Previous studies have indicated similar relationships in hydrophilic film systems: amilose films (Rankin *et al.*, 1958), starch films (Bertuzzi *et al.*, 2007) protein and other edible films (Gennadios *et al.*, 1994a, Longares, Monahan, O'Sullivan, & O'Riordan, 2004, McHugh *et al.*, 1993, Park & Chinnan, 1995, Park, Weller, Vergano & Testin, 1993, Sobral, 2000). Several explanations for the effects of thickness on edible film properties have been reported. McHugh *et al.*, 1993 observed that as film thickness increased, the film provided an increased resistance to mass transfer and this provoked an increase in the equilibrium water vapour partial pressure at the inner film surface. Other authors attributed the thickness effect to film swelling as a result of attractive forces between polymer and water (Park *et al.*, 1993). Film swelling could result in varying film structures.

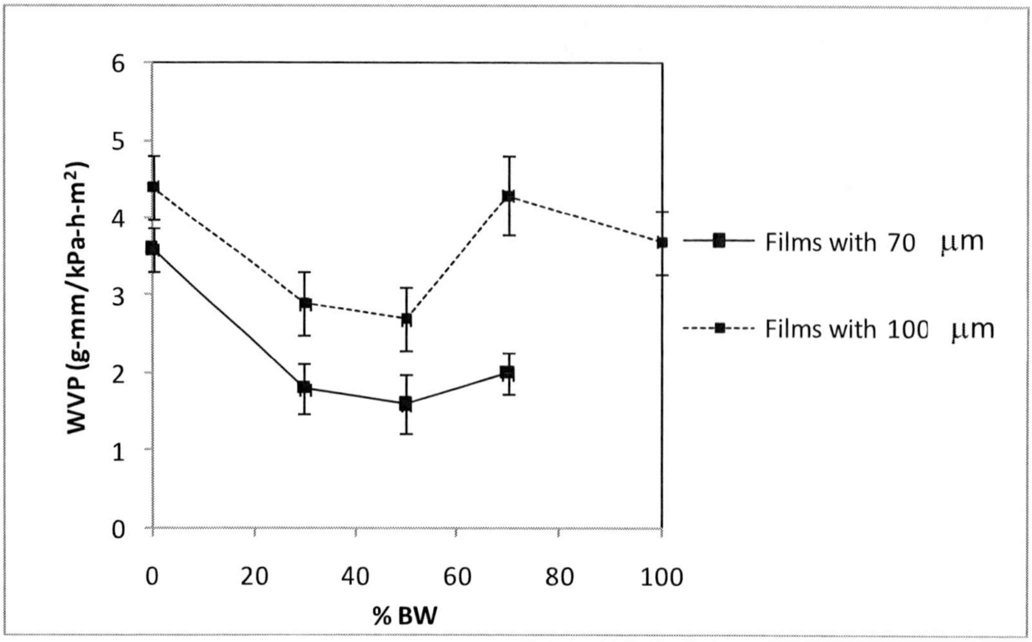

Figure 2. Mean values (and LSD intervals) for water vapor permeability as a function of the BW percentage in the lipid mixture for films prepared with different thicknesses.

2.2. Aroma Permeability

Aroma components are present in food at low levels, but their loss greatly affects sensory properties by reducing flavor intensity or by changing the aroma of the food. Most edible films are quite moisture sensitive, but this inherent hydrophilicity makes them excellent barriers to nonpolar substances such as oxygen and some aroma compounds. Barrier properties of sodium caseinate based films are of interest in order to control moisture, oxygen and aroma compounds (Khawaldia *et al.*, 2004a, Chen, 2002, Fabra *et al.*, 2008a) and Miller & Krochta, 1998 pointed out that the aroma permeability of polymer films is often related to their oxygen permeability. In a recent work (Fabra *et al.*, 2008c), the mass transport of 6 different aroma compounds (ethyl acetate, ethyl butyrate, ethyl hexanoate, 2-hexanone, 1-hexanol and cis-3-hexenol) through sodium caseinate based films with different oleic acid (OA):beeswax (BW) ratios has been studied. The affinity of volatile compounds with the film (defined by molecular interactions) seems to be the main factor governing the transfer. In this sense, lipid-free sodium caseinate based films are mainly hydrophilic and the permeability of aroma compounds was low. Nevertheless, the incorporation of lipids notably changed the aroma barrier properties. In this sense, oleic acid was less efficient than beeswax in reducing aroma permeability, which can be attributed mainly to its physical state. Sorption and diffusion mechanisms can be easier through liquid oleic acid. A marked increase in aroma permeability was observed when OA increased in the lipid mixture; this effect being more pronounced for 1-hexanol, ethyl hexanoate and ethyl butyrate which are more hydrophobic. On the contrary, beeswax is highly crystalline which limits the diffusion of aroma compounds and the aroma sorption phenomena can only occur on the globule surface of the beeswax. So, preferential sorption and diffusion of aroma compounds occurs through oleic acid instead of the caseinate matrix and/or beeswax particles.

The permeability of ethyl esters (ethyl acetate, ethyl butyrate and ethyl hexanoate) in lipid-free sodium caseinate films was not correlated with the carbon chain length and differences were explained on the basis of the mechanisms implied in aroma transfer and the hydrophobicity of the compounds. While the sorption mechanism controls the transfer of small molecules (maximum 5 carbon atoms) without interactions, the diffusion mechanism plays an important role in volatile molecule transfers with high molar volume. So, ethyl acetate and ethyl butyrate transfers were mainly governed by the sorption mechanism and the greater permeability observed for ethyl butyrate could be attributed to its greater hydrophobicity. On the contrary, the greater hydrophobic nature of ethyl hexanoate (log P = 3.62) limited its transfer through control films due to the lower affinity for this aroma compound with sodium caseinate. However, a greater increase of ethyl hexanoate permeability was observed for films prepared with high oleic acid content as compared to ethyl acetate and ethyl butyrate permeability. This fact can be explained by its better solubility in the lipid phase. Similar results were observed for gluten based films (Quezada-Gallo, 1999).

The transfer of 1-hexanol through sodium caseinate-OA-BW films induces interactions between polar groups leading to a more plasticized structure, especially at high contents of oleic acid. A synergistic effect between 1-hexanol and oleic acid could be produced, giving rise to a more plasticized structure with poor aroma barrier properties. Cis-3-hexenol permeability showed significantly lower permeability values than 1-hexanol. These differences have been attributed to the chain unsaturation that implies higher molecular volume of cis-3-hexenol and less flexibility of the molecule. Thus, intrinsic mobility of this aroma compound could be reduced as compared to linear aroma compounds with the same chain length.

3. Tensile Properties

Sodium caseinate films are brittle without the addition of plasticizers and have limited practical use (Chen, 1995). Tomasula *et al.*, 1998 reported that a minimal glycerol concentration of 20% was necessary to prepare films with suitable tensile properties. So, the use of plasticizers is necessary to obtain NaCas based films. Film properties are affected differently depending on the plasticizer. In this sense, glycerol and sorbitol were used effectively due to their ability to reduce intermolecular hydrogen bonding while increasing intermolecular spacing and reducing tensile strength of the films (Siew, Heilmann, Easteal & Cooney, 1999, Audic & Chaufer, 2005, Cheuk-Hang, Kim & Lee, 2005). Table 1 compares tensile properties of NaCas films containing glycerol and sorbitol, as a function of the plasticizer: protein ratios. Tensile strength (TS) and elastic modulus (EM) decreased exponentially and elongation at break (E) increased when the content of these compounds increased. Nevertheless, differences in tensile parameters were observed at each plasticizer-protein ratio depending on whether glycerol or sorbitol was used. Glycerol plasticized NaCas films had lower TS and EM values and higher E values than sorbitol plasticized films. Thus, the films prepared with glycerol were more flexible and more stretchable than films prepared with sorbitol, indicating the greater plasticizing effect of glycerol on these films. This effect can be attributed to the higher hydrophilic nature of glycerol compared to sorbitol (glycerol adsorbs more moisture, which serves as an additional plasticizer) and to the fact that the glycerol molecule is much smaller than the sorbitol molecule, which makes it easier to insert between the protein strands through hydrogen bonds, thus reducing intermolecular protein interactions more efficiently and increasing intermolecular spacing. Similar results were observed by other authors for egg white (Gennadios *et al.,* 1994b), whey protein (McHugh & Krochta, 1994c), gelatin and soluble starch (Arvanitoyannis, Psomiadou, Nakayama, Aiba, & Yamamoto, 1997), fish myofibrillar protein (Cuq, Gontard, Aymard, & Guilbert, 1997), sodium caseinate and soluble starch mixtures (Arvanitoyannis & Biliaderis, 1998), casein (Chick & Ustunol, 1998) and soy protein (Audic & Chaufer, 2005).

Lipids are also commonly used to reduce water vapor permeability due to their non-polar nature or hydrophobicity which provides a good barrier against water vapor migration. In some cases, incorporating lipids, mainly fatty acids, in protein- or polysaccharide- based films may produce a plasticizing effect, leading to a reduction of film strength and an increase of film flexibility, as seen in milk protein (Shellhammer & Krochta, 1997, Fabra *et al.*, 2008a) and wheat gluten films (Gontard *et al.*, 1994). In this section, the effect of OA and BW on the tensile properties of sodium caseinate based films was also analyzed. The results of the tensile tests performed on samples equilibrated at 33% and 58% RH are shown in table 2. In general, OA imparted interesting tensile properties to the films, improving their flexibility and deformability as compared to control films, mainly at 58% relative humidity. On the contrary, BW gave rise to films that were less deformable and flexible than the control ($p<0.05$), at both RH values. The effect of lipid concentrations depends on the lipid used. While EM and TS decreased and E increased as OA content increased, the increase in BW concentration did not provoke a marked effect on tensile properties. These differences can be explained by the fact that beeswax is solid at room temperature, which limits the deformation of the dispersed phase of the film when being stretched (Morillon *et al.,* 2002, Fabra *et al.*, 2008a). On the contrary, OA is liquid and can flow during the film stretching at the temperatures considered. Likewise the amphiphilic nature of OA contributes to the establishment of molecular interactions with protein molecules which modify the properties of the film matrix.

The effect of saturated fatty acids (lauric acid, palmitic acid and stearic acid) on caseinate matrices was also studied and compared with the beeswax and oleic acid effect (Fabra *et al.*, 2009a). Similarly to beeswax and regardless of the aliphatic chain length, saturated fatty acids increased EM and TS and reduced E significantly ($p< 0.05$), as compared with the control film. This was explained by the solid state of these lipids at room temperature and the formation of crystalline structures in sodium caseinate matrices, giving rise to very rigid non-flexible films. In spite of this, saturated fatty acids produced less elastic and flexible films than beeswax, though no marked differences in elongation at break were observed. So, while oleic acid had a plasticizing effect on the films, increasing their elasticity, flexibility and strechability, non-flexible films were obtained when lauric, plamitic or stearic acids were used. As a result, though saturated fatty acids considerably improved water vapor permeability, these films had poor handling properties (Fabra *et al.*, 2009a).

Taking into account that oleic acid provides flexibility to the films while beeswax is more effective at improving water vapor barrier properties, mixtures of both in sodium caseinate films could be an interesting alternative when designing film formulation (Morillon *et al.,*

2002, Fabra *et al.*, 2008a). To this end, both lipids were combined in different OA:BW ratios (100:0, 70:30, 50:50, 30:70, 0:100) when they were incorporated into caseinate films with 1:0.3:0.5 protein:glycerol:lipid ratios. The effect of relative humidity and temperature on the tensile properties of these films was analyzed. Figure 3 shows the obtained values of elastic modulus (EM), tensile strength (TS) and elongation at break (E).

The addition of oleic acid, beeswax or OA:BW mixtures led to a decrease of EM and TS at the three relative humidity levels studied, although this effect is especially marked at the lowest relative humidity value. This effect can be attributed to the discontinuities in the polymer network induced by the lipid addition, decreasing TS because of preferential break points (Pérez-Gagó & Krochta, 2000, Shellhammer & Krochta, 1997, Bravin *et al.*, 2004). As a result, more deformable and flexible films were obtained. Similar behavior was observed by other authors working with whey protein and carnauba wax, candelilla wax, beeswax or a high-melting industrial milkfat fraction (Shellhammer & Krochta, 1997), zein and carnauba wax (Weller, Gennadios & Sanaiva 1998), whey protein and beeswax (Pérez-Gagó & Krochta, 2000), whey protein and soya oil (Shaw, Monahan, O' Riordan & O' Sullivan, 2002) or sodium caseinate and anhydrous milk fat (Khawladia *et al.,* 2004b). As the beeswax content increased in the lipid mixture, EM and TS values increased, which could be explained by the solid state of beeswax which made the lipid particles less deformable when stretching. This effect is more marked when no oleic acid is present in the lipid mixture. OA can modify the melting properties of beeswax while contributing to the integration of BW in the matrix due to its amphiphilic nature and its ability to adsorb at the lipid protein interface.

It is remarkable that the influence of the RH on the EM and TS of the films is greatly increased when no lipids are present in the films. A sharp decrease in both EM and TS is observed in the control film when comparing those stored at 33% and 58% RH. This seems to indicate that a phase transition (glassy to rubbery) occurs in the protein matrix between the equilibrium moisture levels corresponding to these RH values. In the case of the film containing only BW, this effect is still evident, although it is not observed when different proportions of OA are present in the matrix. In turn, the lack of influence of moisture content on the EM of the films when pure OA is present in formulation is remarkable. This seems to indicate that OA, apart from introducing discontinuities in the matrix, acts as a strong plasticizer inhibiting the effect of water content. Specific interactions of OA and protein chains can contribute to the intense plasticizing role of OA due to its amphiphilic nature (Rhim, Wu, Weller & Schneph, 1999). This can be also deduced from the values of elongation at break. The stretchability of the films increased when pure OA was present and decreased progressively when the BW ratio increased in the lipid mixtures, reaching the lowest values when pure BW was used. No significant effect ($p>0.05$) of relative humidity on the film stretchability was observed.

In an effort to improve physical properties of sodium caseinate based films, some authors claim that the combinations of hydrocolloids could improve functional properties of edible films. However, little information exists on the film-forming ability of such combinations of polysaccharides and proteins (Cho *et al.*, 2002, Longares *et al.*, 2005, Fabra *et al.,* 2008b). In a recent work (Fabra *et al.*, 2008b), the effect of λ-carrageenan and alginate on tensile properties of sodium caseinate based films has been analysed. In lipid-free films, polysaccharides improved the tensile properties of the films. Nevertheless, when lipids (oleic acid and beeswax mixtures) were present, λ-carrageenan slightly modified the film properties whereas the addition of alginate produced less flexible and less stretchable films.

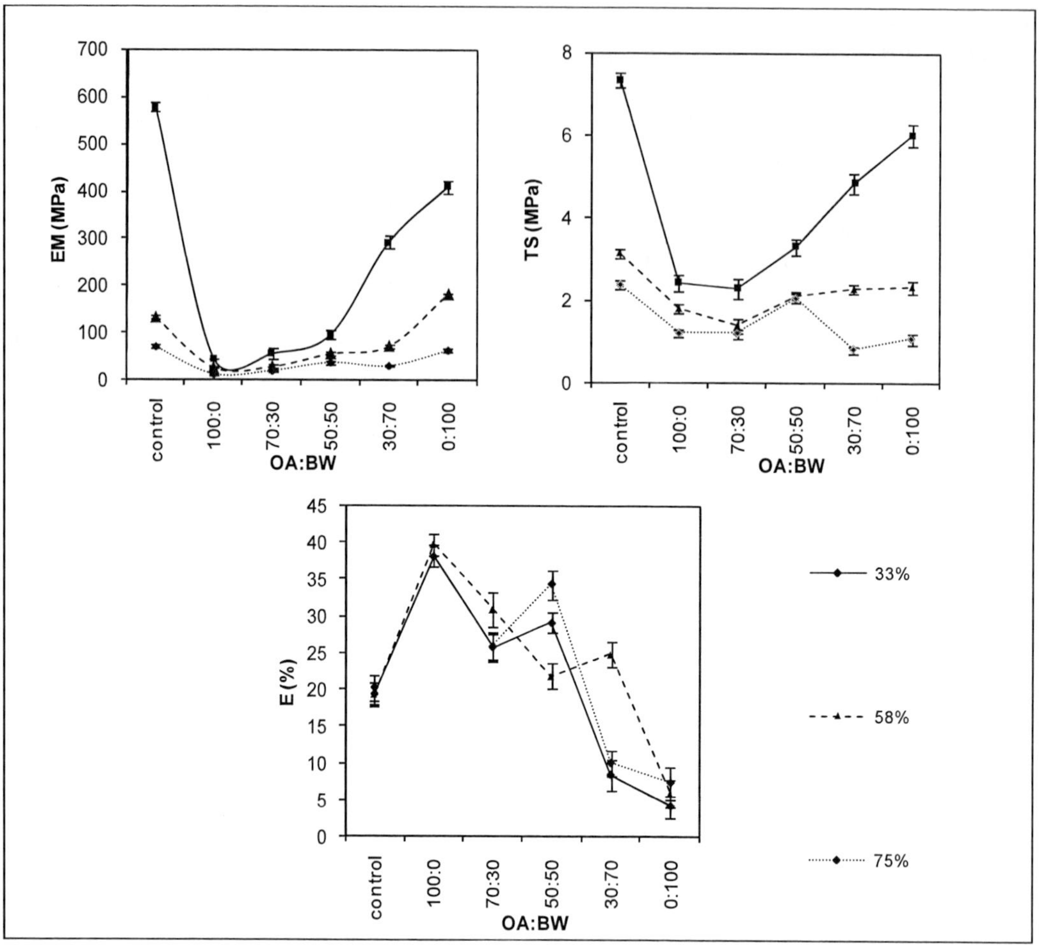

Figure 3. Mean values (and LSD intervals) of tensile parameters of films (100μm) prepared with 1:0.5 protein:lipid ratio and OA:BW mixtures.

The effect of calcium caseinate on sodium caseinate matrices is also discussed in this chapter. As shown in figure 4, calcium caseinate showed greater EM and TS values but similar deformability values (E) to sodium caseinate, suggesting that calcium provides a less flexible structure which contributes to an increase in stiffness and tensile strength. This can be related to the different electrostatic interactions and calcium cross-linking commented on above when water vapour permeability is discussed. In sodium-calcium caseinate matrices, the calcium reinforcement effect is also evident. When lipids (OA:BW) were incorporated, similar effects to those described for sodium caseinate based films were observed, although OA did not show such a notable plasticizing effect on pure calcium caseinate matrices. Very small changes in tensile properties and deformability were detected when pure OA was incorporated. This could indicate that calcium blocks the active points for protein-OA interactions.

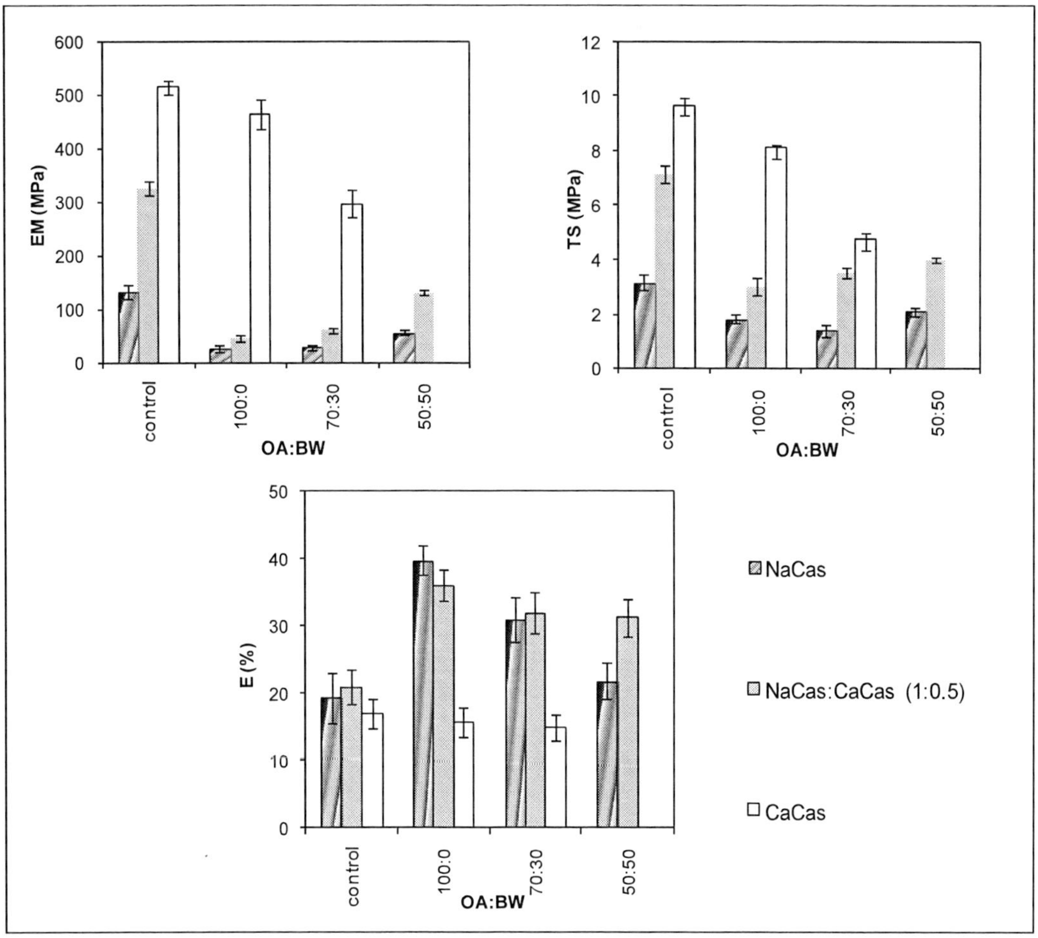

Figure 4. Mean values (and LSD intervals) of tensile parameters of films (100 μm)prepared with different NaCas:CaCas ratios.

Taking into account water vapor permeability and tensile properties, the substitution of sodium caseinate for calcium caseinate in film matrices imparts interesting properties to the films, giving rise to more flexible films than those obtained with calcium caseinate but with better water barrier properties than sodium caseinate based films. This effect was improved even more by the addition of OA:BW mixtures.

As occurred with water vapor permeability, the mechanical properties of sodium caseinate based films were substantially influenced by the film thickness. Figure 5 compares tensile parameters of sodium caseinate edible films of different thicknesses at 20°C and 58% RH. Elastic modulus (EM) and tensile strength (TS) decreased when thickness of the control film (without lipids) increased, whereas differences were not significant for films containing lipids. This can be explained by the fact that discontinuities in the matrix have a greater influence on mechanical behaviour than film thickness. Moreover, a more stretchable matrix was formed in thicker films, which can be explained by the greater cross sectional area permitting greater extension under stress than in the thinner films. Thus, the thinner film reached a limit in their ability to stretch (break point) more quickly than thicker films. Similar results were observed by Longares *et al.*, 2004 when working with whey protein isolate. So, thinner films were stiffer and less stretchable. In contrast, Cuq *et al.*, 1996 and Sobral, 2000 did not find any correlation between % E and film thickness in a puncture test.

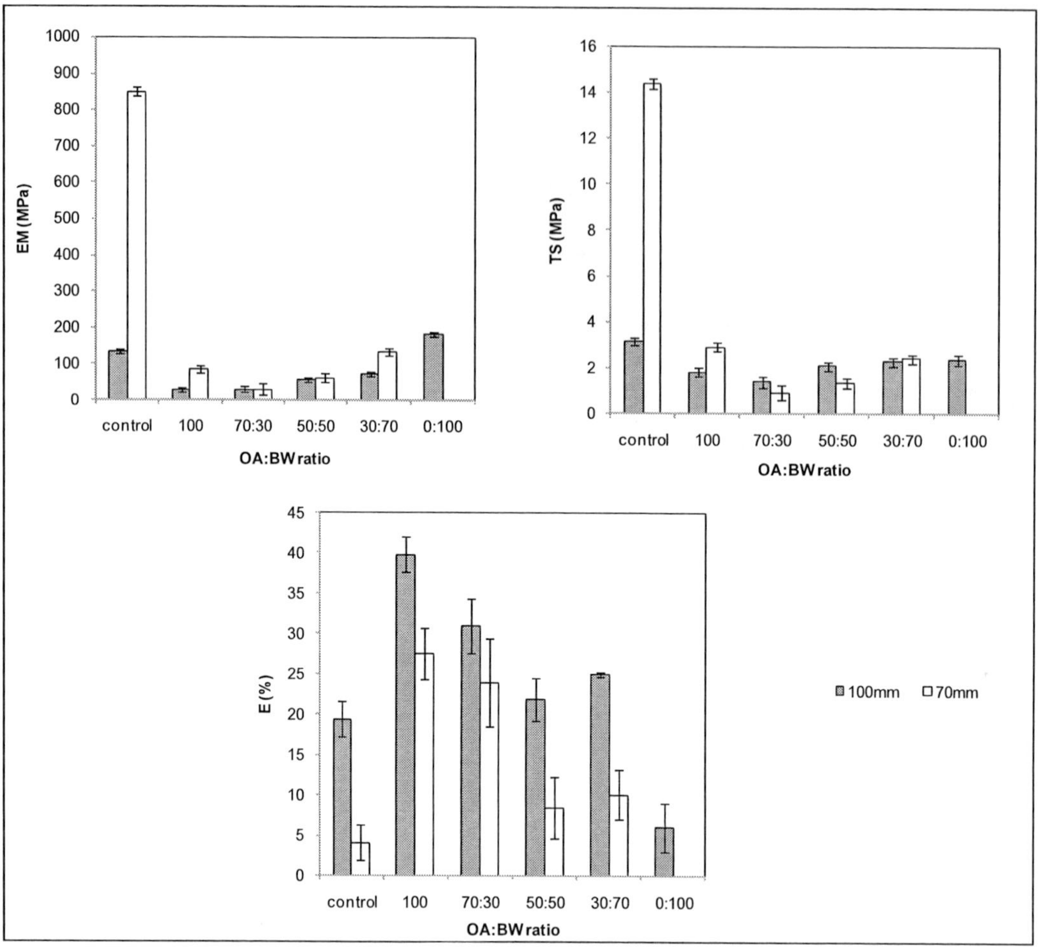

Figure 5. Mean values (and LSD intervals) of tensile parameters (EM, TS and E) of films (100μm) prepared with different thicknesses.

4. Structural Properties

The resulting structure of the films is affected by the distribution of the different components in the film forming system, and their interactions while drying. The microstructural analysis of the films gives relevant information about the arrangement of such components, and helps to understand the mass transference mechanisms through the films, as well as their interaction with light that determines optical properties. For instance, in composite films containing immiscible components, the particle size of the dispersed phase and the heterogeneity of its size distribution and the differences between the refractive indexes of the phases affect the film transparency and gloss (Villalobos *et al.* 2005, Fabra *et al.*, 2009a, b). Scanning Electron Microscopy (SEM) and Atomic Force Microscopy (AFM) are two techniques that can be used to characterize the internal and surface structure of edible films. As reported in previous works (Fabra *et al.*, 2009a, b), the control film has a smooth, homogeneous structure while lipid addition imparts structural discontinuities due to the lack of miscibility of components. Different structures were reached depending on the lipid used; while saturated fatty acids imparted a laminate-like structure, films containing oleic acid and beeswax were not laminar in structure due to the absence of bilayer self-association of the molecules in the film forming emulsions (Fabra *et al.*, 2009a, b). Figure 6 gives an example of cross section images obtained for control films (without lipids) and films prepared with stearic acid $_{(C18:0)}$, oleic acid $_{(C18:1)}$ or beeswax. The laminate-like structure produced by saturated fatty acids can be associated to the growth of the molecular aggregation in the bilayer self-association of the lipid molecules during the film drying and the final crystallization of fatty acids in the protein matrix. Fabra *et al.*, 2009a pointed out that the length of lipid layers in the matrix was greater for lauric acid and decreased as the chain length increased. This indicates that the growth of molecule aggregates during the film drying was more effective when the fatty acid chain was shorter, probably due to the greater molecular mobility. These successive lipid layers in the film matrix greatly contributed to reduce water vapour permeability, thus improving the water barrier properties; the longer the lipid layers, the lower the WVP. This explains the greater water vapour permeability of films containing stearic acid, despite its greater hydrophobic nature (Fabra *et al.*, 2009a).

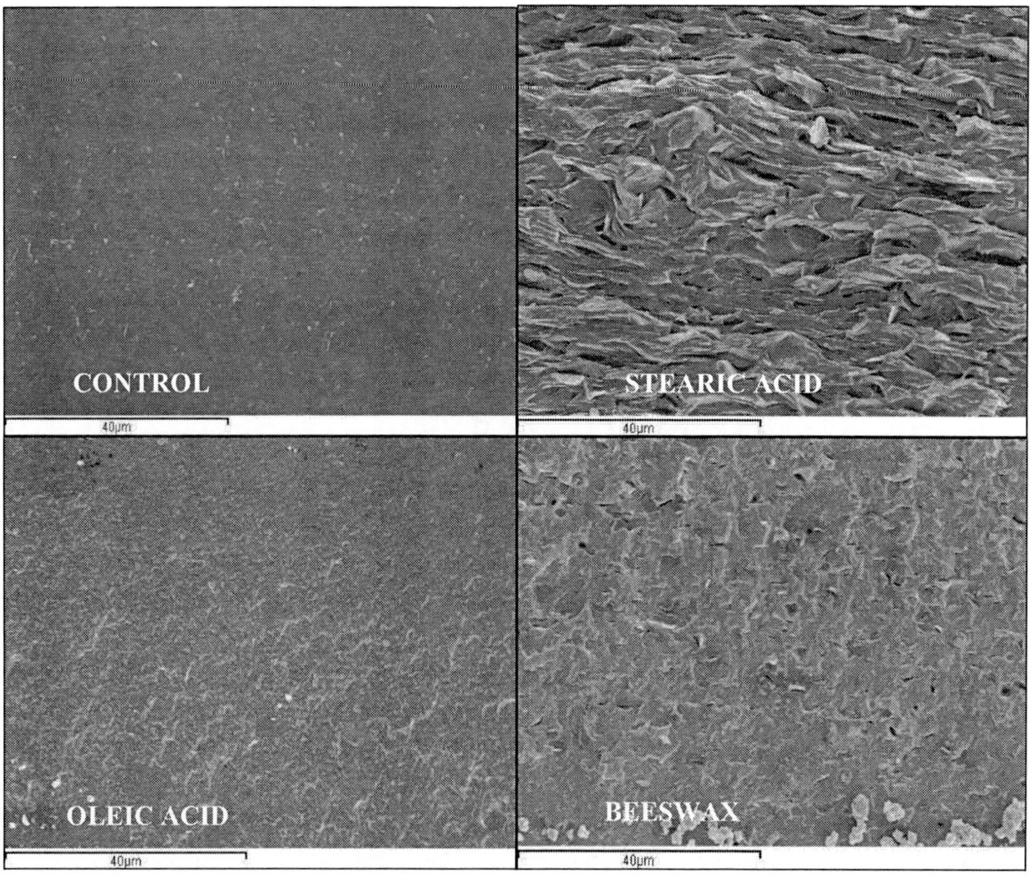

Figure 6. SEM observations of cross-sections of control film (without lipid) and films containing stearic acid $_{(C18:0)}$, oleic acid $_{(C18:1)}$ or beeswax.

On the contrary, in films containing OA and BW lipid droplets were homogeneously dispersed in the protein matrix; more finely distributed for OA and with greater aggregates for BW. This structure did not constitute such an effective barrier to water transport as a laminar structure, which was reflected in the WVP values.

The presence of lipid droplets or laminate-like structures in the film increases its internal heterogeneity and surface roughness due, in part, to the coalescence and creaming phenomena that occur during film drying. The effect of different lipids on film surface topography and roughness was quantified in previous works (Fabra *et al.*, 2009 a, b) through atomic force microscopy (AFM) observations. This roughness can be described by three parameters: average roughness (R_a), which is the average of the absolute value of the height deviations from a mean surface, root mean square roughness (R_q), which is the root mean square average of height deviations taken from the mean data plane, and roughness factor (*r*) which is the ratio between the three-dimensional and two-dimensional surface area produced by projecting the surface onto the threshold plane.

Generally, lipids produced a sharp increase in roughness as compared to control film, which increased as lipid content increased from 1:0.15 to 1:0.5 (protein-lipid ratio) (table 5), mainly for films containing beeswax or saturated fatty acids. The lesser effect of oleic acid on the increase in roughness can be attributed to the fact that oleic acid aggregates are smaller and better dispersed and to its liquid state, which favours its flow over the surface, filling the small voids, thus contributing to obtain a smoother surface. For a determined protein:lipid ratio (1:0.15), saturated fatty acids imparted the highest R_a and R_q values which were similar for lauric $_{(C12:0)}$, palmitic$_{(C16:0)}$ and stearic acids$_{(C18:0)}$, while according to r(%), only stearic acid imparted the maximum roughness. This seems to indicate that the surface of the films containing stearic acid showed a great number of smaller irregularities, so the ratio between the real and the projected surfaces (r) is the greatest. This coincides with the small size of the laminar particles observed in this film from SEM micrographs (Fabra *et al.*, 2009a).

Table 5. Roughness parameters obtained from atomic force microscopy (AFM) images: Ra, Rq and r. (Three images were analyzed in each case.)

NaCas:lipid ratio	Type of lipid	R_a (nm)	R_q (nm)	r (%)
control		9.6 (1.4)^a	12.5 (1.5)^a	0.039 (0.001)^a
1:0.15	OA	84.4 (15.5)^b	128 (18)^{bc}	1.2 (0.4)^b
1:0.5		69 (9)^b	103 (12)^b	2.98 (0.92)^c
1:0.15	BW	219 (45)^c	173 (30)^c	2.4 (0.4)^c
1:0.5		414 (17)^d	525 (25)^d	3.31 (0.97)^c
1:0.15	LA	193 (9)^c	245 (19)^e	0.6 (0.3)^a
	PA	359 (16)^e	351 (29)^f	1.1 (0.2)^{a,b}
	SA	256 (73)^c	293 (37)^{e,f}	3.4 (0.7)^c

Mean values (standard deviation)

$^{a-f:}$ Different superscripts within a column indicate significant differences among formulations (p<underline 0.05)

The effect of oleic acid-beeswax mixtures on surface roughness has also been described by Fabra *et al.*, 2009b for 1:0.3:0.5 protein:glycerol:lipid films. R_a and R_q values showed a sharp increase when BW was present in the lipid mixture, these increasing as the BW ratio increased.

6. Optical Properties

Gloss and transparency are properties which are greatly relevant to the film's functionality due to their great impact on the appearance of the coated products. In composite films containing lipids, these properties can change greatly as compared with the pure polymer films due to the presence of a dispersed phase in the matrix with a different refractive index from that of the continuous polymeric phase. Film transparency evaluated through the Kubelka-Munk coefficients K/S and internal transmittance Ti increases when either K/S or Ti increases. In this sense, high K/S and Ti values are related to homogeneous and transparent films, while a decrease in these values implies film matrices with greater opacity. The different transparency of the films is related with their internal structure developed during the film drying (Villalobos *et al.*, 2005, Fabra *et al.*, 2009b) which, in turn, is greatly affected by the initial microstructure of the film forming emulsions (volume fraction of the dispersed lipids and size of lipid particles) and its development throughout the film drying process. During water removal, flocculation, coalescence and creaming phenomena can occur, depending on the drying conditions, which determine the final film microstructure.

The gloss of edible coatings as affected by surfactants, lipids, relative humidity and storage time has been studied by Trezza & Krochta, 2000. They found that for lipid/surfactant dispersion coatings, the increase in the surfactant level greatly decreased gloss with respect to the pure hydrocolloid film, depending on the surfactant type and the particle size in the dispersed phase. An increase in the particle size implied a loss of glossiness. The gloss of films containing surfactants/lipids has been correlated with the surface roughness determined by AFM (Villalobos *et al.*, 2005, Fabra *et al.*, 2009a); the greater the surface roughness, the lower the gloss. Nevertheless, the angle of incident light and the intrinsic properties (refractive index) of the materials also affect gloss (Trezza & Krochta, 2001). Different studies have been carried out to improve mechanical and barrier properties of sodium caseinate based films (Avena-Bustillo & Krochta, 1993, Banerjee & Chen, 1995, Arvanitoyannis, Psomiadou, & Nakayama, 1996, Chick & Hernández, 2002, Barreto *et al.*, 2003, Khawaldia, Banon, Desobry & Hardy, 2004b), but few have focussed on optical properties such as gloss and transparency, which in turn are very important since they greatly affect the appearance of the coated product. In this chapter, the effect of lipid type and mixtures (OA and BW) and the substitution of sodium caseinate by calcium caseinate on gloss and transparency are discussed.

Table 6 shows the values of Ti at 550 nm as a function of the film composition, where the effect on the film transparency of lipid mixtures (OA:BW) and the substitution of sodium caseinate by calcium caseinate can be clearly observed. As commented on above, sodium caseinate control films exhibited the greatest transparency (higher Ti and K/S values), whereas the presence of lipids increased the film opacity, especially when beeswax increased in the lipid mixture.

Table 6. Internal transmittance (T_i) at 550 nm and gloss values (60°) of the films as a function of NaCas:CaCas ratios and lipid composition

NaCas:CaCas ratio	Protein:lipid ratio	Type of lipid	Ti 550nm	Gloss 60°
1:0	Control		82.8	44 (3)^{a}
	1:0.15	Lauric acid	69.5	29 (1)^{a}
		Palmitic acid	82.0	11 (1)^{b}
		Stearic acid	83.0	13.6 (1.2)^{b}
		Oleic acid	85.2	37 (7)^{a}
		Beeswax	83.4	7.3 (0.8)^{b}
	1:0.5	100:0 (OA:BW)	82.6	29 (8)^{b}
		70:30 (OA:BW)	81.3	17 (3)^{c}
		50:50 (OA:BW)	81.7	14 (2)^{cd}
		30:70 (OA:BW)	80.6	13 (2)^{d}
		0:100 (OA:BW)	78.2	6.3 (0.2)^{e}
1:0.5	Control		81.3	21 (1)^{bc}
	1:0.5	100:0 (OA:BW)	81.1	16 (1)^{cd}
		70:30 (OA:BW)	80.2	16.0 (0.9)^{cd}
		50:50 (OA:BW)	80.3	13 (1)^{d}
0:1	Control		79	46 (3)^{a}
	1:0.5	100:0 (OA:BW)	79.5	30 (2)^{b}
		70:30 (OA:BW)	77.3	26 (2)^{b}

a-e: Different superscripts within a each NaCas:CaCas ratio and protein:lipid ratio indicate significant differences among formulations ($p < 0.05$).

The effect of saturated fatty acids (lauric acid, palmitic acid and stearic acid) on sodium caseinate based films was also analyzed by comparing it to the oleic acid effect for films with 0.15 lipid-protein ratio (Fabra *et al.*, 2009a). Saturated fatty acids implied lower values of K/S and T_i than OA and the film with lauric acid was the one which showed the highest opacity. Saturated fatty acids conferred more heterogeneity on the film structure, which led to light dispersion, thus increasing the opacity. Fabra *et al.*, 2009a attributed the lowest transparency reduction of OA to its liquid state and to the smaller particle size of lipid droplets in the film matrix whereas saturated fatty acids give rise to films with large laminar structures of lipid which greatly contribute to the internal light scattering, decreasing transparency.

As shown in table 6, films containing caseinate mixtures showed higher opacity (less internal transmittance) than that observed for pure sodium caseinate and close to that shown by calcium caseinate. The influence of calcium on the matrix opacity must be attributed to the greater degree of protein chain aggregation promoted by this ion and the greater compactness of the structure. This will imply higher values of the refractive index and a subsequent increase in the reflected light. In calcium caseinate films and matrices prepared with mixtures of both caseinates, the effect of OA was not significant, but BW incorporation reduced the film transparency.

Table 6 shows the gloss values at 60º. A decrease in gloss was observed when lipids were added in the sodium caseinate matrices and this effect became more significant as beeswax content increased. This behaviour is parallel to that described for internal transmittance and is related to the films' heterogeneous structure developed during drying. Surface gloss is closely related with the smoothness or absence of irregularities on the surface. The loss of gloss must be attributed to the presence of irregularities on the film surface; the rougher the surface, the lower the gloss (Ward & Nussinovitch, 1996). In fact, the development of gloss in the studied films coincided with the development of the previously described surface roughness. In a recent work (Fabra *et al.*, 2009a y b), significant correlations between the film surface roughness and optical (transparency and gloss) parameters could be established.

Table 7 shows the Pearson matrix correlation for roughness parameters and gloss values at 60º and 20º for sodium caseinate films containing 1:0.5 and 1:0.15 protein-lipid ratios. Lipid phase was composed by different lipids (saturated fatty acids, OA, BW and OA-BW mixtures). The statistically significant correlations can be observed.

Lipid introduction in the protein matrix implied a decrease in transparency and gloss, especially when there was a high content of beeswax or saturated fatty acids. This can be attributed to the formation of greater lipid aggregates in both the internal and surface parts of the film. In the case of OA, creaming did not imply notable irregularities on the surface because it is liquid and flows, filling the surface voids generated during drying. Taking into account the influence of calcium, calcium caseinate based films are glossier than those obtained with sodium caseinate. This could be attributed to the greater film compactness promoted by calcium which will imply an increase in light reflection.

Table 7. Pearson correlation coefficients for gloss and roughness parameters

	R_a (nm)	R_q (nm)	r (%)	Gloss 20º	Gloss 60º
R_a (nm)	1	.930$^{(**)}$	0.514	-.784$^{(**)}$	-.879$^{(**)}$
R_q (nm)		1	.636$^{(+)}$	-.710$^{(*)}$	-.756**
r (%)			1	-.646$^{(*)}$	-.680*
Gloss 20º				1	.862$^{(**)}$
Gloss 60º					1

(**) Significant correlations ($p<0.01$)
(*). Significant correlations ($p<0.05$)

7. References

[1] Arvanitoyannis, I. & Biliaderis, C. G. (1998). Physical properties of polyol-plasticized edible films made from sodium caseinate and soluble starch blends. *Food Chemistry,* 62 (3), 333-342.

[2] Arvanitoyannis, I., Psomiadou, E. & Nakayama, A. (1996) Edible Films made from Sodium Caseinate, Starches, Sugars or Glycerol. Part I, *Carbohydrate Polymer*, 31, 179-192.

[3] Arvanitoyannis, I., Psomiadou, E., Nakayama, A., Aiba, S. & Yamamoto, N. (1997). Edible films made from gelatin, soluble starch and polyols. Part III. *Food Chemistry*, 60, 593-604.
[4] Avena-Bustillos, R. J. & Krochta, J. M. (1993). Water vapor permeability of caseinate-based edible films as affected by pH, calcium cross-linking and lipid content. *Journal of Food Science,* 58, 904-907.
[5] Audic, J. L & Chaufer, B. (2005). Influence of plasticizers and crosslinking on the properties of biodegradable films made from sodium caseinate. *European Polymer Journal*, 41, 1934-1942.
[6] Banerjee, R. & Chen. (1995). Impact of microfluidization on the mechanical and water vapor barrier properties of milk protein based edible films. *Journal Dairy Science,* 77 (suppl.1), 24.
[7] Barreto, P. L. M., Roeder, J., Crespo, J. S., Maciel, G. R., Terenzi, H., Pires, A. T. N. & Soldi, V. (2003). Effect of concentration, temperature and plasticizer content on rheological properties of sodium caseinate and sodium caseinate/sorbitol solutions and glass transition of their films. *Food Chemistry,* 82, 425-431
[8] Bertuzzi, M. A., Castro Vidaurre, E. F., Armada, M. & Gottifredi, J. C. (2007). Water vapor permeability of edible starch based films. *Journal of Food Engineering,* 80, 972-978.
[9] Brault, D., D'Aprano G. & Lacroix, M. (1997). Formation of freestanding sterilized edible films from irradiated caseinates. *Journal of Agricultural Food Chemistry*, 45, 2964–2969.
[10] Bravin, B., Peressini, D. & Sensidoni, A. (2004). Influence of emulsifier type and content on functional properties of polysaccharide lipid-based edible films. *Journal of Agricultural and Food Chemistry* 52(21), 6448-6455.
[11] Chen, H. (1995) Functional properties and applications of edible films made of milk proteins. *Journal of Dairy Science,* 78, 2563-2583.
[12] Chen, H. (2002). Formation and properties of casein films and coatings. In: A. Gennadios (Ed.), *Protein-based films and coatings* (pp. 181-211). New York: CRC PRESS.
[13] Cheuk-Hang Wan, V., Kim, M. S. & Lee, S. Y. (2005). Water vapor permeability and mechanical properties of soy protein isolate edible films composed of different plasticizer combinations. *Journal of Food Science*, 70(6), 387-391.
[14] Chick, J. & Hernández, R. J. (2002). Physical, thermal and barrier characterization of casein–wax-based edible films. *Journal of Food Science*, 67(3), 1073-1079
[15] Chick, J. & Ustunol, Z. (1998). Mechanical and barrier properties of lactic acid and rennet precipitated casein-based edible films. *Journal of Food Science*, 63, 1024–1027.
[16] Cho, S.Y., Park, J.W. & Rhee, C. (2002). Properties of laminated films from whey powder and sodium caseinate mixtures and zein layers. *Lebensmittel-Wissenschaft und-Technologie*, 35(2), 135-139.
[17] Coughlan, K., Shaw, N.B., Kerry, J.F. & Kerry, J.P. (2004). Combined effects of proteins and polysaccharides on physical properties of whey protein concentrate-based edible films. *Journal of Food Science*, 69 (6), 271-275.
[18] Cuq, B.; Gontard, N.; Aymard, C. & Guilbert, S. (1997). Relative humidity and temperature effects on mechanical and water vapor barrier properties of myofibrillar protein-based films. *Polymer Gel Networks*, 5 (1), 1-15.

[19] Cuq, B., Gontard, N., Cuq, J. L. & Guilbert, S. (1996). Functional properties of myofibrillar protein-basedbiopackagin g as affected by film thickness. *Journal of Food Science*, 61, 580–584.
[20] Debeaufort, F., Quezada-Gallo, J. A., Delporte, B. & Voilley, A. (2000). Lipid hydrophobicity and physical state effects on the properties of bilayer edible films. Journal of Membrane Science, 180, 47-55.
[21] Fabra, M.J., Talens, P. & Chiralt, A. (2008a). Tensile properties and water vapour permeability of sodium caseinate films containing oleic acid-beeswax mixtures. *Journal of Food Engineering*, 85, 393-400.
[22] Fabra, M.J., Talens, P. & Chiralt, A. (2008b). Effect of alginate and λ-carrageenan on tensile properties and water vapour permeability of sodium caseinate-lipid based films. *Carbohydrates Polymers,* 74 (3,4), 419-426.
[23] Fabra, M.J., Hambleton, A., Talens, P., Debeuafort, F., Chiralt, A. & Voilley, A. (2008c). Aroma barrier properties of sodium caseinate-based films. *Biomacromolecules,* 9, 1406-1410.
[24] Fabra, M.J., Jiménez, A., Atarés, L., Talens, P. & Chiralt, A. (2009a). Effect of fatty acids and beeswax addition on properties of sodium caseinate dispersions and films. *Biomacromolecules,* (In press).
[25] Fabra, M.J., Talens, P. & Chiralt, A. (2009b). Microstructure and optical properties of sodium caseinate films containing oleic acid-beeswax mixtures. *Food Hydrocolloids,* 23(3), 676-683.
[26] Gennadios, A., Bradenburg, A. H., Park, J. W., Weller, C. L. & Tcstin, R. F. (1994a). Water vapor permeability of wheat gluten and soy protein isolates films. *Industrial Crops and Products*, 2, 189–195.
[27] Gennadios, A., McHugh, T. H., Weller, C. L. & Krochta, J. M. (1994b). Edible coatings and films based on proteins. In J. M. Krochta, E. A. Baldwin, & M.O. Nísperos-Carriedo (Eds.), *Edible Coatings and Films to Improve Food Quality* (pp.201-277). Technomic Publishing: Lancaster, PA.
[28] Gennadios, A., Park, H. J. & Weller, C. L. (1993). Relative humidity and temperature effects on tensile strength of edible protein and cellulose ether films. *Transactions of the American Society of Agricultural Engineers*, 36: 1867-1872.
[29] Greener, I. & Fennema, O. (1992). The effect of relative humidity gradient on wáter vapor permeance of lipid and lipid-hydrocolloid bilayer films. *Journal of the American Oil Chemists'Society,* 69(11), 1081-1087.
[30] Gontard, N.; Guilbert, S. & Cuq, J. L. (1993). Water and glycerol as plasticizers affect mechanical and water vapour barrier properties of an edible wheat gluten film. *Journal of Food Science,* 58 (1), 206-211.
[31] Gontard, N., Duchez, C., Cuq, J.L. & Guilbert, S. (1994). Edible composite films of wheat gluten and lipids: water vapour permeability and other physical properties. *International Journal of Food Science and Technology*, 29, 39-50.
[32] Ho, B. (1992). Water vapor permeabilities and structural characteristics of casein films and casein-lipid emulsion films. M. S. Thesis, University of California, Davis, CA, USA.
[33] Khwaldia, K.; Perez, C.; Banon, S.; Desobry, S. & Hardy, J. (2004a). Milk proteins for edible films and coatings. *Critical Review in Food Science and Nutrition*, 44(4), 239-251.

[34] Khwaldia, K.; Banon, S.; Desobry, S. & Hardy, J. (2004b). Mechanical and barrier properties of sodium caseinate-anhydrous milk fat edible films. *International Journal of Food Science and Technology,* 39, 403-411.

[35] Kristo E., Biliaderis, C.G. & Zampraka, A. (2007). Water vapour barrier and tensile properties of composite caseinate-pullulan films: biopolimer composition effects and impact of beeswax lamination. *Food Chemistry*, 101, 753-764.

[36] Lim, L. T., Mine, Y. & Tung, M. A. (1999) Barrier and tensile properties of transglutaminase cross-linked gelatin films as affected by relative humidity, temperature and glycerol content. *Journal of Food Science*, 64 (4), 616-622.

[37] Longares, A. Monahan, F.J., O´Riordan, E. D. & O´Sullivan, M. (2005). Physical properties of edible films made from mixtures of sodium caseinate and WPI. *International Dairy Journal,* 15(12), 1255-1260.

[38] Longares, A., Monahan, F. J., O'Sullivan, M. & O'Riordan, E. D. (2004). Physical properties and sensory evaluation of WPI films of varying thickness. *Lebensmittel-Wissenschaft und-Technologie*, 37(5), 545–5

[39] McHugh, T. H., Avena-Bustillos, R. & Krochta, J. M. (1993). Hydrophilic edible films: modified procedure for water vapor permeability and explanation of thickness effects. *Journal of Food Science*, 58(4), 899–903.

[40] Mc Hugh, T. H. & Krochta, J. M. (1994a). Milk protein based films and coatings. *Food Technology,* 48 (1), 97-103.

[41] Mc Hugh, T. H. & Krochta, J. M. (1994b).Water vapour permeability properties of edible whey protein-lipid emulsion films. *Journal of the American Oil Chemists Society,* 71, 307-312.

[42] Mc Hugh, T. H. & Krochta, J. M. (1994c). Sorbitol vs. glycerol plasticized whey protein edible film: integrated oxygen permeability and tensile property evaluation. *Journal of Agricultural and Food Chemistry*, 42 (4), 841-845.

[43] Miller, K. S., & Krochta, J.M. (1997). Oxygen and aroma barrier properties of edible films: A review. *Trends in Food Science and Technology*, 8, 228-237.

[44] Monedero, F. M., Fabra, M.J., Talens, P. & Chiralt, A. (2009). Effect of oleica cid-beeswax mixtures on mechanical, optical and wáter barrier properties of soy protein isolate based films. *Journal of Food Engineering*, 91 (4), 509-515.

[45] Morillon, V., Debeaufort, F., Blond, G., Capelle, M. & Voilley, A. (2002). Factors affecting the moisture permeablity of lipid-based edible films: a review. *Critical Review in Food Science and Nutrition*, 42(1), 67-89.

[46] Park, H. J. & Chinnan, M. S. (1995). Gas and water vapor barrier properties of edible films from protein and cellulosic materials. *Journal of Food Engineering*, 25, 497–507.

[47] Park, H. J., Weller, C. L., Vergano, P. J. & Testin, R. F. (1993). Permeability and mechanical properties of cellulose-based edible materials. *Journal of Food Science*, 58(6), 1361–1364, 1370.

[48] Park, H. J., Testin, R. F., Park, H. J., Vergano, P. J. & Weller, C. L. (1994). Fatty acid concentration effect on tensile strength, elongation, and water vapor permeability of laminated edible films. *Journal of Food Science*, 59, 916-919.

[49] Parris N., Coffin, D.R., Joubran, R.F. & Pessen, H. (1995). Composition factors affecting the water vapor permeability and tensile properties of hydrophilic flms. *Journal of Agricultural and Food Chemistry*, 43, 1432-1435.

[50] Pérez-Gago, M. B. & Krochta, J. M. (2000). Drying temperature effect on water vapor permeability and mechanical properties of whey protein-lipid emulsion films. *Journal of Agricultural and Food Chemistry*, 48 (7), 2687-2692.

[51] Pérez-Gagó, M. B. & Krochta, J. M. (2001). Lipid particle size effect on water vapor permeability and mechanical properties of whey protein/beeswax emulsion films. *Journal of Agricultural and Food Chemistry*, 49, 996-1002.

[52] Quezada-Gallo, J. A. (1999) Influence de la structure et de la composition de réseaux macromoléculaire pour les transferts de volatiles (eau et arômes). Applications aux emballages comestibles et plastiques. Ph.D. Thesis, ENSBANA, Université de Bourgogne, Dijon.

[53] Rankin, J. C., Wolff, A., Davis, H. A. & Rist, C. E. (1958). Permeability of amylose film to moisture vapor, selected organic vapors and the common gases. *Industrial and Engineering Chemistry*, 3, 120–123.

[54] Rhim, J. W. Wu, Y., Weller, C. & Schnepf, M. (1999). Physical characteristics of emulsified soy protein-fatty acid composite films. *Science Aliments*, 19, 57-71.

[55] Roos, Y. H. (1995). *Phase transitions in Food*. San Diego, CA: Academic Press.

[56] Sabato, S. F., Ouattara, B., Yu, H.; D'Aprano, G., Le Tien, C., Mateescu, M. A. & Lacrois, M. (2001). Mechanical and barrier properties of cross-linked soy and whey protein based films. *Journal of Agricultural and Food Chemistry,* 49, 1397-1403.

[57] Shaw, N. B.; Monahan, F. J., O' Riordan, E. D. & O' Sullivan, M. (2002). Effect of soya oil and glycerol on physical properties of composite WPI films. *Journal of Food Engineering*, 51, 299-304.

[58] Shellhammer, T. H. & Krochta, J. M. (1997). Whey protein emulsion film performance as affected by lipid type and amount. *Journal of Food Science,* 62 (2), 390-394.

[59] Siew, D. C. W., Heilmann, C., Easteal, A. J. & Cooney, R. P. (1999). Solution and Film Properties of Sodium Caseinate/ Glycerol and Sodium Caseinate/Polyetylene Glycol Edible Coating Systems. *Journal of Agricultural and Food Chemistry,* 47, 3432-3440.

[60] Sobral, P. J. D. (2000). Influencia da espessura de biofilmes feitos a base de proteínas miofibrilares sobre suas propiedades funcionais. Pesquisa Agropecuária Brasileira, 35(6), 1251–1259.

[61] Schwartzberg, H. G. (1986). Modeling of gas and vapor transport through hydrophilic films. In M. Mathlouthi (Ed.), *Food packaging and preservation. Theory and practice* (pp. 115–136). London: Elsevier Applied Science Pub.

[62] Tanaka, M., Ishizaki, S., Suzuki, T. & Takai, R. (2001). Water vapour permeability of edible films prepared from fish water soluble proteins as affected by lipid type. *Journal of Tokyo University of Fisheries*. 87, 31-37.

[63] Tomasula, P. M., Parris, N., Yee, W. & Coffin, D. (1998). Properties of films made from CO_2-precipitated casein. *Journal of Agricultural and Food Chemistry*, 46, 4470-4474.

[64] Trezza, T. A. & Krochta, J. M. (2000). The gloss of edible coatings as affected by surfactants, lipids, relative humidity and time. *Journal of Food Science*, 65 (4), 658-662.

[65] Trezza, T. A. & Krochta, J. M. (2001). Specular reflection, gloss, roughness and surface heterogeneity of biopolymer coatings. *Journal of Applied Polymer Science*, 79 (12), 2221

[66] Turgeon, S. L. & Beaulieu, M. (2001). Improvement and modification of whey protein gel texture using polysaccharides. *Food Hydrocolloids*, 15, 583-591.

[67] Villalobos, R., Chanona, J., Hernández, P., Gutiérrez, G. & Chiralt, A. (2005). Gloss and transparency of hydroxypropyl methylcellulose films containing surfactants as affected by their microstructure. *Food Hydrocolloids,* 19, 53-61.

[68] Ward, G. & Nussinovitch, A. (1996). Gloss properties and surface morphology relationships of fruits. *Journal of Food Science*, 61(5), 973-977.

[69] Weller, C. L., Gennadios, A. & Sanaiva, R. A. (1998). Edible bilayer films from zein and grain sorghum wax or cannauba wax. *Lebensmittel-Wissenschaft und-Technologie* 31, 279-285.

[70] Yang, L. & Paulson, A. T. 2000. Effects of lipids on mechanical and moisture barrier properties of edible gellan film. *Food Research International,* 33, 571-578.

[71] Yu, H., Sabato, S. F., D´Aprano, G. & Lacroix, M. (2004). Effect of the addition of CMC on the aggregation behaviour of proteins. *Radiation Physics and Chemistry*, 71, 129-133.

[72] Zaleska, H., Ring, S. & Tomasik, P. (2000). Apple pectin complexes with whey protein isolate. *Food Hydrocolloid*, 14, 377-382.

In: Food Hydrocolloids: Characteristics, Properties... ISBN: 978-1-60876-222-4
Editor: C. S. Hollingworth, pp. 145-177

Chapter 5

Molecular Dynamics in Food Hydrocolloids Observed by NMR

Shingo Matsukawa*[*], *Makoto Shimizu, Bona Dai, and Ayaka Shinsho
Department of Food Science and Technology, Tokyo University of Marine Science and Technology, Tokyo 108-8477, Japan

1. Introduction

Many solutions of polysaccharides and proteins exist as a solution at high temperature and become a gel at low temperature, and they are widely used in food industries to make a gel. Rheological properties drastically change when the solution becomes a gel. This change is based on the structural change of the polysaccharides and proteins, such as that from random coil to helix structure. The structural change causes the change of tumbling motion and translational motion in the solutions. NMR is a powerful tool for investigating the molecular motion. Relaxation times for the macromolecules reflect the molecular motion, and relaxation times for water reflect the motion of water molecules and also macromolecules through the chemical exchanging between water proton and labile proton on the macromolecules. The diffusion coefficients of molecules in the hydrocolloid system give the information about the mobility of molecules and the structure of the gel network.

In this chapter, we introduce the basic theories and techniques of NMR for measurements of relaxation times and the diffusion coefficient in section 2, and some experimental results for solutions of gelatin [1], gellan [2,3], carrageenans [4] and agar [3] in section 3.

* Address:
4-5-7 Konan, Minato-ku, Tokyo 108-8477, Japan
Tel & Fax: 81-3-5463-0581
E-mail: matsukaw@kaiyodai.ac.jp

2. MEASUREMENTS OF MOLECULAR DYNAMICS BY NMR

In this section, we introduce some NMR techniques that play an important role in characterizing the structure and dynamics of food hydrocolloid systems. Measurements of relaxation times and diffusion coefficient are important and will be described in detail. The relaxation times give information on the tumbling motion of molecules such as polymer chain and water in food hydrocolloid systems. The relation between the motional change and structural change with the network formation is investigated through the relaxation measurements. On the other hand, the diffusion coefficient gives information about the translational motion of molecules, which is useful for the structural changes of gel structure in hydrocolloid systems.

2.1. Basic NMR Techniques

Pulse NMR Method

In the pulse NMR experiment, high-power rf pulses are applied to a sample for rotating the equilibrium longitudinal magnetization, and subsequently, the free-introduction decay (FID) of the transverse component is detected as a time-dependent function. The Fourier transform converts FID into the corresponding frequency spectrum, of which the chemical shift gives information about chemical structures. Measurements for relaxation times and diffusion give longitudinal (spin-lattice) relaxation time T_1, transverse (spin-spin) relaxation time T_2, and diffusion coefficient D. The T_1 and T_2 are closely related to the correlation time in molecular motion by the BRP theory [5] when they are governed by dipolar interactions, and the D is a parameter for a measurement of the velocity of the translational motion. Therefore, pulse NMR is a very useful method for the elucidation of molecular dynamics in polymer systems.

Excitation by Pulses

In the pulse NMR experiments for weakly interacting spins, such as nuclei in solution, the motion of the ensemble of spins is described by using the macroscopic magnetization vector in a classical manner, although a single spin is considered to obey quantum mechanics. This classical formalism is helpful to understand the pulse NMR experiments, and the behavior of the magnetization **M** in the magnetic field **H** during experiments is described by the Bloch equation as follows [6]:

$$d\mathbf{M}/dt=\gamma\mathbf{M}\times\mathbf{H}-(M'_x\mathbf{i}'+M'_y\mathbf{j}')/T_2-(M'_z-M_0)\mathbf{k}'/T_1 \quad (1)$$

where **i'**, **j'**, and **k'** are the unit vectors in the laboratory frame. The M'_x, M'_y, and M'_z components are along **I'**, **j'**, and **k'**, respectively, and M_0 is the equilibrium value of M'_z. In Bloch equations, T_1 and T_2 are used as parameters for the decay of longitudinal magnetization and transverse magnetization, respectively. The application of an rf pulse of the same frequency with the Lamor frequency ω_0 for the static field $\mathbf{H}_0$ causes the rf magnetic field $\mathbf{H}_1$ on the x-axis in the rotating frame of ω_0. Then, **M** is rotated around the x-axis into the y-z plane (Figure 1). For the rf pulse much shorter than T_1 and T_2, the time rate of change of **M** in the rotating flame of ω_0 is given by

$$d\mathbf{M}/dt=\gamma\mathbf{M}\times\mathbf{H}_1 \quad (2)$$

and the angle of the rotation θ is given by

$$\theta=\gamma H_1 t_p \quad (3)$$

where t_p is the pulse width, which is the length of time for the application of the rf pulse. In many pulse NMR experiments, $\pi/2$ and π pulse, for which θ=90°and 180°, respectively, are frequently used and made by the corresponding t_p for the H_1.

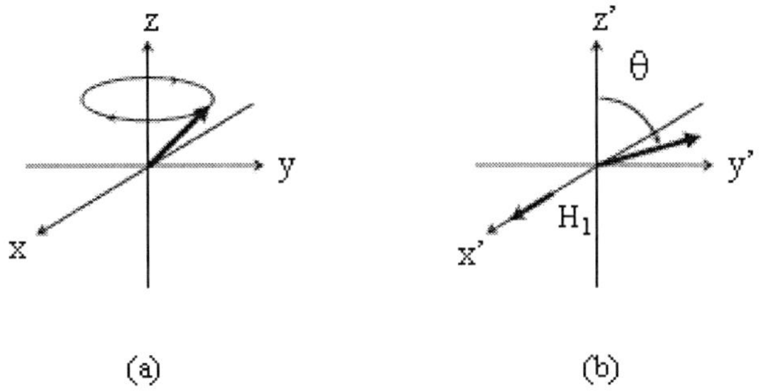

Figure 1. Rotation of the magnetization vector **M** around the x-axis in the y-z plane in the rotating flame of the Lamor frequency ω_0 for the static field H_0 by the rf pulse of ω_0, which induces the magnetic field H_1 in the rotating flame.

2.2. NMR Relaxation Times

T_1 Measurements

Spin-lattice relaxation is a process through which the equilibrium population in spin states is reached after a perturbation through the energy transfer from an excited spin to the lattice, the surrounding thermal reservoir. On the basis of the phenomenological description, the z component of the magnetization M_z obeys the equation

$$dM_z/dt=-(M_z-M_0)/T_1 \quad (4)$$

where M_0 is the equilibrium value of the magnetization. The inversion recovery (IR) method is frequently used to measure the T_1 value. Figure 2 shows the π_x pulse to invert the magnetization vector lying along the z-axis to that lying along the - z-axis(a), the inverted magnetization reverts to the equilibrium condition (b), $\pi_x/2$ pulse is used as a second pulse after delay time t to inspect the z component of the magnetization M_z. When t=0, $M_z=-M_0$; therefore, Eq.4 gives

$$M_z=M_0(1-2\exp(-t/T_1)) \quad (5)$$

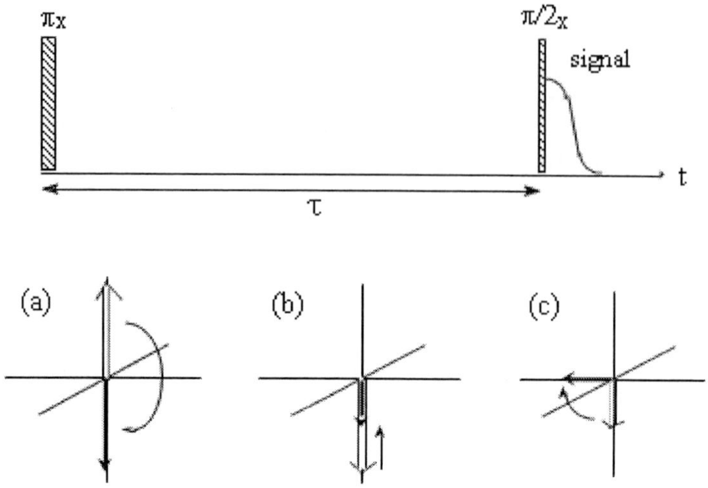

Figure 2. Pulse sequence and magnetization trajectories for the inversion recovery method: (a) the π_x pulse inverts the magnetization vector lying along the z-axis to that lying along the -z-axis, (b) the inverted magnetization reverts to the equilibrium condition, and (c) the $\pi_x/2$ pulse is used as a second pulse after delay time τ to inspect the z component of the magnetization M_z.

From a series of measurements varying t, the plot of M_z against t is obtained and the T_1 value is given by a least-squares fitting to the experimental data by using Eq.5. Before the first π_x pulse, every experiment has to be done after enough retention time, at least five times T_1, to reach the equilibrium condition.

T_2 Measuremtns

Spin-spin relaxation is a process of decay in the net transverse magnetization owing to the loss of phase coherence by the mutual exchange of spin energies after a pulse to make the transverse magnetization. The phase coherence also decreases by the effect of magnetic field inhomogeneity, therefore, this effect should be removed for the exact T_2 measurement. In the pulse NMR experiments with the spin echo sequence, the phase coherence is refocused at some time after rf pulses. This phenomenon is described by the classical formalism using the macroscopic magnetization vector.

Spin Echo Method

After the rf pulse is switched off, the time rate of change of the net transverse magnetization $M_+=M_x+iM_y$, where M_x and M_y are the x and y components of **M**, respectively, is derived from Eq.1 as follows:

$$dM_+/dt = -i\gamma H_0 M_+ - M_+/T_2 = i\omega_0 M_+ - M_+/T_2 \quad (6)$$

where H_0 is the z component of H_0. The time rate of change of the net transverse magnetization in the rotating frame M_+ introduced by $M_+ = M_+ \exp(i\omega_0 t)$ is written as

$$dM_+/dt = -M_+/T_2 \quad (7)$$

In Eq.7, the T_2 emerges as a time constant for the decay where the perturbed M_+ returns to the equilibrium state; $M_+ = 0$. When there is an imhomogeneity of the static magnetic field, the z component of the total external magnetic field H_z at a point **r** is expressed by using field gradient **G**, which described the imhomogeneity in the first approximation:

$$H_z = H_0 + \mathbf{G} \cdot \mathbf{r}. \quad (8)$$

Then, **M** in Eq.6 is rewritten as a time- and space-dependent function:

$$\partial M_+/\partial t = -i\gamma(H_0 + \mathbf{G}\cdot\mathbf{r})M_+ - M_+/T_2 = i\omega_0 M_+ - i\gamma(\mathbf{G}\cdot\mathbf{r})M_+ - M_+/T_2 \quad (9)$$

and M_+ is given by

$$\partial M_+/\partial t = -i\gamma(\mathbf{G}\cdot\mathbf{r})M_+ - M_+/T_2 \quad = -M_+/[1/T_2 + i\gamma(\mathbf{G}\cdot\mathbf{r})]^{-1} \quad (10)$$

In Eq.10, the $T_2^* = [1/T_2 + i\gamma(\mathbf{G}\cdot\mathbf{r})]^{-1}$ emerges as a time constant for the decay under the existence of an inhomogeneteiy of the static magnetic field. The FID after a pulse decays with the time constant of T_2^*.

With $\Psi(\mathbf{r},t)$ introduced by $M_+ = \Psi(\mathbf{r},t)\exp(-t/T_2)$, Eq.10 is rewritten as

$$\partial\Psi(\mathbf{r},t)/\partial t = -i\gamma(\mathbf{G}\cdot\mathbf{r})\Psi(\mathbf{r},t) \quad (11)$$

This indicates that spin vectors precise in the rotating flame with the individual rates of $-\gamma(\mathbf{G}\cdot\mathbf{r})$ corresponding to the locations and that the spin isochromat (the macroscopic magnetic vector at individual **r**) dephases, which causes a decay of the net transverse magnetization.

In the spin echo experiments [7], this rotation is canceled at the echo point; the vectors for spin isochromat precess to become the angle from the x-axis $\alpha = -\gamma(\mathbf{G}\cdot\mathbf{r})\tau$ during the τ period; the π_x pulse (the π pulse along the x-axis) flips the vectors to become $\alpha = \pi - \{-\gamma(\mathbf{G}\cdot\mathbf{r})\tau\}$; and the vectors precess $-\gamma(\mathbf{G}\cdot\mathbf{r})\tau$ during the next τ period and are aligned on the —y-axis (Figure 3). The solution for $\Psi(\mathbf{r},t)$ after the π pulse is given by

$$\Psi(\mathbf{r},t) = -A(t)\exp[-i\gamma(\mathbf{G}\cdot\mathbf{r})(t-2\tau)] \quad (12)$$

where $A(t)$ is a space-independent function. Eq.12 shows that $\Psi(\mathbf{r},t)$ for the individual spins in the whole of the sample is identical with $A(t)$ at $t=2\tau$; therefore, the net transverse magnetization becomes a maximum. Similarity, in the spin echo experiment with the $\pi/2_x$

pulse and π_y pulse (the π pulse along y-axis), the spin vectors in the whole of the sample are aligned on the y-axis, and $\Psi(\mathbf{r},t)$ is given by the expression with the opposite sign of the right-hand side of Eq.12.

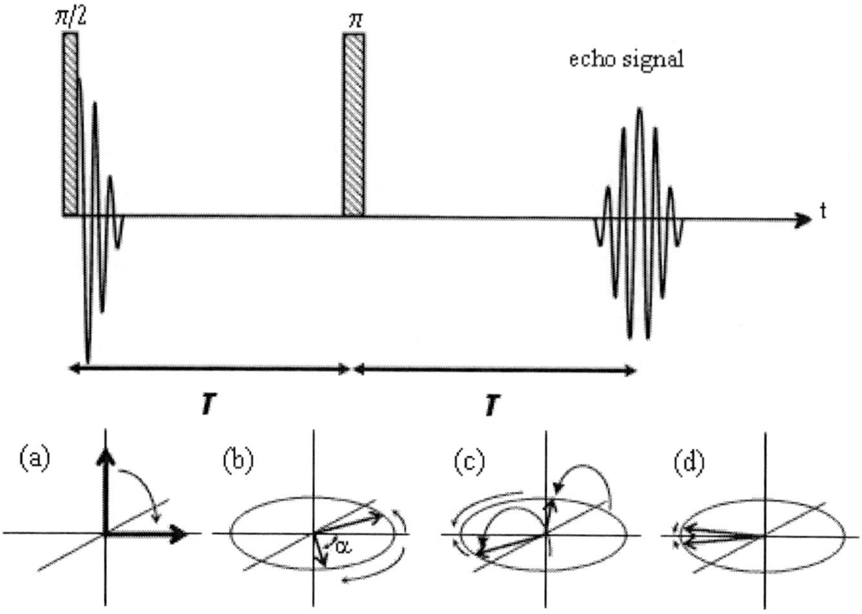

Figure 3. Pulse sequence and magnetization trajectories for the spin echo experiment: a) the magnetizations are aligned along the y axis by an r.f. $\pi/2$ puls, b) the individual spin isochromat presses to be the angle from the x-axis α=-γ(G · **r**) τ during the τ period corresponding to the locations, c) the π_x pulse flips spin isochromat to be α=-{- γ (G · **r**) τ}, d) the individual spin isochromat presses –γ(G · **r**) τ during the next τ period and are aligned on the —y-axis.

Carr-Purcell-Meiboom-Gill (CPMG) Method

In the Carr-Purcell-Meiboom-Gill (CPMG) method [8,9,10], a $\pi/2_x$ pulse and a series of π_y pulses are used to form an echo train (Figure 4); the vector of individual spin isochromat induced by the $\pi/2_x$ pulse precess to become $\alpha=-\gamma(\mathbf{G}\cdot\mathbf{r})\tau$ during the τ period, flip to become $\alpha=-[\gamma(\mathbf{G}\cdot\mathbf{r})\tau]$ after the π_y pulse, precess to become $-\gamma(\mathbf{G}\cdot\mathbf{r})\tau$ during the next τ period and align on the y-axis to make an echo, and flip at $t=2n\tau$ and align on the y-axis at $t=(2n+1)\tau$ in a similar manner. The angle error of the π pulse is compensated every two π pulses by using the π pulse along the y-axis. The solution for $\Psi(\mathbf{r},t)$ between $t=(2n-1)\tau$ and $t=(2n+1)\tau$ is given by

$$\Psi(\mathbf{r},t)=A(\mathrm{t})\exp[-\mathrm{i}\gamma(\mathbf{G}\cdot\mathbf{r})(\mathrm{t}-2n\tau)]. \quad (13)$$

Considering the diffusion term $D\nabla^2\Psi(\mathbf{r},t)=-D\gamma^2G^2(t-2n\tau)^2\Psi(\mathbf{r},t)$, we have

$$\mathrm{d}A(t)/\mathrm{d}t=-D\gamma^2G^2(t-2n\tau)^2A(t). \quad (14)$$

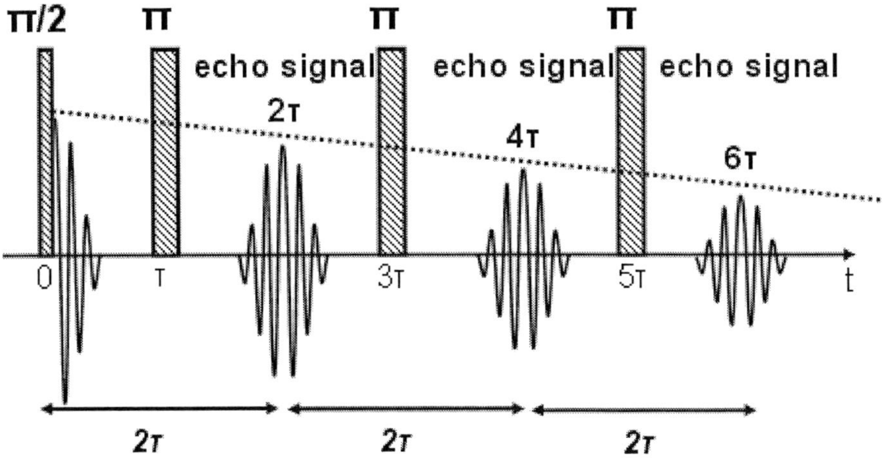

Figure 4. Pulse sequence and magnetization trajection the individual spin isochromat or the Carr-Purcell-Meiboom-Gill method.

The integration from t=0 to t=τ from t=(2m−1)τ to t=(2m+1)τ, where m=1,2,…,n−1, and from t=(2n−1)τ to 2τ gives

$$A(2n\tau)=A(0)\exp(-2/3D\gamma^2G^2\tau^3n). \qquad (15)$$

And $M_+(t)$ at t= 2$n\tau$ is given by

$$M_+(t)=M_+(0)\exp(-2n\tau/T_2-2/3D\gamma^2G^2\tau^3n) =M_+(0)\exp(-t/T_2-tD\gamma^2G^2\tau^2/3). \qquad (16)$$

As indicated in Eq.16, it is possible to make the effect of diffusion negligible by employing a small τ value with large n in the CPMG experiment, and thus we have

$$M_+(t)=M_+(0)\exp(-t/T_2). \qquad (17)$$

Under the condition that the effect of diffusion is not negligible, the FID in the CPMG experiment decays single exponentially with the time constant of T_2'=$\{1/T_2-D\gamma^2G^2\tau^2/3\}^{-1}$, whereas the FID in spin echo decays with the stronger time dependence, which is indicated below, rather than with exponential decay.

2.3. Diffusion Coefficient

A field gradient attaches spatial information to an NMR signal; therefore, it can produce the spatial distribution of a nucleus, that is, NMR imaging. When two field gradients for dephasing and rephasing are applied, the NMR signal decays due to the displacement of the nucleus during the period between the two field gradients. This gives the diffusion coefficient for the Fickian diffusion in free space. The field gradient should be applied linearly in space and well controlled in time schedule of a pulse sequence. Recently, the Field gradient system has been greatly improved and it is possible to apply very strong field gradients under precise control. This advanced the applications for the investigation of macromolecular diffusion in gels [11]. Some basic theory are described below.

Effect of Diffusion on Spin Echo

When a nucleus in a minute area at **r** diffuses to another area following Fick's law, the decay of $\Psi(\mathbf{r},t)$ by the diffusion is given by [12]

$$\partial\Psi(\mathbf{r},t)/\partial t = D^2\Psi(\mathbf{r},t) \tag{18}$$

where D is the diffusion coefficient. The nuclear mean square displacement in the z direction $<z^2>$ is given by

$$\langle z^2 \rangle = 2DT_d \tag{19}$$

where T_d is the time allowing diffusion. With the effect of the diffusion, taken into account Eq.11 is rewritten as

$$d\Psi(\mathbf{r},t)/dt = -i\gamma(\mathbf{G}\cdot\mathbf{r})\Psi(\mathbf{r},t) + D\nabla^2\Psi(\mathbf{r},t). \tag{20}$$

From Eq.12, the diffusion term is given as

$$D\nabla^2\Psi(\mathbf{r},t) = -D\gamma^2G^2(t-2\tau)^2\Psi(\mathbf{r},t). \tag{21}$$

On the basis of Eqs.12 and 21, Eq. 20 is rewritten as

$$dA(t)/dt = -d\gamma^2G^2(t-2\tau)^2A(t). \tag{22}$$

From the result of the integration of Eq.22 from t=0to $t=\tau$ and from $t=\tau$ to t=2τ, $A(2\tau)$ is given by

$$A(2\tau) = -A(0)\exp(-2/3D\gamma^2G^2\tau^3). \tag{23}$$

Therefore, $M_+(2\tau)$, the net transverse magnetization at t=2τ, is given by Eq.23 as

$$M_+(2\tau) = -M_+(0)\exp(-2\tau/T_2 - 2/3D\gamma^2G^2\tau^3). \tag{24}$$

Equation 24 shows that the spin echo intensity decays by the T_2 relaxation process and also by the effect of the diffusion, which becomes greatly pronounced in large τ measurements.

Diffusion Coefficient Measurements

The Larmor precession frequency depends on the magnetic field experienced by a nucleus. Therefore, it has spatial dependence under a field gradient. The spatially dependent Larmor frequency $\omega(\mathbf{r})$ at position **r** under a spatially linear field gradient **g** is expressed as follows [13]:

$$\omega(\mathbf{r}) = \gamma(H_0+\mathbf{g}\ \mathbf{r}) = \omega_0 + \gamma\mathbf{g}\ \mathbf{r} \tag{25}$$

where H_0 is the externally applied magnetic field and $\mathbf{g}\ \mathbf{r} = 0$ at position $\mathbf{r} = 0$. When the duration of the field gradient is δ, the difference of the phase angle $\phi(\mathbf{r})$ at r from that at $\mathbf{r} = 0$ is

$$\phi(\mathbf{r}) = \gamma\ \mathbf{g}\ \mathbf{r}\ \delta \tag{26}$$

The distance in the direction of g where $\phi(\mathbf{r}) = 2\pi$ is

$$q^{-1} = 2\pi/\ \gamma g\delta \tag{27}$$

q^{1} is the length scale with the field gradient. For example, q^{-1} is 235 μm when $g = 10$ G/cm and $\delta = 1$ ms. When the sample size, or the size of the detection area, is several times larger than q^{-1}, the total signal intensity vanishes because of the dephasing.

For the diffusion coefficient measurements, a second field gradient is applied in order to diphase the dephased magnetization. Figure 5 shows a typical pulse sequence with two pulsed field gradients (PFGs) having a rectangular shape along the z-axis and the dephasing and rephasing behavior of the magnetization when the individual spins did not change their positions at interval Δ between the two PFGs. a) The vectors of spin isochromat (the macroscopic magnetic vector at individual **r**) are aligned along the y-axis by an rf $\pi/2$ pulse. b) Under the first PFG, the vectors precess at the angular velocity of $\gamma\mathbf{g}\ \mathbf{r}$ corresponding to the individual positions in the rotating frame. c) At the end of the first PFG, the vectors are spirally twisted by a pitch of q^{-1}. d) The application of an rf π pulse along the y-axis rotates the individual vectors along the y-axis by 180 degrees, creating a mirror-symmetrical arrangement of the vectors with respect to the y-z plane. e) Under the second PFG, the individual vectors precess at the same angular velocity as that under the first PFG. f) At the end of the second PFG, the vectors are aligned along the y-axis. When the nucleus has a displacement of Δz in the z direction during Δ, it has a phase angular shift

$$\phi(\Delta z) = 2\pi\Delta z/q^{-1} = \gamma g\delta\Delta z \tag{28}$$

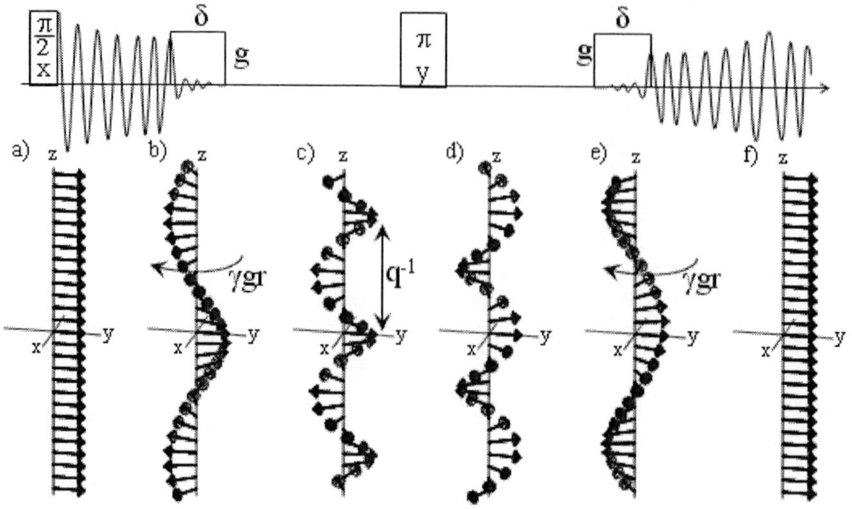

Figure 5. A typical pulse sequence with two pulsed field gradients of rectangular shape and the dephasing and rephasing behavior of the magnetization; a) The vectors of spin isochromat (the macroscopic magnetic vector at individual **r**) are aligned along the y-axis by an rf π/2 pulse. b) Under the first PFG, the vectors precess at the angular velocity of γ**g r** corresponding to the individual positions in the rotating frame. c) At the end of the first PFG, the vectors are spirally twisted by a pitch of q^{-1}. d) The application of an rf π pulse along the y-axis rotates the individual vectors along the y-axis by 180 degrees, creating a mirror-symmetrical arrangement of the vectors with respect to the y-z plane. e) Under the second PFG, the individual vectors precess at the same angular velocity as that under the first PFG. f) At the end of the second PFG, the vectors are aligned along the y-axis.

The echo signal intensity $I(2\tau,g\delta)$ at 2τ is proportional to the vector sum of the vectors of the spin isochromats in the sample, and is therefore expressed as follows:

$$I(2\tau,\mathrm{g}\delta) = I(2\tau,0)\iint \cos(\phi(\Delta z))\times\rho(\mathbf{r})p(\mathbf{r},\Delta z)d\mathbf{r}d\Delta z \quad (29)$$

where $\rho(\mathbf{r})$ is the density of the nucleus and is constant for a homogeneous sample, $p(\mathbf{r},\Delta z)$ is the probability of the displacement during Δ for the nucleus at **r**, and $I(2\tau,0)$ is the total signal intensity without PFG, which is expressed as follows:

$$I(2\tau,0) = I(0,0)\exp(-2\tau/T_2) \quad (30)$$

where $I(0,0)$ is the initial signal intensity immediately after the rf π/2 pulse. For the free diffusion in an isotropic medium, $p(r,\Delta z)$ becomes the Gaussian distribution

$$p(r,\Delta z)=(4\pi D\Delta)^{-1/2}\exp(-\Delta z^2/4D\Delta) \quad (31)$$

where D is the diffusion coefficient. Taking the diffusion during δ into account, $I(2\tau,g\delta)$ is rewritten as follows

$$I(2\tau,\mathrm{g}\delta)=I(0,0)\exp(-2\tau/T_2-(\gamma g\delta)^2D(\Delta-\delta/3)) \quad (32)$$

In common measurements of the free diffusion, $g\delta$ is varied under constant Δ. For the diffusion in a restricted space, the D value obtained by Eq.32 is the apparent diffusion constant

$$D_{app}(\Delta)=<\Delta z^2>(\Delta)/2\Delta \quad (33)$$

where $<\Delta z^2>(\Delta)$ is the mean square of Δz^2 during Δ. $<\Delta z^2>(\Delta)$ is proportional to Δ for the free diffusion; however, becomes smaller than the proportional value due to spatial restriction, which gives the information of the space size of the restriction.

The signal decay caused by the displacement under the field gradient can be used for removing peaks of small molecules from a mixture spectrum of small and large molecules. Conversely, the spectrum of peaks for small molecules can be obtained by subtracting the decayed spectrum for large molecules from the mixture spectrum. The mixture spectrum is composed of peaks contained in different molecules that exponentially decay with the square of $\gamma g\delta$ according to individual diffusivity. By changing Eq.32 into a sum of each component, the peak intensity at each frequency in the mixture spectrum is expressed as follows

$$I\ (2\tau, g\delta)=I(0,0)\times\Sigma\ f_i \exp\ ((-2\tau)\ /\ T_{2,i} - (\gamma g\delta)\ 2\ D\ (\Delta-3/\delta)) \quad (34)$$

where f_i and $T_{2,\,i}$ are fraction and T_2 for the component with the diffusion coefficient D_i. By using the T_2 weighted fraction of the i-th component

$$F_i(2\tau)=f_i \exp(-2\tau/T_{2,i}) \quad (35)$$

Equation 34 is rewritten as follows

$$I\ (2\tau, g\delta)=I\ (0,0)\times\Sigma\ f_i \exp\ (-(\gamma g\delta)^2\ D_i\ (\Delta-3/\delta) \quad (36)$$

When F is expressed as a continuous function of D, Eq.36 is rewritten as follows:

$$I(2\tau,g\delta)=I(0,0)\int F(D_i)\exp(-KD)dD \quad (37)$$

where $K = (\gamma g\delta)^2(\Delta-\delta/3)$. $F(D)$ is a T_2 enhanced distribution of D. Eq.37 shows that $I(2\tau,g\delta)$ is the Laplace transformation of $F(D)$ and therefore, an inverse Laplace transformation of $I(2\tau,g\delta)$ will give $F(D)$. The 2D spectrum with dimensions of frequency and D, or diffusion-ordered NMR spectroscopy (DOSY), gives a separate spectrum for each species in a mixture on the diffusivity and the distribution of the diffusivity for each species [14].

3. Molecular Dynamics in Food Hydrocolloid Systems

In this section, some experimental results for Agars, Carrageenans, Gellan Gum and Gelatin solutions are introduced and discussed. Dynamics of polymer chains in the hydrocolloid solutions is discussed based on the changes of signal intensities and diffusion coefficient obtained by NMR measurements. The relaxation time of water reflects the change of rigidity in polymer chain through the chemical exchange between water proton and labile proton on the polymer. Theoretical calculations are presented to understand the experimental results. Further, the mesh sizes in hydrocolloids gels are estimated from the restriction of probe molecule in translational motion.

3.1. Dynamics of Polymer Chain

Carrageenans are a group of sulphated galactan polysaccharides extracted from marine red algae. There are three main commercial classes of carrageenans: κ-, ι- and λ-carrageenan. They are widely used in the food industry as gelling, stabilizing and thickening agents[15]. Their gelation ability is different due to chemical structures [16]. The high-resolution ^{1}H NMR spectra of 1% κ- and ι-carrageenan were measured at various temperatures [4]. The signal intensities of polysaccharide start to decrease obviously around each T_{sg}, which were determined by rheologic measurements as to be 25 and 54°C respectively. This observation suggests the generation of immobile component with short T_2, i.e., the signal decays during the dead time before the detection in high-resolution ^{1}H NMR experiments. The stronger peak at chemical shift of 3.8 ppm for κ- and ι-carrageenan was chosen for following quantitative analysis, which corresponds to position 5 and 6 protons of the D-galactose residue according to reported assignment [17,18]. Temperature dependence of the signal intensity is demonstrated in Figure 6 for both carrageenans[4]. The intensity nearly keeps constant at temperature above the respective T_{sg}, indicating all protons at 5 and 6 positions are observable when polysaccharide chain exists in random coil. On further cooling, however, the temperature dependence became quite different between the two cases. For κ-carrageenan, the intensity decreased steeply in a range of 10°C starting from a temperature slightly above T_{sg}, and only ca. 10% of the signal remained at temperature lower than T_{sg}, i.e., ca. 90% of polysaccharide chains are transformed into solid-like immobile component by cooling. On the contrary, the signal intensity of ι-carrageenan exhibited a gradual decrease in a range of 30 °C starting at around T_{sg} and about 20% of the signal remained at the lowest temperature investigated. These results are coordinated with the gradual increase of G' over a wide temperature range and one order lower G' value for ι-carrageenan as contrasted with κ-carrageenan at the same concentration (data not shown).

Gellan gum is a microbial polysaccharide derived from Pseudomonas elodea. The polymer exists as a random coil in solution at high temperature and makes a conformational transition to a double helix with decreasing temperature. In the presence of cations, the conformational transition is accompanied by network formation, producing a gel[19]. Figure 7 shows the temperature dependence of the signal intensity at 1.3 ppm for 1% gellan gum together with the thermal scanning rheological measurements [3]. The peak intensity of gellan gum became to decrease at around 40 °C, which is higher than T_{sg} and corresponds to the temperature at which G' and G'' started to increase. This finding suggests that the gellan gum chains began to form aggregates above T_{sg}. Below T_{sg}, a portion of gellan chains remained as solute within the network; these chains decreased in number with progressive involvement in network formation during decreasing temperature. The peak intensities of pullulan increased gradually with decreasing temperature, probably in response to increasing magnetization recovery (during the repetition interval of 1s) with decreasing T_1. This indicates that pullulan was not involved in the aggregation of gellan chains; consequently, pullulan is interpreted to have no intermolecular interactions with gellan chains other than hydrodynamic interaction, which reflects the diffusion coefficient, D, of pullulan, and is used for the estimation of network size as shown below.

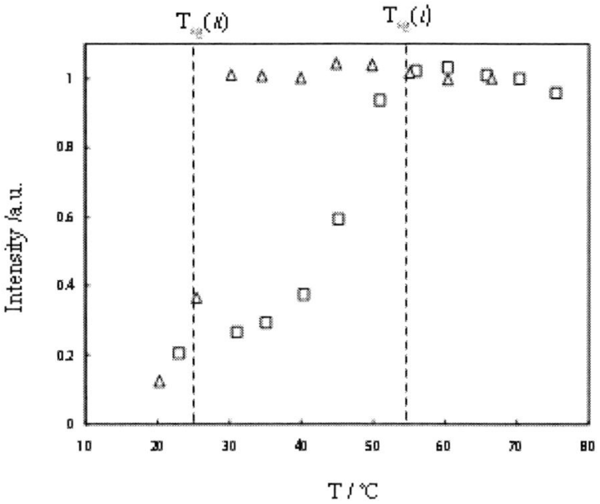

Data from ref.4.

Figure 6. Temperature dependence of intensity of proton signal for 1% κ-carrageenan (triangle) and ι-carrageenan (square) on cooling. $T_{sg}(\kappa)$ and T_{sg} (ι)denote the gelling points for 1% κ and ι-carrageenan, respectively.

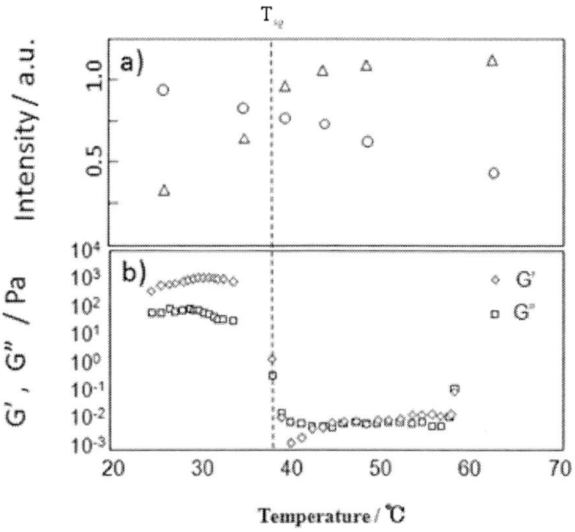

Data from ref.3.

Figure 7. Temperature dependence of NMR signal intensity for gellan (triangles) and pullulan (circles) (a), and G' and G'' on cooling (b) for a 1 wt% gellan gel containing 5 mM $CaCl_2$.

Figure 8 shows the temperature dependence of *G'* and *G"* of the mixed solution of high-acyl (HA) and low-acyl (LA) gellan[2]. Although HA gellan concentration was 0.5% in the solution, *G'* and *G"* steeply increase in the same temperature range and showed the same T_{sg} as that for 1% HA gellan solution. This indicates that HA gellan chains in the mixed solution interact with each other to form double helices and finally the gel, to almost the same extent as those in the 1% HA gellan solution. It is considered that the existence of LA gellan chains decrease the chain mobility of HA gellan, and as a result, the interaction between the HA gellan chains is enhanced. A similar behavior was found in the mixed gellan solution containing 50 mM NaCl[20]. However, *G'* and *G"* at temperature lower than T_{sg} are smaller than those of 1% HA gellan solution, indicating that the network structure in the mixed solution is weaker and involves fewer chains than that in the 1% HA gellan solution. The smaller increase of *G'* and *G"* at ca. 30°C in the mixed solution are thought to be due to the transition from the random coil to the double helix (C-H transition) of the LA gellan chains, which imparts stiffness and provides reinforcement to the network. For a clear understanding, CD spectra were measured and $[\theta]_{203}$ and $[\theta]_{218}$ are plotted against temperature in Figure 9[2]. As temperatures were decreased, $[\theta]_{203}$ showed an abrupt decrease at 25 °C for LA gellan solution whereas $[\theta]_{218}$ showed an increase at 76 °C for HA gellan solution, both of which corresponded to the drastic changes in the CD spectra. On the other hand, $[\theta]_{203}$ was almost unchanged for HA gellan solution at ca. 25 °C and $[\theta]_{218}$ was almost constant for LA gellan solution at ca. 76 °C. By contrast, both $[\theta]_{203}$ and $[\theta]_{218}$ were changed for the mixed solution.

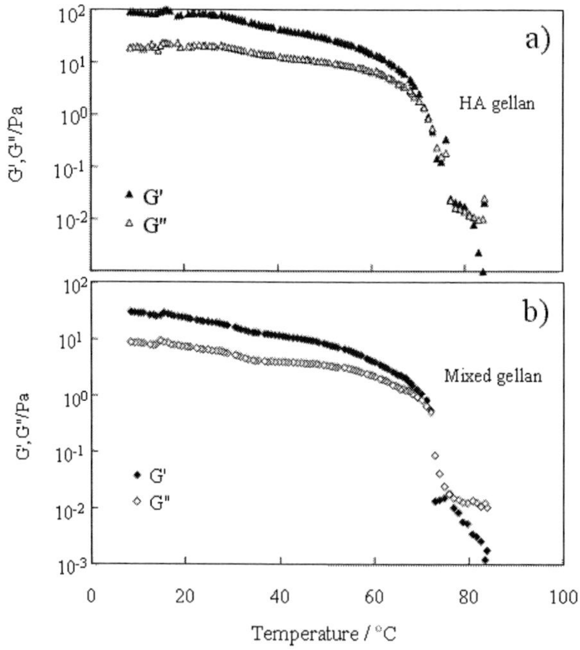

Data from ref. 2.

Figure 8. Temperature dependence of storage modulus *G'* (solid symbols) and loss modulus *G"* (open symbols) on cooling at 0.5 °C/min of 1% HA gellan solution a) and mixed solution b). The mixed solution contains 0.5% HA gellan and 0.5% LA gellan.

The results clearly indicate that the C-H transitions of LA and HA gellan chains occur independently in the mixed solution. Moreover, the C-H transitions occur at temperatures corresponding to the C-H transition temperatures for 1% LA gellan or HA gellan solution despite the fact that the concentrations of LA and HA gellan were half those in their respective solutions. This indicates that HA gellan chains in the mixed solution interact with each other to almost the same extent as the HA gellan chains in the 1% HA gellan solution. The same goes for LA gellan chains in the mixed solution and those in the 1% LA gellan solution. The existence of LA gellan chains probably contributes to the decrease in chain mobility of HA gellan and vice versa. This is similar to the rheological properties, where *G'* and *G"* for the 1% mixed solution increase steeply in the same temperature range and show the same T_{sg} as that for the 1% HA gellan solution.

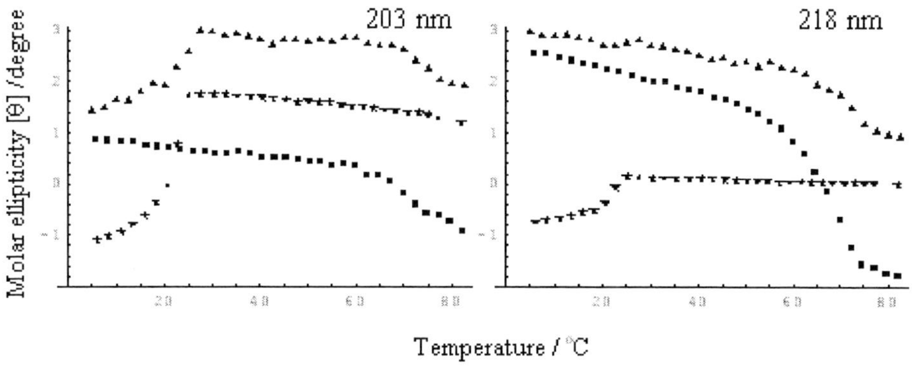

Data from ref. 2.

Figure 9. Temperature dependence of [θ] at 203 nm and 218 nm of 1% LA gellan solution (stars), 1% HA gellan solution (squares), and the mixed solution (triangles).

Figure 10 shows the NMR spectra of the mixed solution measured with the pulsed-field-gradient spin-echo sequence using two field gradients (strength = 3.0 T/m, duration = 1 ms) at various temperatures in the cooling process [2]. The signal assignable to water at 4.7 ppm was completely eliminated due to diffusion and only the signals assignable to gellan remained. Signal decay due to the diffusion of gellan chains was negligible. With decreasing temperature, the signals indicative of gellan became smaller although the linewidths were almost the same. This indicates that there are two fractions having different chain mobility: one has strongly restricted mobility with a short T_2 relaxation time, which causes complete decay of signals during the echo time of 20 ms, and the other has certain mobility with a long T_2 relaxation time, which leads to the preservation of signals with narrow line widths. The signals at 1.3 and 2.1 ppm are assigned to the methyl protons of rhamnose and the protons of the acetate group on (1, 3)β-D-glucose, respectively. The former exists on both LA and HA gellan chains and the latter exists only on the HA gellan chain. Therefore, the change of the chain mobility of HA gellan is reflected by the intensity of the two signals at 1.3 and 2.1 ppm; on the other hand, that of LA gellan chains is reflected by the intensity of the signal at 1.3 ppm.

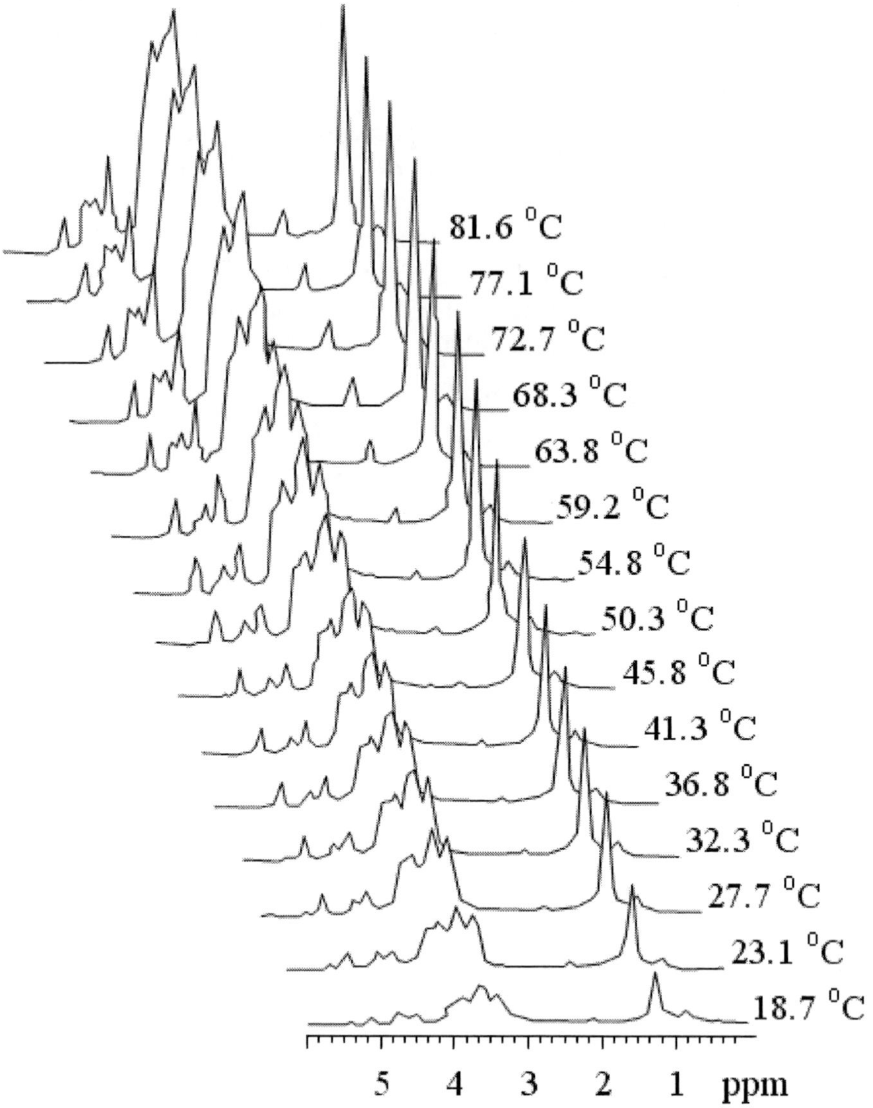

Data from ref. 2

Figure 10. NMR spectra of mixed solution measured with the pulsed-field-gradient spin-echo sequence using two field gradients (strength = 3.0 T/m, duration = 1 ms) at various temperatures in the cooling process.

The temperature dependence of $I_{1.3}$ and $I_{2.1}$ (signal intensities at 1.3 and 2.1 ppm normalized to those at 80°C are unity, respectively) for 1% HA gellan solution is shown in Figure 11a and that of $I_{1.3}$ and $I_{2.1}$ for the mixed solution is shown in Figure 11b[2]. For the 1% HA gellan solution, both $I_{1.3}$ and $I_{2.1}$ were decreased abruptly at temperatures ranging from 72 to 60°C, and this was followed by a gradual decrease with further decrease of the temperature (Figure 11a), which is comparable to the abrupt changes and the subsequent and gradual changes of *G'* and *G"*(Figure 8a) and $[\theta]_{218}$ (Figure 9) in the above-mentioned temperature range. Therefore, it is considered that the fraction of gellan with the strongly restricted chain mobility is attributed to the double helix structure and the subsequently formed aggregates. The increase of $I_{1.3}$ and $I_{2.1}$ at temperatures ranging from 85 to 72 °C is probably due to the increase in signal recovery during the repetition time due to the short T_1. For the mixed solution, both $I_{1.3}$ and $I_{2.1}$ started to decrease at 72 °C; however, the extent of decrease of $I_{1.3}$ is half that of $I_{2.1}$ (Figure 11b). $I_{2.1}$ originates in the HA gellan chain and $I_{1.3}$ originates in both HA and LA gellan chains, therefore, it is considered that only the HA gellan chain is involved in such high-order structures as the double helix structure and the aggregates, in this temperature range. Following a gradual decrease, $I_{1.3}$ shows a steep decrease down to ca. 25 °C. This is attributable to the formation of the double helix structure of the LA gellan chain. $I_{2.1}$ also bends downward at ca. 25 °C, which is probably caused not by the formation of high-order structures by HA gellan chains, but by the formation of a double helix by the LA gellan chains, which decreases the mobility of the HA gellan chains that remain as random coils.

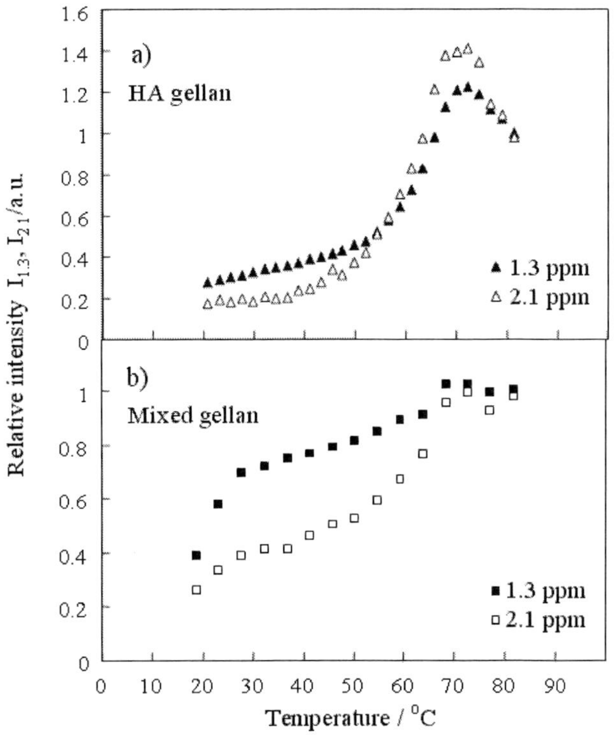

Data from ref. 2

Figure 11. Temperature dependence of signal intensities at 1.3 ppm (solid symbols) and 2.1 ppm (open symbols) of 1% HA gellan solution a) and mixed solution b). The signal intensities are normalized to those at 80 °C and are unity.

3.2 Relaxation Time of Water

The water proton spin-spin relaxation time (^{1}H T_2) is used to detect directly the coexistence of water molecules with different mobility [21] as well as indirectly the change of polymer mobility through the chemical exchange between water proton and exchangeable proton on polymer [22,23]. Hills et al. showed the effect of chemical exchange to water ^{1}H T_2 at given temperature by simulating the water ^{1}H T_2 as a function of pulse space (the duration between $\pi/2$ and π pulse in CPMG sequence) [24]. For the temperature-induced gelation systems, it is often demonstrated that the observed water ^{1}H T_2 rapidly decreases and subsequently increases, accompanying a local minimum of water ^{1}H T_2 during the gelation [25]. For the melting process of carrageenan systems, Lewis et al. reported the influence of ion species and ion concentration on water ^{1}H T_2, and found that the local minimum of water^{1}H T_2 has close relation with the onset of the gel melting determined by ORD [26]. These reports suggest that water ^{1}H T_2 is useable as a general indicator for macroscopic structural change of polysaccharide chains in the sol and the gel state as well as during the gelation process, if we know the exact relation between water ^{1}H T_2 relaxation mechanisms and the structure of the chain in various states.

The structural change of gelatin molecule in solution with the temperature change was investigated by measuring viscoelasticity, circular dichroism and water ^{1}H T_2 [1]. Figure 12 (a), (b) and (c) show the temperature dependence of storage modulus G' and loss modulus G'', specific ellipticity at 233 nm $[\theta]_{233}$ and ^{1}H T_2 for 8 % gelatin solution, respectively, on cooling and heating process at 0.5 °C/min. In the viscoelasticity measurements on cooling (Figure 12(a)), G' and G'' steeply increasedafter G' crossed over G'' at 27 °C, which is attributed to the sol-to-gel transition temperature T_{sg}. On the other hand, a specific ellipticity at 233 nm $[\theta]_{233}$, which is considered to reflect the content of optically active chemical structure of the carboxyl group in the helix form of gelatin, increased more slowly comparing to G' and G'' at the temperature range (Figure 12(b)). These facts indicate that the effective crosslinking points, which form three-dimensional networks, are generated with a small portion of helix structure at T_{sg}, and then the helix structure continuously grows with decreasing in temperature to 0 °C. The temperature dependence of ^{1}H T_2 for water showed a downward curvature at temperature lower than T_{sg} [Figure 12(c)]. The observed change of the ^{1}H T_2 is caused not by the mobility change of water itself, but by the mobility change of gelatin through the chemical proton exchange between water protons and labile protons in gelatin and through the motional exchange between free water and bound water [4]. This is experimentally proved by the observation that water in gelatin solution showed a similar temperature dependence of the diffusion coefficient with that of pure water (data not shown). Therefore, it is considered that the mobility of gelatin polymer chain decreased with the formation of the gel structure and the helix structure. The curves for the temperature dependence of G', G'' and $[\theta]_{233}$ on heating shift to higher temperature than those on cooling. Yoon et al. presented that no temperature hysteresis is found in measurement of the diffusion coefficient D by a forced Rayleigh scattering method for 10% gelatin solution, i.e., the value of D was identical in temperature ascending side and descending sides after the aging for more than 24 h [27]. Therefore, it is said that there exists a delay for structural change which is reflected in viscoelasticity and CD measurements on cooling and heating at 0.5 °C/min. The delay in ^{1}H T_2 measurement seems to be smaller than that in viscoelasticity and CD measurements. This is probably due to that the structural change reflected in ^{1}H T_2 is rapid as compared with the structural change reflected in viscoelasticity and CD measurements, such as the sol-gel transition and the coil-helix transition.

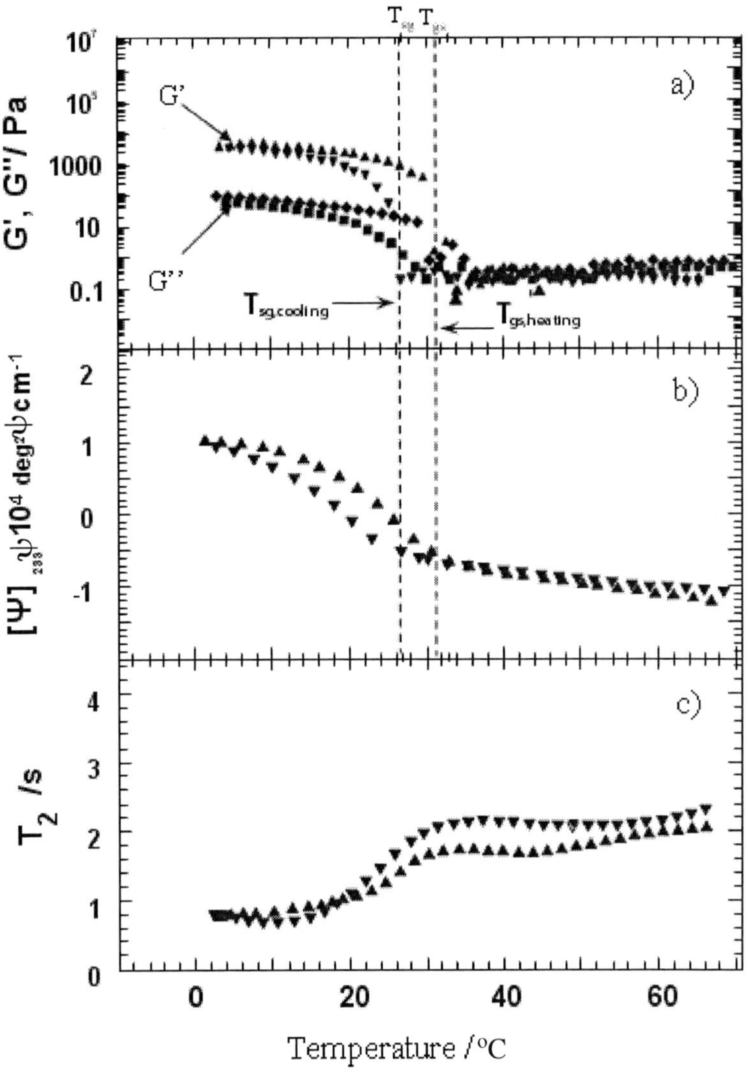

Data from ref. 1.

Figure 12. Temperature dependencies of viscoelasticity (a; storage modulus G' and loss modulus G''), specific ellipticity at 233 nm $[\Psi]_{233}$ (b) and ^{1}H T_2 (c) for 8% gelatin solution, respectively, on cooling process (▼, ■) and heating process (▲, ◆) at 0.5 °C/min.

Theoretical approaches for water ^{1}H T_2 relaxation behavior based on proton exchange between water and polysaccharides undergoing structural change will give us a clear physical picture about the structural change. ^{1}H T_2 relaxation rate aqueous system of polysaccharides where random coils, ordered chains (helices) and ordered chains in aggregation (aggregated helices) coexist at a ratio which varies with temperature. ^{1}H T_2 relaxation of water molecule is predominantly caused by the molecular dipole-dipole interaction modulated by the rotational and the diffusive motion of water molecule. In case of proton with sufficient mobility, such as water proton in pure water or extremely dilute polysaccharide solution, ^{1}H T_2 value is given by

$$T_2 \sim 1/\tau_c \sim \exp(-E/RT), \quad (38)$$

where τ_c is a correlation time for random motion of water molecule, E an activation energy for the motion, R the gas constant and T the absolute temperature. When E is constant, i.e., water molecule does not change the motional mode in certain temperature range, the plot of $\ln T_2$ against $1/T$ gives a straight line with a slope of $-E/$R. In dilute polysaccharide solution, most water molecules exist as free water (i.e., bulk water) although a trace of water molecules is motionally perturbed by macromolecules. These two kinds of water molecules undergo a very fast motional (or diffusive) exchange, which makes the mobility of water molecules well averaged. The water ^{1}H T_2 (T_{2w}) is approximately regarded as the ^{1}H T_2 of pure water in the following theoretical analysis. The observed ^{1}H T_2 value for water proton in polysaccharide solutions is often influenced by a chemical exchange between water proton and exchangeable proton (i.e., labile proton) on polysaccharide. As we know, the sol-gel transition is accompanied with structural change of polysaccharide chain. Therefore, we should separately examine the water ^{1}H T_2 under three different conditions(Figure 14), namely, the existence of the random coil or the ordered chain (case A), the coexistence of the random coil and the ordered chain without aggregation (case B), and the coexistence of the random coil, the ordered chain with/without aggregation (case C).

Case A

This is the general case for the chemical exchange between water proton and labile proton. Here, F_w and F_p are defined as the fraction of water proton and labile proton, respectively, and $F_w + F_p = 1$. We treat the condition in which F_p is sufficiently small. At lower resonance frequency, the differences of the resonant frequency ($\delta\omega$) between water proton and the labile protons are small enough as compared with $1/T_{2p}$ and $(1/T_{2p}\tau_p)^{1/2}$ and the dependence of ^{1}H T_2 on pulse space can be omitted theoretically:

$$F_p << 1,\ \delta\omega << 1 / T_{2p},\ \delta\omega << (1 / T_{2p}\tau_p)^{1/2}, \quad (39)$$

where T_{2p} and τ_p are the T_2 and the mean residence lifetime of the labile protons, respectively. Then, the observed water ^{1}H T_2 (T_{2obs}) is expressed by [28]

$$1/T_{2obs} = F_w / T_{2w} + F_p / (T_{2p} + \tau_p). \quad (40)$$

When the chemical exchange rate is extremely fast so that $\tau_p << T_{2p}$, Eq. 40 is simplified as

$$1/T_{2obs,ext\text{-}fast} = F_w / T_{2w} + F_p / T_{2p}. \quad (41)$$

When the chemical exchange rate is extremely slow so that $T_{2w} < \tau_p$, the labile protons have no contribution to water ^{1}H T_{2obs}:

$$1/T_{2obs,ext\text{-}slow} = 1 \,/\, T_{2w} \,. \tag{42}$$

When T_{2w}, T_{2p} and τ_p are similarly expressed with Eq.38, each of them is formulated by

$$T_{2w} \sim \exp(-E_w/RT), \quad T_{2\pi} \sim \exp(-E_p/RT), \quad \tau_p \sim \exp(E_{ex,p}/RT) \tag{43}$$

where E_w, E_p and $E_{ex,p}$ are activation energies for the motion of water proton, for the motion of labile proton, and for the chemical exchange between water protons and labile protons, respectively. In addition, the mean residence lifetime of water proton τ_w is expressed as $\tau_w = F_w\tau_p/F_p$ in an equilibrium state. By using Eqs. 39-43, we can evaluate the temperature dependence of T_{2obs}, T_{2w}, T_{2p}, $T_{2obs,ext\text{-}fast}$, $T_{2obs,ext\text{-}slow}$, τ_p and τ_w. Assuming appropriate values for E_w, E_p, $E_{ex,p}$ and F_p as representative values for the system investigated, the Arrhenius plots are obtained as shown in Figure 13. In the range of extremely fast proton exchange ($\tau_p << T_{2p}$), T_{2obs} follows with the straight line of $T_{2obs,ext\text{-}fast}$ (dash-dotted line) and decreases with decreasing temperature. On the other hand, in the range of extremely slow proton exchange ($T_{2w} < \tau_p$), T_{2obs} almost follows the straight line of T_{2w} (upper broken line) and decreases with decreasing temperature. In course from the extremely fast to the extremely slow proton exchange, the T_{2obs} gradually deviates from $T_{2obs,ext\text{-}fast}$ in the range of fast exchange, and increases with decreasing temperature under the condition of $\tau_p \geq T_{2p}$ and $\tau_w \leq T_{2w}$ in the range of intermediate exchange, and finally approaches to the straight line of T_{2w} because T_{2obs} is gradually free from the contribution of the labile proton through moderately slow exchange. In this series of change of ^{1}H T_{2obs}, the inverse proportion of ^{1}H T_{2obs} to temperature at the intermediate exchange region is quite characteristic and often presents in the gel state.

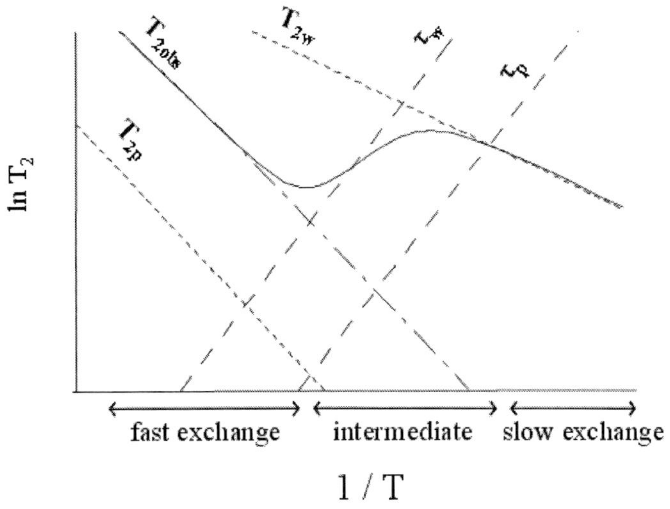

Figure 13. Schematic Arrhenius plot of T_{2obs} (solid line), T_{2w} (upper broken line), T_{2p} (lower broken line), $T_{2obs,ext\text{-}fast}$ (dash-dotted line), τ_w (upper dashed line) and τ_p (lower dashed line) calculated by using Eqs. 39-43.

For polysaccharide aqueous systems, the T_{2p} can not be detected directly and is estimated by the ^{1}H T_2 of inert proton (unexchangeable proton) on polysaccharide (T_{2inert}). The T_{2inert} is decreased more than 1 order with the conformational transition from random coil to helix. Therefore, the relative rate of chemical exchange ($1/\tau_p$) to the relaxation rate ($1/T_{2p}$) for labile protons might be different between on the random coil and on the helix. In the temperature range investigated, as demonstrated below, the chemical exchange of water proton with labile proton on the random coil can be approximated to be in the fast range, and that on the ordered chain to be in the intermediate range. Therefore, the water ^{1}H T_2 modified by the labile protons on the random coil ($T_{2,water/coil}$) and on the ordered chain ($T_{2,water/order}$) is expressed by Eq.44 and Eq.45, respectively:

$$1 / T_{2,water/coil} = F_w / T_{2w} + F_p / (T_{2coil} + \tau_{coil}), \tag{44}$$

$$1/ T_{2,water/order} = F_w / T_{2w} + F_p / (T_{2order} + \tau_{order}), \tag{45}$$

where T_{2coil} and τ_{coil} denote T_2 and the mean residence lifetime of labile protons on random coil, T_2,order and τ_{order}, represent T_2 and the mean residence lifetime of labile protons on ordered chain, respectively.

Case B

In the case undergoing the structural change such as random coil-helix transition, the random coil coexists with the ordered chain in specific temperature range. Therefore, water protons experience both type of the chemical exchange with labile proton on the random coil and the ordered chain as shown in Figure 14. Then, the observed water ^{1}H T_{2obs} is given by the following equation:

$$1/T_{2obs} = F_w / T_{2w} + F_p f_{coil} / (T_{2coil} + \tau_{coil}) + F_p f_{order} / (T_{2order} + \tau_{order}), \tag{46}$$

where f_{coil} and f_{order} presents the fraction of proton on the random coil and the ordered chain to the whole polysaccharide protons, respectively.

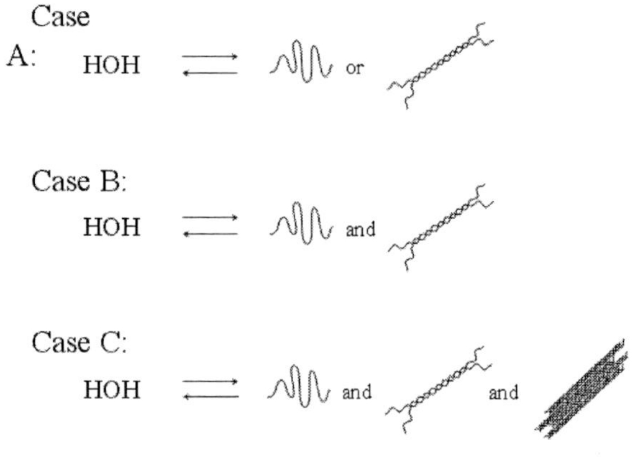

Figure 14. Schematic drawing for chemical exchange in biopolymer solutions.

Case C

It has been reported for some polysaccharide systems, such as κ-carrageenan and agarose, that rigidly aggregated junction zones (the ordered chains in aggregation) are formed in the gel state [29] and the T_2 of proton on such ordered chains became shorter by two orders. Therefore, the labile protons existing at the interior of the junction zones might lose their opportunity to exchange proton with bulk water, and hence have no effect on the bulk water ^{1}H T_{2obs}. In this situation, the fraction of the labile protons effectively working on chemical exchange with water is reduced. In order to incorporate this fact in the analysis of water ^{1}H T_{2obs}, a parameter, f_{agg}, should be introduced as a fraction of ineffective labile protons to the chemical exchange. Labile protons on exterior of the junction zones are possible to exchange protons with bulk water. Therefore, as shown in Figure 14, we must estimate T_{2obs} under chemical exchanges with labile protons existing on the random coils, the ordered chains and the exterior of the junction zones. In this analysis, simply, we assume that the latter two types of labile protons can be classified together as labile proton on the ordered chain, and f_{coil} + f_{order} = 1 - f_{agg}. Then, T_{2obs} is expressed by

$$1/T_{2obs} = \{F_w/T_{2w} + F_p f_{coil}/(T_{2coil} + \tau_{coil}) + F_p f_{order}/(T_{2order} + \tau_{order})\} / (1 - f_{agg}). \qquad (47)$$

Summarily, we can propose the locus of T_{2obs} expected in the temperature-induced sol-gel transition process of polysaccharides in Figs.15(a), (b) and (c). Here, we introduced some parameters for simulation, such as E_p for the labile proton on the random coil (E_p(coil)) and on the ordered chain (E_p(order)) and $E_{ex,p}$ between water proton and the labile proton on the random coil ($E_{ex,p}$(coil)) and on the ordered chain ($E_{ex,p}$(order)) as well as F_p and E_w. They were given in the caption of Figure 15. If the transition is immediately completed at T_{sg} without forming aggregated structure, T_{2obs} vertically jumps from the dashed line for the case of random coil to the broken line for the case of helix as shown in Figure 15(a). Before starting the transition, almost all labile protons exist on the random coils and T_{2obs} follows the dashed line in Figure 15(a) induced by Eq.40 on cooling. At temperature lower than T_{sg}, T_{2obs} follows the broken line induced by Eq.46 on cooling. On the other hand, if the transition from random coil to ordered chain without aggregation is gradually completed in a broader temperature range around T_{sg}, we have to take account of the decrease of f_{coil} as a promotion of the temperature-induced gelation. The T_{2obs} continuously and gradually pursues the points determined by varying f_{coil} with temperature as shown in the bold line in Figure 15(b). After reaching the final f_{coil} in the gel state, for example 0.1, the T_{2obs} follows the water ^{1}H T_2 curve for f_{coil} = 0.1 on cooling. If aggregation occurs synchronously during random coil-helix transition, the fraction of aggregated structure will have influence on the water ^{1}H T_{2obs} according to Eq.47. The effect of f_{agg} was clearly exhibited in Figure 15(c), which is based on a theoretical simulation with tentatively assumed f_{agg} = 0 (i.e., without aggregation) and f_{agg} = (1-f_{coil})/2 (i.e., half of the ordered chains are included into the interior of junction zones) for the 4% κ-carrageenan (F_p=0.003). At the same f_{coil}, the larger the f_{agg} is, the longer the T_{2obs} becomes.

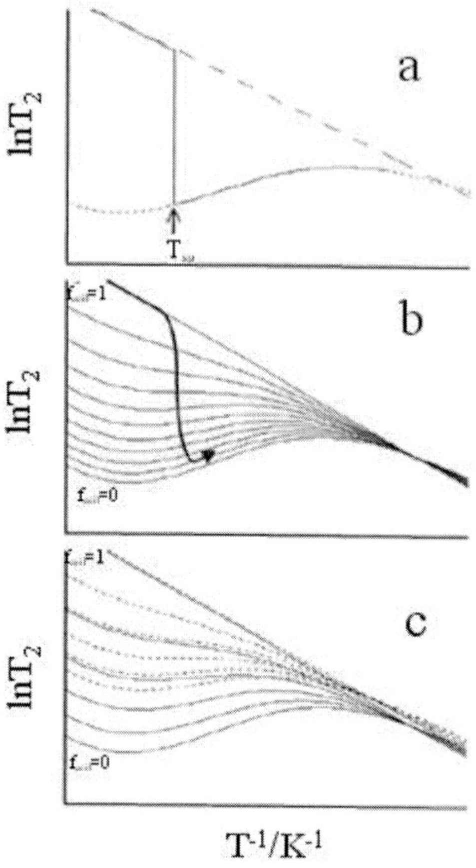

Figure 15. Arrhenius plot of T_{2obs} calculated using Eqs. (9) and (10) on cooling. The parameters, F_p=0.003, E_w=15kJ/mol, $E_{p(coil)}$=20kJ/mol, $E_{p(order)}$=30kJ/mol, $E_{ex,p(coil)}$=$E_{ex,p(order)}$=30kJ/mol are used for the simulation. (a) The gelation happens immediately with f_{agg}=0. f_{coil}=0.1 in the gel state. Dashed line, ^{1}H T_2, water/coil; broken line, ^{1}H T_2 calculated with f_{coil}=0.1; solid line, expected trace of T_{2obs}. (b) The gelation gradually happens with f_{agg}=0. Solid line, the water ^{1}H T_2 calculated using various f_{coil} (from lower to upper at an interval of 0.1); bold line, expected trace of ^{1}H T_{2obs} on cooling. (c) The effect of aggregation on T_{2obs} during gelation. The water ^{1}H T_2 calculated using various f_{coil} (from lower to upper at an interval of 0.2 for both solid and broken line): solid line, f_{agg}=0; broken line, f_{agg}=0.5(1-f_{coil}).

The temperature dependence of water ^{1}H T_{2obs} is plotted in Figure 16 for 4% κ, ι and λ-carrageenan systems. The T_{2obs} in 4% κ-carrageenan started a gradual decrease from the temperature above T_{sg} on cooling, and then drastically decreased in a narrow temperature range (from 43 to 33 °C), followed by a gradual increase with further decreasing temperature. In the heating process, the T_{2obs} did not trace the route in the cooling process back, but decreased continuously and then turned to increase at about 50 °C to return back to the original value for the sol state at about 70 °C. An obvious thermal hysteresis was exhibited in the T_{2obs} profiles between the cooling and the heating process. For 4% ι- carrageenan, the change of T_{2obs} almost reversibly occurred from 65 to ac. 50 °C except a little longer T_{2obs} on heating. For 4% λ-carrageenan, the T_{2obs} kept a uniformly gradual and reversible change in the whole temperature range investigated, suggesting no sol-gel transition.

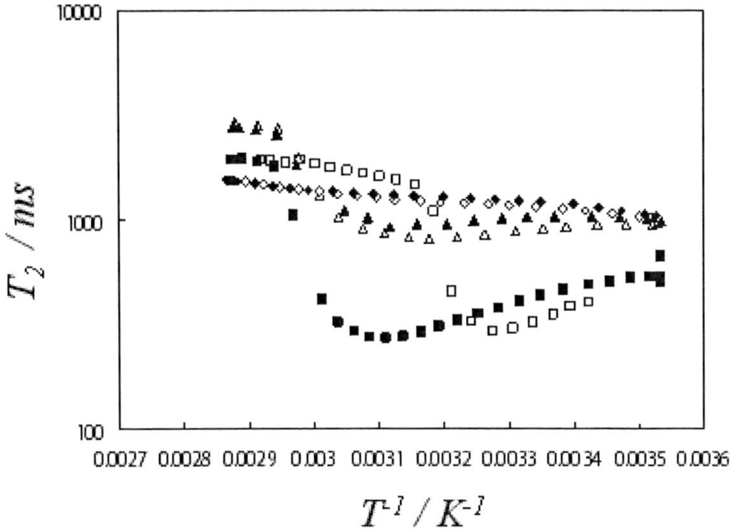

Data from ref. 4.

Figure 16. Temperature dependence of water ^{1}H T_{2obs} of 4% carrageenan solutions. □/■, Δ/▲ and ◇/◆denote κ-, ι- and λ- carrageenan in cooling/heating process, respectively.

The Chemical Exchange in the Sol State

In Figure 17, water ^{1}H T_{2obs} for κ-carrageenan system at different concentration (1% and 4%) and water (as reference) are demonstrated as a plot of ln(T_{2obs}) against 1/T. In the sol state, where polysaccharide chain exists in random coil, the T_{2obs} exhibits obvious dependence on concentration of polysaccharide. The shorter ^{1}H T_{2obs} at higher concentration can be explained by Eq.40 with a higher F_P. Comparing the slopes of the three cases in Figure 17, the slopes become a little gentler for κ-carrageenan solutions than for water. This fact indicates that the chemical exchange of water proton with the labile proton on random coil is in the fast exchange range with $\tau_p \leq T_{2p}$ on the base of the theoretical scheme shown in Figure 13. The same result can be obtained for ι- and λ-carrageenan from Figure 16.

Water ^1H T_{2obs} during the Sol-Gel Transition - Contribution of f_{order} and f_{agg}

The T_{2obs} values shifted to shorter at the vicinity of T_{sg} and reached a minimum value, as shown in Figures16 and 17. As already discussed in the theoretical approach and demonstrated in Figure 15(b) and (c), this observation can be explained by the increase in f_{order} and f_{agg} during the promotion of gelation. The T_{2obs} consecutively follows the points determined by f_{coil} and temperature to reach to the curve for f_{coil} = 0.1, and then follows the f_{coil} = 0.1 curve in Figure 15(b) on further cooling, referred to the only 10% of NMR signal remained at about 20 °C (Figure 6). κ- carrageenan is known to form the aggregated junction zone in the gel state [25]. With due regard to the contribution of the aggregated structure to the T_{2obs}, the curve of T_{2obs} should shift upwards as simulated in Figure 15(c). If we compare the traces of T_{2obs} in the cooling and the heating process for 4% κ-carrageenan in Figure 17, this expectation has been clearly recognized. Namely, the T_{2obs} in the heating process traces

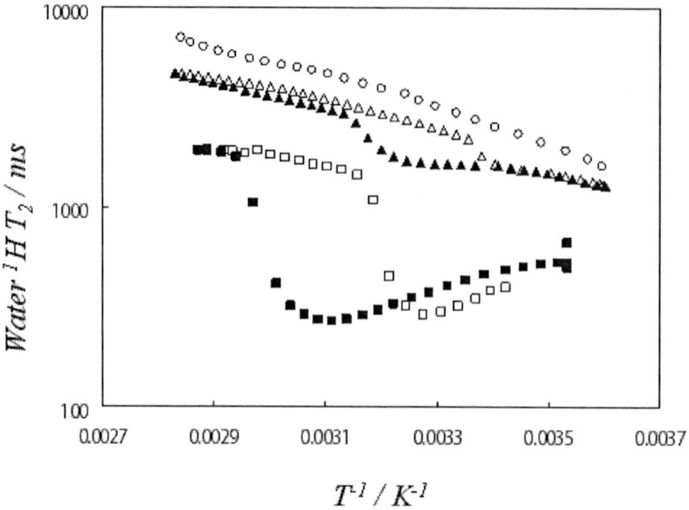

Data from ref. 4.

Figure 17. The Arrhenius plot of water ^{1}H T_{2obs} of pure water (○), 1% (△/▲) and 4% (□/■) κ -carrageenan solutions. The open marks denote the cooling and the heating process, respectively.

the upper line, i.e., a longer T_{2obs} value as compared with that in the cooling process. As the aggregated helices are much more thermally stable than the free ordered chains, they can notreversibly dissociate in the heating process. Therefore, the actual percentage of f_{agg} to f_{order} becomes larger in the heating process than in the cooling process. This higher f_{agg} is partly caused by keeping the sample at lower temperature (10 °C) before heating as the experimental condition. The thermal hysteresis is also observed in the profile of T_{2obs} for 1% κ-carrageenan, but the upward shift of the T_{2obs} in the heating process is not clearly distinguished. It may be due to the smaller F_p at lower concentration of polysaccharide, or the aggregation is only formed at around T_{sg}. For 4 % ι-carrageenan, the fact that the T_{gs} equals to the T_{gs} obtained by the rheological measurement can explain the reason why no thermal hysteresis was exhibited in the profile of T_{2obs}. Under this situation, the upward shift of T_{2obs} in the heating process as compared with the cooling process is still observed in the gel state. It demonstrates that a loose aggregation or association structure exists in 4% ι-carrageenan system, which has dissociated at about 55 °C and has no influence to T_{gs}.

Chemical Exchange in the Gel State – Effect of the Intermediate Exchange

In the gel state, we often detect a characteristic phenomenon, i.e., the increase of water 1H T2obs with decreasing temperature. According to Figure 15, such temperature dependence of T2obs implies that the chemical exchange between water proton and labile proton on the ordered chain happens in the intermediate range, where the relationship τorder ≥ T2order and τw ≤ T2w is satisfied. With further decreasing temperature, the chemical exchange in the slow range and then the extremely slow range could be expected in the gel state, but not detected in the present experiments because of freezing of the sample. The intermediate exchange was also observed up to the Tgs in the heating process of κ- and ι-carrageenan systems.

3.3 Diffusion Coefficients of Solute Molecule

The gelation of polysaccharides is induced by the aggregation of polysaccharide chains. Polysaccharide chains that aggregate to form a network have a strongly restricted molecular motion, commonly resulting in very short ^{1}H T_2; consequently, the echo signal for the polysaccharide chains of the aggregates disappears because of T_2 decay. Figure 18 shows changes in the ^{1}H NMR spectra of the 1.3 wt% agar gel containing 0.1 wt% dendrimer as the temperature was decreased from 58 °C to 20 °C[3]. The peak intensities of the agar chains started to decrease at around 40 °C, indicating that a portion of the agar chains aggregated at the gelling temperature, T_{sg}, to form the network and that another portion of the agar chains remained as solute within the network at T_{sg}; these chains gradually aggregated to form the network as the temperature decreased further.

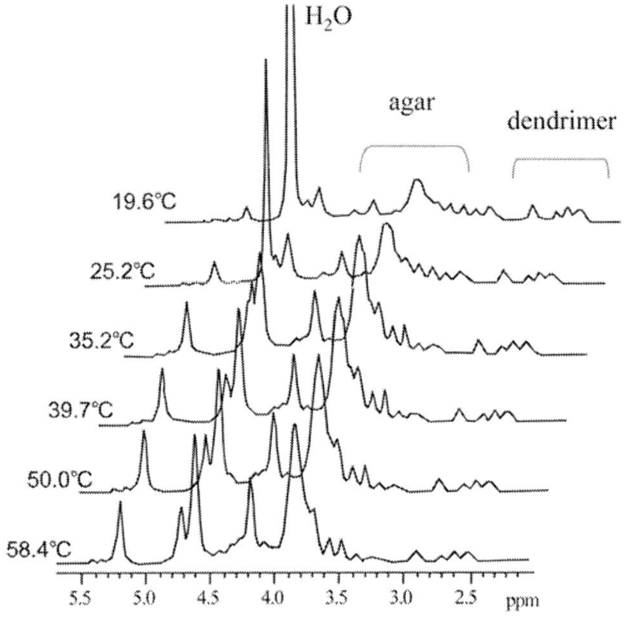

Data from ref. 3.

Figure 18. Progressive changes in ^{1}H NMR spectra of a 1.3 wt% agar gel containing 0.1 wt% dendrimer with decreasing temperature.

The peaks of the dendrimer did not decrease with decreasing temperature, indicating that the dendrimer was not involved in the aggregation of agar chains. We therefore interpret that the dendrimer has no intermolecular interaction with agar chains other than hydrodynamic interaction.

Hydrodynamic interaction would be reflected by the diffusion coefficient of the dendrimer, D_{dend}, in the agar solution. Figure 19 shows the temperature dependence of D_{dend} in the agar solution. D_{dend} in dilute solution, $D_{dend,0}$, was also measured at various temperatures. $D_{dend,0}$ showed a dependence on diffusion time, Δ, at each temperature, probably related to convection in the dilute solution. This dependence was pronounced at high temperatures (data not shown). When the directions of convection at each part in the sample solution negligibly change during Δ, the mean square of displacement by convection in the z direction (field gradient direction), $\langle z^2 \rangle_{conv}$, is given by the square of $\langle v_z \rangle \Delta$, where $\langle v_z \rangle$ is the average velocity of the convection in the z direction. The apparent D observed at Δ, $D_{app}(\Delta)$, is therefore expressed as

$$D_{app}(\Delta) = \frac{\langle Z^2 \rangle_{diff} + \langle Z^2 \rangle_{conv}}{\Delta} = \frac{D_0 \Delta + (\langle V_z \rangle \Delta)^2}{\Delta} = D_0 + \langle V_z \rangle^2 \Delta \qquad (48)$$

where $\langle z^2 \rangle_{diff}$ is the mean square of displacement by diffusion. At lower temperatures, $D_{app}(\Delta)$ shows a linear dependence on Δ; therefore, $D_{dend,0}$ at 30 °C was obtained using Eq.48. From $D_{dend,0}$ at 30 °C, the hydrodynamic radius R_H of the dendrimer at 30 °C was calculated by

$$D_{dend,0} = k_B T / (6\pi \eta_s R_H), \qquad (49)$$

where k_B, T, and η_s are the Boltzmann constant, absolute temperature, and viscosity of the solvent, respectively. Because R_H of the dendrimer is thought to be constant with changing temperature, $D_{dend,0}$ at various temperatures was calculated using Eq.49 (as shown in Figure 19).

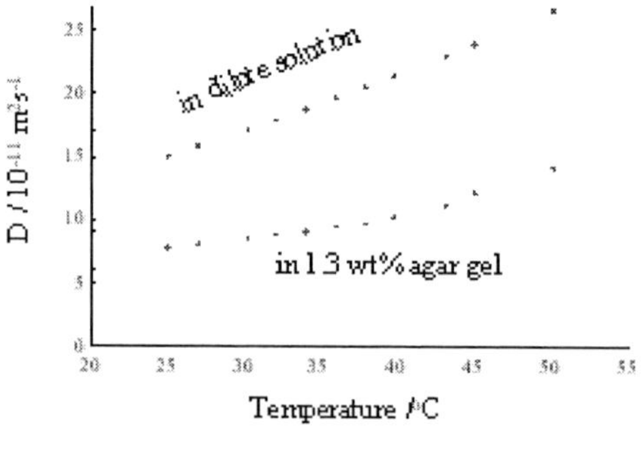

Data from ref. 3.

Figure 19. Temperature dependence of the diffusion coefficient (D) of a dendrimer. D in solution was calculated from the theoretical temperature dependence of the dilute solution.

As seen in Figure 19, D_{dend} is smaller than $D_{dend,0}$ and was observed to decrease with decreasing temperature. As a parameter that describes the restriction of diffusion in gel, D/D_0 is plotted against temperature in Figure 20. D/D_0 decreased with decreasing temperature above T_{sg} (ca. 40 °C). The agar used in this study is a mixture of agarose and agaropectin with variable chemical structure and molecular weight. Therefore, it is considered that a portion of the agar chains forms micro-aggregates that do not induce gelation, but restrict diffusion of the dendrimer. Below T_{sg}, D/D_0 increases with decreasing temperature. When D/D_0 is changed by hydrodynamic interaction with the agar network, D/D_0 is expressed as follows [30,31]:

$$D/D_0 = \exp(-R_H/\xi), \tag{50}$$

where ξ is the dynamic shielding length, considered to represent the hydrodynamic mesh size. The values of ξ corresponding to the D/D_0 values calculated using Eq.50 are indicated on the right-hand axis in Figure 20, where an increase in D/D_0 indicates an increase in mesh size. As shown in Figure 18, solute agar chains are involved in aggregation below T_{sg}, thereby accounting for the decrease in peak intensity. This inclusion into the aggregation is considered to reduce the local agar concentration and enlarge the mesh size.

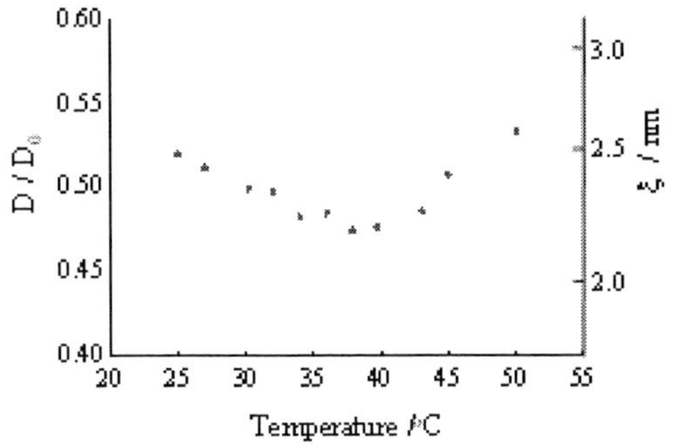

Data from ref.3.

Figure 20. Temperature dependence of the D/D_0 of a dendrimer in 1.3 wt% agar gel. The corresponding ξ, calculated using $D/D_0 = \exp(-R_h/\xi)$, is indicated on the right-hand axis. (Data from ref.3)

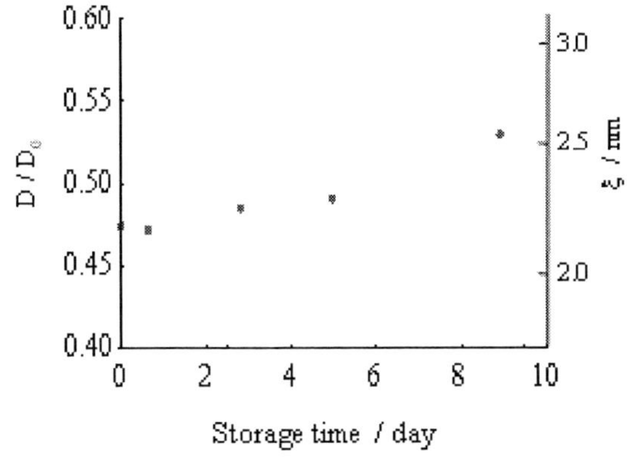

Data from ref.3.

Figure 21. Storage-time dependence of the D/D_0 value of dendrimer in 1.3 wt% agar gel. The corresponding ξ is indicated on the right-hand axis. (Data from ref.3)

The rate of water exudation from the agar gel under compression increased after storage for 3 days, suggesting an increase in mesh size during storage (data not shown). Figure 21 shows the dependence on storage time of D/D_0 values for the dendrimer in 1.3 wt% agar gel after sudden cooling from 58 °C to 25 °C. The values of ξ corresponding to the D/D_0 values calculated using Eq.50 are indicated on the right-hand axis. The initial value of ξ is smaller than that at 25 °C (see Figure 20), suggesting that sudden cooling creates a network structure with a smaller mesh size than that achieved by slow cooling. However, ξ became large (exceeding 2.5 nm) on Day 9 of the experiment, suggesting that during storage the mesh size gradually increases and the solute agar chains become progressively involved in the aggregation.

Figure 22 shows the temperature dependence of ξ of the 1 wt% gellan gum containing 5 mM $CaCl_2$ and 0.1 wt% pullulan. ξ was calculated from the $D_{pul}/D_{pul,0}$ value of pullulan using Eq.50, where D_{pul} was measured by pulsed field-gradient spin-echo (PFGSE) NMR with Δ of 10 ms, and $D_{pul,0}$ was calculated using Eq.49 and the following equation[32]

$$R_H = 2.96\ 10^{-9}\ Mw^{0.49}, \tag{51}$$

where Mw is the molecular weight of pullulan. It is clear that ξ increased with decreasing temperature below a T_{gel} value of approximately 42 °C. From this result, it is considered that the solute gellan is involved in aggregation at temperatures below T_{gel} to enlarge the interspaces among the network chains, in the same manner as that observed for agar gel. Here, ξ was largely constant during the heating process. The 1 wt% gellan gel containing 5 mM $CaCl_2$ showed a high level of stress and retained its highly ordered structure during reheating after setting at low temperature [33]. The result shown in Figure 22 indicates that the microscopic structure of the gel network is retained during reheating [34].

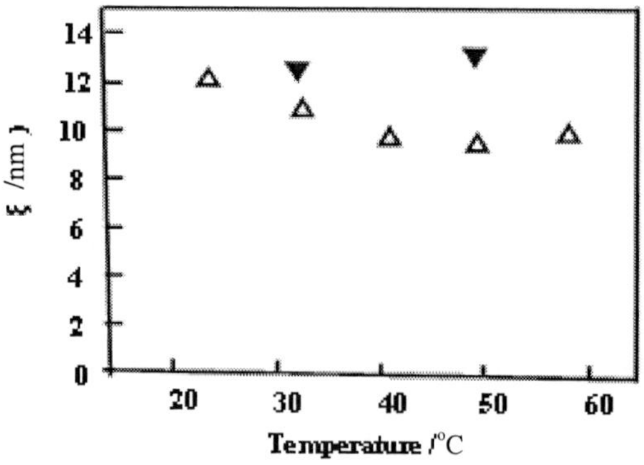

Data from ref.3.

Figure 22. Temperature dependence of the ξ value of a 1 wt% gellan gel containing 5 mM $CaCl_2$ on cooling (open triangles) and heating (filled triangles)

4. Conclusion

On the basis of some experimental results for food hydrocolloid systems, it has been demonstrated that NMR is a useful tool for the investigation of the structure and dynamics of hydrocolloids. Relaxation times for the macromolecules and water gives the information about tumbling motions of molecules in the hydrocolloids. The diffusion coefficients of molecules in the hydrocolloids system give the information about the mobility of molecules and the structure of the gel network. These experimental results have been compared with other macroscopic measurements and discussed regarding the relations between the macroscopic properties and the molecular motions and structures obtained by the NMR measurements.

NMR techniques have been elevated and widely spread in many fields, and many are expected to take advantage of NMR for the investigation of molecular dynamics and structure of food hydrocolloids. In particular, micro NMR imaging and NOE (Nuclear Overhauser Effect) measurements would become prevalent methods, the former gives the micro structure and the latter the interaction between molecules in hydrocolloids, as well as the measurements of relaxation times and diffusion coefficients.

References

[1] Abe, K., Matsukawa, S., Abe, M. and Watanabe, T. (2001). Kinetics for structural change of gelatin studied by viscoelasticity, CD and 1H-NMR measurements. *Trans.MRS-J,* 26, 589-592.

[2] Matsukawa, S., & Watanabe, T. (2007). Gelation mechanism and network structure of mixed solution of low- and high-acyl gellan studied by dynamic viscoelasticity, CD and NMR measurements. *Food Hydrocolloids,* 21, 1355-1361.

[3] Matsukawa, S., Sagae, D. & Mogi, A. (2009) Molecular diffusion in polysaccharide gel systems as observed by NMR. *Progress in Colloid and Polymer Science.* In Press.

[4] Zhang, Q.J., Matsukawa, S., & Watanabe, T. (2004). Theoretical analysis of water 1H T2 based on chemical exchange snd polysaccharide mobility during gelation. *Food Hydrocolloids,* 18, 441-449.

[5] Bleombergen, N., Purcell, E.M. and Pound, R.V. (1948). *Phys.Rev,* 73, 679.

[6] (a) Farrar, T.C. and Becker, E.D. (1971). Pulse and Fourier Transform NMR. Introduction to Theory and Methods. New York: Academic Press. (b) Bloch, F. (1946). *Phys. Rev,* 70, 460.

[7] Harn, E.L. (1950). *Phys. Rev,* 80, 580.

[8] Carr, H.Y. and Percell, E.M. (1954). *Phys.Rev,* 94,630.

[9] Meiboom, S. and Gill, D. (1958). *Rev. Sci. Instrum.* 29,688.

[10] Frarrar, T.C. and Becker, E.D. (1971). Pulse and Fourier Transform NMR. New York: Academic Press.

[11] (a) Matsukawa S, Ando I. Macromolecules 1996;29:7136;(b) 1997;30:8310; (c) 1999;32:1865; (d) Matsukawa S, Yasunaga H, Zhao C, Kuroki S, Kurosu H, Ando I. *Prog. Polym. Sci.* 1999;24:995.

[12] Abragham, A., (1954). Principles of Nuclear Magnetic Resonance, Oxford University Press, Oxford.

[13] (a) Tanner, J.E. and Stejskal, E.O., (1968). J. Chem. Phys, 49, 1768. (b) J.Karger, H.Pfeier, and W.Heink. (1988). Adv. Magn. Reson, 12,1. (c) W.S.Prince. Annual Report on NMR Spectroscopy; G.A.Webb, Ed. London: Academic Press, 35,140. (d) T.Nose. (1993). ibid., 27,217.

[14] (a) Morris, K. and Johnson, C.S.,Jr, J. Am. Chem. Soc.(1992),114,3139-3141. (b)Johnson, C.S.,Jr.,(1999),Diffusion ordered nuclear magnetic resonance spectroscopy: principles and applications, *Prog. Nuc. Magn. Reson. Spect.,* 34, 203-256.

[15] De Ruiter, D. A., & Rudolph, B. (1997). Carrageenan biotechnology. *Trends in Food Science and Technology,* 8, 389-395.

[16] Viebke, C., Piculell, L., & Nilsson, S.(1994). On the mechanism of gelation of helix-forming biopolymer. *Macromolecules,* 27, 4160-4166.

[17] Knutsen, S. H., & Grasdalen, H. (1992). The use of neocarrabiose oligosaccharides with different length and sulphate substitution as model compounds for 1H-NMR spectroscopy. *Carbohydrate Research,* 229, 233-244.

[18] Van de Velde, F., Peppelman, H. A., Rollema, H. S., & Tromp, R. H. (2001). On the structure of κ/ι-hybrid carrageenans. *Carbohydrate Research,* 331, 271-283.

[19] Jansson, P., Lindberg, B., & Sandford, P. A. (1983). Structural studies of gellan gum, an extracellular polysaccharide elaborated by Pseudomonas elodea. *Carbohydrate Research,* 54, 517-522.

[20] Kasapis, S., Giannouli, P., Hember, M.W. N., Evageliou, V., Poulard, C., Tort-Bourgeois, B., et al. (1999). Structural aspect and phase behaviour in deacylated and high acyl gellan systems. *Carbohydrate Polymers,* 38, 145-154.

[21] Richardson, S. J., & Steinberg, M.P. (1987). Application of nuclear magnetic resonance. In L.B. Rockland, & L.R. Beuchat (Eds.*), Water activity: theory and applications to food* (pp.235-285). New York: Marcel Dekker

[22] Okada, R., Matsukawa, S., & Watanabe, T. (2002). Hydration structure and dynamics in pullulan aqueous solution based on 1H NMR relaxation time. *Journal of Molecular structure,* 602-603, 473-483.

[23] Zhang, H., Nishinari, K., Williams, M. A. k., Foster, T. J., & Norton, I. T. (2002). A molecular description of the gelation mechanism of curdlan. *International Journal of Biological Macromolecules*, 30, 7-16.

[24] Hills, B. P., Cano, C., & Belton, P. S. (1991). Proton NMR relaxation studies of aqueous polysaccharide systems. *Macromolecules,* 24, 2944-2950.

[25] Abeltt, S., Lillford, P. J., Baghdadi, S. M. A., & Derbyshire, W. (1976). NMR relaxation in polysaccharide gels and film. *American Chemical Society Symposium Series,* 34, 344-359.

[26] Lewis, g. P., Derbyshire, W., Abeltt, S., Lillford, P. J., & Norton, I. T. (1987). Investigations of the N.M.R. relaxation of aqueous gels of the carrageenan family and of the effect of ionic content and character. *Carbohydrate Research,* 160, 397-410.

[27] Yoon, H., Kim, H. and Yu, H.(1989)..*Macromolecules,* 22,848-852.

[28] McConville, P., & Pope, J. M. (2001). 1H NMR T2 relaxation in contact lens hydrogels as a probe of water mobility. *Polymer,* 42, 3559-3568.

[29] Borgström, J., Piculell, L., Viebke, C., & Talmon, Y. (1996). On the structure of aggregated kappa-carrageenan helices. A study by cryo-TEM, optical rotation and viscometry. *International Journal of Biological Macromolecules,* 18, 223-229.

[30] de Gennens,P.-G. (1976). *Macromolecules,* 9, 594-8.

[31] Cukier, R. (1984). *Macromolecules,* 17. 252-5.

[32] Nishinari, K., Kohyama, K., Williams, P.A., Phillips, G.O., Burchard, W., & Ogino, K. (1991). *Macromolecules,* 24, 5590-5593.

[33] Miyoshi, E., & Nishinari, K. (1999). *Prog. Coll. Polym. Sci.* 114, 68-82.

[34] Matsukawa, S., Tang, Z., & Watanabe, T. (1999) *Prog. Coll. Polym. Sci.* 114, 15-24.

In: Food Hydrocolloids: Characteristics, Properties... ISBN: 978-1-60876-222-4
Editor: C. S. Hollingworth, pp. 179-203

Chapter 6

THE INFLUENCE OF STRUCTURAL PROPERTIES AND BIOLOGICAL FUNCTIONS OF SWEET CASSAVA AND SEAWEED POLYSACCHARIDES

Albert Linton Charles[1*], Jan-Ying Yeh[2], Tzou-Chi Huang[3], and Bor-Rung Ou[4]

[1] Department of Tropical Agriculture and International Cooperation, National Pingtung University of Science and Technology (NPUST), Pingtung 91201, Taiwan

[2]Department of Biotechnology, Asia University, Taichung County 413, Taiwan

[3]Department of Food Science, NPUST, Pingtung 912, Taiwan

[4]Department of Animal Science and Biotechnology, Tunghai University, Taichung, 407 Taiwan,

DRAFT ABSTRACT

There is now new evidence concerning the participation of plant-derived polysaccharides that may play an important role in reactive oxygen species in the etiology and physiopathology of human diseases. Ethanol-extract and sulfated polysaccharides of sweet cassava and seaweed, respectively, were tested for their structural, antiproliferative, and antioxidative properties. Animal cell cultures, including hepatoma HepG2, glioma C6, and Chang liver cell lines, were used as models to evaluate the antiproliferative, DPPH, and hydrogen peroxide radical scavenging properties of the extracts. The antioxidative properties of seaweed were screened and investigated for its hepatoprotective activity in rat. Sweet cassava polysaccharide is mainly composed of the major neutral sugars sucrose, fructose, glucose, galactose, and arabinose. ^{1}H NMR and ^{13}CNMR showed that the cassava polysaccharides was composed of 1, 3-linked glucan chains with 1, 6-linked galactan side chains attached to position 6. Both sweet cassava polysaccharides and a seaweed diet were independently found to significantly increase UGT1A1 and UGT1A6 (phase II enzymes) mRNA levels and to decrease CYP1A1 and

* Corresponding author. Tel.: +886 8 774 0286, fax. +886 8 774 0446.
E-mail address: alcharles@mail.npust.edu.tw (A. L. Charles)

CYP1A2 (Phase I) mRNA levels in chemically-injured rat liver microsomes. *In vitro* and intracellular antioxidative enzyme activity assays demonstrated and confirmed the potential of cassava root and seaweed extracts and were attributed to the structural properties of the polysaccharide compounds. A mechanism relating the antioxidative properties of the polysaccharide glucan chains to stimulate surface cellular receptors and induce antioxidant response enzymes is proposed in this study, as the underlying factor for the immune-stimulating properties of polysaccharides.

Keywords: sweet cassava, seaweed, polysaccharide, phase II enzymes, antioxidation.

INTRODUCTION

Primary plant cell walls, composed of a matrix of polysaccharides which provide structural strength, are central to the control of growth and provide a shield against pathogen attack (Western *et al.*, 2000). Plant-derived polysaccharides are increasingly studied for their antioxidant properties and biological activities. These polysaccharides are involved in biological responses and can activate complement system, which plays an important role in primary host defense mechanisms. Recently, there have been ongoing studies of BRMs (biological response modifiers) because they can replace chemical drugs against human disease. With the view of developing new drugs without adverse effects, many researchers have been seeking biologically active substances. Commercially available polysaccharides, such as lentinan, larch arabinogalactan, krestin and so on, were the results of these efforts, which were applied in cancer therapy. Immune stimulating substances, such as anticomplementary polysaccharide, macrophage-activating polysaccharide, B-cell proliferation activating materials and anti-cancer substances have been studied. Production efficiency, including yield and quality of starch from cassava is markedly influenced by environmental conditions, especially water-stressed plants during early plant development and immediately before root harvest. Evidently, the root tuber in its response to water-stressed conditions synthesizes a mucilaginous compound, a secondary metabolite with high absorption capabilities. Presumably, this appears to be the plant's defense mechanism to adapt to water stressed conditions. The extent of defense of the mucilaginous substance for the plant is largely unknown and is worth investigating. A few researchers have hypothesized that the mucilage is assumed to have protective functions such as preventing desiccation and infection of the wounded tuber, which indicates special structural conformation and biological properties of plant mucilaginous compound.

CASSAVA

Cassava is a member of the family, Euphorbiaceae, tribe Manihotae, subfamily Crotonoideae (Cock, 1985). Early literature on cassava described the genus with two edible species, *M. ultissima* Phol or sweet and *M. aipi* Phol, delineating species which have high and low cyanogenic glucoside concentrations respectively. More recently, cassava was classified as all being the same species *M. esculenta*. It is the only one of 98 species in its family that is widely cultivated for food production. Cassava is grown chiefly as a food – is a perennial vegetable propagated shrub grown throughout the lowland tropics for its starchy, thickened roots. It is one of the most important sources of metabolic energy for 500 million people in the tropics and it also has several significant industrial uses. This crop also called manioc, tapioca or yuca, is one of the most important food crops in the humid tropics, being particularly suited to conditions of low nutrient availability and able to survive drought. It was initially domesticated in Central and South America but has become an economic cash crop in many regions of the world as a nonconventional food source. Cassava tubers contain only about 1-2 % of protein and have a starch content of approximately 85% on a dry weight basis with a low content of sulfur-containing amino acids (Charles *et al.*, 2005a). The tubers are a source of flour used in composite formulations, and starch for industrial and food use.

Cassava will grow in many soils (Figure 1) and under a wide range of climatic conditions, but like most crops, it yields best in fertile soils with little or no drought stress.

Figure 1. Photograph of cassava root tubers taken in Pingtung, Taiwan.

Seaweeds

Seaweeds are extensively used as food by coastal people, particularly in East Asia, e.g. Japan, China, Korea, Taiwan, and Vietnam. Seaweeds are also harvested or cultivated for the extraction of alginate, agar and carrageenan, gelatinous substances collectively known as hydrocolloids or phycocolloids. Hydrocolloids have attained commercial significance, especially in food production as food additives. The food industry exploits the gelling, water-retention, emulsifying and other physical properties of these hydrocolloids. Agar is used in foods such as confectionery, meats and poultry products, desserts and beverages and moulded foods. Carrageenan is used in preparation of salad dressings and sauces, dietetic foods, and as a preservative in meat and fish products, dairy items and baked goods.

Seaweeds, traditionally used as food, are a rich source of dietary fiber, minerals, and proteins and are considered to be a rich source of antioxidants with the ability to scavenge reaction oxygen species and free radicals (Charles *et al.*, 2007). Studies have revealed that seaweeds decreased serum total cholesterol in the seaweed-based diet groups compared with those of the control rat group and various seaweed species, screened for their hepatoprotective activity using CCl_4-induced liver injury in the rat as a model, were reported to exhibit significant hepatoprotection by reducing the acute increase of glutamic pyruvic transaminase and glutamic oxoaloacetic transaminase. Seaweeds administration enhanced dioxin excretion and reduced dioxin absorption in rats administered the dioxin mixture.

Cell wells from marine algae contain sulfated polysaccharides which are not found in land plants. Functionalities of sulfated polysaccharides were found to be anticoagulant, antiviral, antioxidative, immuno-modulating and antitumor. Similarly, active sulfated-polysaccharides, which are referred to as heparinoids, are used as substitutes and are well known to have remarkable anticoagulant activity. However, it is unknown whether seaweed species such as *M. nitidum* could induce important detoxifying enzymes referred to as Phase I and II enzymes. In the biomedicine and pharmaceutical industries, products from seaweeds including alginates are used in wound dressings; agar is extensively used as culture medium, and Carrageenans, alginates and agaroses, together with other macroalgal polysaccharides, are reported in the literature to have several important biological activities or applications in biomedicine.

Composition of Mucopolysaccharides

Plant roots actively modify the rhizosphere (the soil region influenced by the root) through the production of mucilage which is a complex polysaccharide gel (Osborn *et al.*, 1999). Mucilage, a high molecular weight, water-soluble polysaccharide, containing more than 30,000 different sugars, is common in higher plants. In most cases its physiological function is uncertain; it may serve as reserve substances or contribute to the moisture balance and/or resistance to drying out. Understanding the chemical composition of root mucilage is essential to quantifying the level and nature of carbon deposition from roots into soils and identifying the mechanisms of recognition and interaction between plant roots and other organisms, both beneficial and pathogenic. The sugar composition of the mucilage has been established as consisting of arabinose and galactose, but there is disagreement on the quantity of protein associated with the mucilage. A histochemical study identified arabinogalactan proteins (AGPs) in taro corm mucilage ducts (Jiang & Ramsden, 1999). However, the elucidation of the structures of such polysaccharides (e.g., maize mucilage, rice mucilage, taro mucilage, seaweed sulfated-polysaccharides) is complex and typically requires multi-step strategies. Several techniques are available for the elucidation of polysaccharide structures and common techniques usually begin with enzymatic or total hydrolysis (Charles *et al.*, 2005b). These techniques objectively aim to efficiently release the constituent monosaccharide building blocks from the parent polysaccharide, with minimum decomposition, and to derivatize these fragments to enable their subsequent identification and quantification. Typical derivation techniques include formation of volatile methyl ethers which are easily identifiable by analytical methods including high-performance liquid chromatography (HPLC), high performance anionic exchange chromatography (HPAEC) and gas chromatography and nuclear magnetic resonance (NMR) analyses.

POLYSACCHARIDE STRUCTURE-BIOLOGICAL ACTIVITY OF POLYSACCHARIDES

Polysaccharide is a major component of plant whose functions have been less studied for their complex structures and diversities of their molecular weights than other components. Biologically active polysaccharides from plant extracts are macromolecules and their molecular masses disperse from thousands to a million daltons.

Studies on structure–activity relationship for these complement activating pectic polysaccharides have clarified that β-(1→6)-linked galacto-oligosaccharides, which are attached to rhamnogalacturonan core or to β-(1→3)-galactan backbone, play important roles in the expression of the complement. To investigate the chemical properties of thyme purified polysaccharide, total carbohydrate, uronic acid, protein, and structural sugar contents were examined. According to the purification step, galacturonic acid content was greatly changed and reached up to 44.8 mol% and uronic acid content was to 61.5%. It was concluded that these changes might have possible roles in showing anticomplementary activity and acidic polysaccharides, such as pectic polysaccharides from *Plantago major* (Samuelson *et al.*, 1996). In that report, a biologically active pectin type polysaccharide, PMII, was isolated from the leaves of *Plantago major* L. which is traditionally used in wound healing. PMII consists of 71.1% galacturonic acid, 4.2% rhamnose, 8.8% arabinose and 8% galactose. It has anti-complementary activity and induces production of tumor necroses factor alpha (TNF-α) from human monocytes. Similary, a type of polysaccharide, using a modified acidified extraction step was used for isolating polysaccharides from seaweeds.

Sulfated polysaccharides have a wide range of important biological properties. Marine algae are a rich source of sulfated polysaccharides with novel structures, and these compounds have anticoagulant properties. Various anticoagulant-active polysaccharides, especially from red and brown algae, have been isolated and characterized. They contain a variety of sulfated galactans and sulfated fucans, which are among the most abundant nonmammalian sulfated polysaccharides found in nature. These polysaccharides are potent thrombin and factor Xa inhibitors mediated by anthithrombin or heparin cofactor II. There are, however, fewer reports of anticoagulants from green algae than from brown and red algae. The polysaccharide from *Monostroma nitidum* has greater antithrombin activity than that of heparin (Mao *et al.,* 2006).

Arabinogalactan, a polysaccharide, is classified into three groups and type II is heavily branched and mainly composed of 3-linked and 3, 6-linked galactose and its anticomplementary activities were discovered from *Angelica acutiloba.* Much of the mechanism underlying the inhibitory effect of polysaccharides or polysaccharide–peptide complexes involving complementary and anticomplementary activities remains unclear. However, from the discussion above, there is a strong background of literature and research, already concluded and ongoing that allows for the feasibility of carrying out similar research activities in investigating the chemical structures and antioxidative properties of mucopolysaccharides and investigating how the structural properties of these compounds influence their biological properties.

PLANTS- A SOURCE OF ANTIOXIDANTS

Antioxidants are substances that protect the body against oxidation by scavenging biologically toxic reactive oxygen species (ROS). Further, antioxidants are known to reduce the risk of developing more than 100 diseases including acquired immunodeficiency syndrome (AIDS), heart disease, stroke, artherosclerosis, diabetes, cancer, and to slow the aging process. Consequently, antioxidants in natural products have become an area that has gained much attention.

The use of traditional medicine is widespread, and plants still present a large source of natural antioxidants that might serve as leads for the development of novel drugs. Several anti-inflammatory, digestive, antinecrotic, europrotective, and hepatoprotective drugs have recently been shown to have an antioxidant and/or radical scavenging mechanism as part of their activity. In searching for novel natural antioxidants, some plants have been extensively studied in the past few years for their antioxidant and radical scavenging components. These include *in vitro* antioxidant activities of yam tuber mucilage, antioxidant and antiproliferative activities of sweet potato (Duvivier *et al.*, 2008), oolong tea, antioxidant activity of polysaccharide conjugate from green tea, seaweed (Charles *et al.*, 2007) and wild Korean ginseng leaves. Information on antioxidants in seaweeds is extensive in the literature, whereas, cassava as a source of antioxidant or anticarcinogenic agents has scarce, and neither has the possibility or analysis of biological natural functional products have ever been raised or pursued.

Many natural antioxidants originating from plant sources have since been identified as free radical or active oxygen scavengers. It is thought that the risk of ROS-related diseases is decreased by reinforcement of the defense mechanism against oxidative stress. ROS are highly oxidative substances, which may attack almost all cellular components and materials, especially the proteins, lipids and DNA of healthy cells. It is a well-known fact that ROS, such as superoxide radical (O^-_2), hydroxyl radical and (OH), show toxic effects when in contact with tissues of many organisms. Many people are interested in the antioxidation qualities of plants and their products such as red wine, tea and so on. There is superoxide anion scavenging activity (SOSA) in these plants that induces O_2^- quenching activities, which are similar to those of superoxide dismutase (SOD). Currently, the O_2^- scavenging activities of plant polysaccharides are also being studied (Charles *et al.*, 2009). It is expected that the development of food materials that have high SOD scavenging activity will continue to advance.

ANTIOXIDATIVE PROPERTIES OF POLYSACCHARIDE

Polysaccharides are involved in biological responses and can activate complement system, which plays an important role in primary host defense mechanisms. Recently, there have been ongoing studies of BRMs (biological response modifiers) because they can replace chemical drugs for human disease. With the view of developing new drugs without adverse effects, many researchers have been seeking biologically natural active substances. Commercially available polysaccharides, such as lentinan, larch arabinogalactan, krestin and so on, were the results of these efforts, which were applied in cancer therapies. Especially, immune stimulating substances, such as anticomplementary polysaccharide, macrophage-activating polysaccharide, B-cell proliferation activating materials and anti-cancer substances have been studied.

The analysis of various species of mushrooms has resulted in the identification of a family of high molecular weight, hot-water-soluble polysaccharides (PS) and polysaccharide–peptide complexes (PSPC), which have tested positive for antitumor activities in animal studies (Mizuno *et al.*, 1995). NMR analysis revealed that the antitumor PS is composed of a variety of linear and branched glucans. They appear in various conformations, and some are in the gel state. As a result of the complexity and heterogeneity of PS and PSPC and limited suitable bioassay systems, the mechanisms of action of these PS remain obscure. Until now, research on the antitumor activities is based mainly on the results derived from experiments with implanted Sarcoma180 or chemically induced tumors in mice. Antimitotic tests in tumor cell lines have been, so far, rather inconsistent or negative (Mizuno *et al.*, 1995). Aside from their antitumor effects, mushroom-derived PS and PSPC seem to function as immunomodulators; this was observed in animals as well as in cultured macrophages and T-lymphocytes. It has been postulated that the antitumor effect of PS and PSPC may be mediated through the cytokines released by activated macrophages and T-lymphocytes, instead of through direct cytocidal effects on tumor cells. However, a morc comprehensive understanding is necessary to investigate the antitumor activity of polysaccharides in a wider range of cell systems and relate their structural conformations to their immunological or antioxidative functions.

In vitro Studies of Antioxidant Activity

The role of free radicals and active oxygen is becoming increasingly recognized in the pathogenesis of many human diseases, including cancer, aging, and atherosclerosis. Until now, studies on a variety of plant polysaccharides have mainly focused on the activities, but the activity mechanism has not yet been well understood. In this retrospect, one of the objectives is to investigate the mechanisms of antioxidant activity, as related to the structural properties of chemical properties of polysaccharides of cassava and seaweed.

Since cassava, cultivated as bitter (high HCN) and sweet species (low HCN), is widely used in some countries as a staple dietary and as a composite ingredient form either animals or humans, information about its acute and chronic toxicity is largely available in the literature. However, no information about the presence of polysaccharide mucilage exhibiting biological activity exists, as compared to the vast library of information on seaweeds'. This means that it is urgent to conduct a cytotoxicity assessment of this compound, in order to firstly, dispel the stigma against the crop and secondly to ascertain the biological activity of the tuber material. Since cell proliferation inhibition is a standard phenomenon involved in toxicological processes, the aim of this study was to investigate the *in vitro* cytotoxicity properties of cassava polysaccharides. Collecting reliable information about cassava polysaccharides cytotoxicity in vitro requires a relevant cellular model and sensitive analytical tools. We therefore studied the effect of cassava polysaccharides on a human hepatic cell line (HepG2), and rat glioma C6 cell lines which are quite suitable for cytotoxicity assessment on account of the quality and stability of their enzymic and metabolic background. To ensure the relevance of this study, several cellular targets were investigated as recommended by the European Centre for the Validation of Alternative Methods (ECVAM) (Osseni *et al.,* 2000).

Phase I, II and Antioxidant Enzymes

Phase I enzymes results in the activation of chemicals to a toxic or mutagenic product, including aflatoxins and polycyclic aromatic hydrocarbons. Another potentially damaging effect of phase I enzymes is the production of oxygen free radicals that occurs as a result of cytochrome P450 activity. The balance between the abilities of potential chemoprotective agents, such as polysaccharides, to inhibit P450s and induce phase II enzymes may be critical for their capacity to function as anticarcinogenic agents (Charles *et al.*, 2009).

Studies have indicated that anticarcinogenesis activity is attributed to the concomitant induction of phase II enzymes such as glutathione S-transferase (GST) and DT-diaphorase and antioxidant enzymes such as SOD, catalase, glutathione peroxidase (GPX), and glutathione reductase (GR) in the mouse liver.

SODs are ubiquitous metalloproteins that play a major protective role in living organisms by catalizying the dismutation of superoxide into hydrogen peroxide and molecular oxygen. There are different types of SODs characterized by the redox-active metals at the catalytic site. In mammalian systems, copper/zinc SOD (Cu/Zn-SOD) is located primarily in the cytosol and nucleus whereas manganese SOD (Mn-SOD) exists mainly in mitochondria. Glutatione (L-r-glutamyl-L-cysteinylglycine) (GSH) and GSH-related enzymes, GPX, GR, and GST, are important antioxidants and phase II enzymes that play critical roles in the detoxification or reducing the formation of reactive carcinogenic species. GSH is a biologically active thiol tripeptide that plays an important role in detoxifying exogenous compounds and oxidative insults. GPX is a family of four selenoproteins that incorporates selenium as selenocysteine (Se-Cys) into the amino acid sequence. These enzymes exhibit different substrate specificities and collaborate to provide the main enzyme defense system against oxidative stress in mammalian cells. Cellular GPX (cGPX) prevents the formation of ROS by catalyzing the decomposition of hydrogen peroxide with the concomitant oxidation of GSH to GSSG. GSH reductase is the other important enzyme responsible for the availability of GSH in the GSH redox cycle by catalyzing the reduction of GSSG to GSH utilizing NADPH (nicotinamide adenine dinucleotide phosphate). GST comprises a family of isozymes with various substrate specificities and amino acid sequences which are involved in metabolic detoxification by catalyzing the reaction between GSH and electrophilic substrates, preventing toxic damage in several tissues. Catalase is a widely distributed enzyme that destroys hydrogen peroxide by dismutation (catalatic activity), and also demonstrates peroxidatic activity.

Structural Properties and Biological Functions of Sweet Cassava Polysaccharides

Mucilage or mucopolysaccharide component; a high molecular weight, water-soluble polysaccharide and containing more than 30,000 different sugars, is common in roots and other parts of higher plants. In most cases its physiological function is uncertain; it may serve as reserve substances or contribute to the moisture balance and/or resistance to drying out. The slimy texture (observed in cassava tubers during cooking) of taro corms and taro products has long been attributed to the presence of water-soluble gum, or mucilage. Amin (1956) reported that the taro mucilage contained mainly D-galactose and glucose and Lin and Huang (1993) reported that the taro mucilage also contained arabinose and protein. The sugar composition of the mucilage has been established as consisting of arabinose and galactose but there is disagreement on the quantity of protein associated with the mucilage. A histochemical study identified arabinogalactan proteins (AGPs) in taro corm mucilage ducts (Jiang & Ramsden, 1999). Few studies (Uhumwangho *et al.*, 2005; Charles *et al.*, 2006) have been carried out on this mucilaginous gum material of either cassava or within the scope of seaweed sulfated polysaccharides (Charles *et al.*, 2007). In our work, we investigated the molecular characterization, chemical composition and structural properties of polysaccharide material extracted from root tubers of sweet variety of the *M. esculenta.*

Cassava Mucilage Molecular Weight Distribution

The weight average (Mw) and number average molecular mass (Mn) of pure cassava mucilage extracts is 3.6×10^5 and yields a polydispersity index (Mw/Mn) of 1.5. The great molecular mass found here for pure cassava mucilage may be an indication of the formation of large molecular aggregates that were unable to dissociate during permeation through the gel beads in the column and impossible to detect by differential refractometry.

High Performance Chromatography Analyses of Sweet Cassava Polysaccharides

When the mucilage was analyzed from fresh 1 and 2 years cassava tuber samples, the dominant components identified by HPAEC were glucose and xylose, (Charles *et al.*, 2006). In more recent studies, the mucilage samples isolated from dried cassava flours exhibited higher concentrations of monosaccharide compared with the mucilage extracted from fresh cassava. From this data, the increased monosaccharide concentration indicates the availability of monosaccharide components from dried cassava flour samples. Chromatograms indicate that with maturation of tubers, sugar composition is altered, and glucose is the major peak of the tuber mucilages. Glucose, xylose, arabinose and galactose are identified as the major neutral sugar residues of the hydrosylates. The mucilage hydrolyzed for 10 h also indicated similar chromatographic patterns in that the fractions eluting before 20 min contained only monosoccharides. From the chromatograms, the tuber mucilages contained oligosaccharides which were detected by increasing the NaOH mobile phase concentration. In the analysis, 0.1

mM and 0.5 mM of NaOH were used in order to increase the accuracy of detection and separation of fraction peaks.

The differences in carbohydrate compositions can be mainly attributed to growth as well as the presence of starch components in the gum before purification. Hence, polysaccharides from mature tubers compared to young tubers, better represents the carbohydrate composition in which glucose, galactose and arabinose are the major sugars. The purified mucilage compounds were used in the remaining tests to characterize the properties of cassava mucilage. In contrast to purification procedures for other polysacharide sources, from taro, maize, cassava intrinsic protein content is not in significant amounts (Charles *et al*, 2004; Charles *et al.*, 2006) to contaminate its starch isolates, and hence its mucilage material upon isolation. However, since protein content ranged from 0.7-0.45%, the mucilage was treated with TCA to remove unconjugated proteins and its presence in the resulting purified and crude mucilage was determined by

FTIR Spectroscopy of Cassava Mucilage

Application of FTIR (Fourier Transmission-Infra Red) spectroscopy has been shown to be a useful tool in monitoring structural changes in biopolymers. Structural differences are observed between the crude and purified mucilage patterns (Figure 2). The samples exhibit the characteristic IR absorption of polysaccharide at 1650 and 1250 cm^{-1}. The IR absorption at 800 cm^{-1} of the samples is the characteristic absorption of mannose. Additionally, the polysaccharides showed IR absorption at 850 and 920 cm^{-1}, characteristic of an α-D-glucan. The presence of proteins was detected and illustrated by FT-IR spectra in Figure 2. The strong absorbance peak of the two ester carbonyl stretches at 1550 cm^{-1} and at 1650 cm^{-1} indicated high protein content in the crude cassava mucilage.

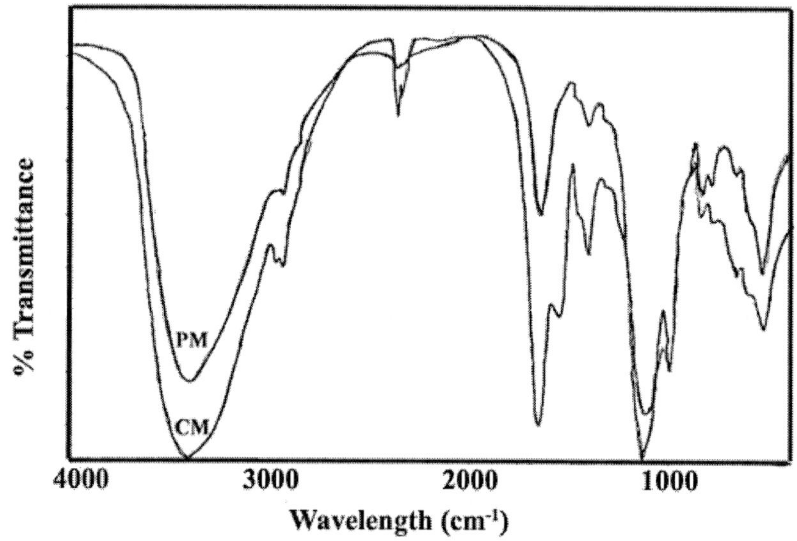

Source: Charles *et al.*, 2008.

Figure 2. FTIR spectra of sweet cassava root tuber mucopolysaccharide PM, purified mucilage, CM, crude mucilage.

These absorbance peaks were greatly reduced perhaps to negligible content in the purified mucilage due to slight shifts of major absorption bands in the purified mucilage compared with those in the crude mucilage. In the mean time the C-O stretch (1150-1000 cm^{-1}) mostly from the carbohydrate structures, became stronger in the purified mucilage, indicating that cassava mucilage is mainly a carbohydrate compound. The (C-O) and ring vibrations ranging at 1200-950 cm^{-1} as well as in the "anomeric region" at 950 - 750 cm^{-1} show differences in the α- and β-glucan spectra. In the case of crude and purified cassava mucilage, the characteristic bands for (1, 3) - and/or (1, 6)-linked β-D-glucan are 1161, 1035 and 885. For α-(1, 4, 1, 6)-linked glycogen, the characteristic bands are at ca. 1153, 1006, 926, 854 and 765. Consequently, the intensity ratios 1161/1153, 1035/1006, and 885/926, 885/854 and 885/765 cm^{-1} could be applied to ratios of α and β -linkage determination.

β-Glucan appears to be an unbranched polysaccharide composed of (1→4)-linked glucopyranosyl (Glcp) residues (70%) and (1→3)-linked-Glcp residues (30%). The (1→3) - and (1→4)-linkages are distributed neither statistically randomly nor in a strictly repeating sequence. Longer sequences of (1→4)-linkages have been identified in rye and these sequences are believed to be responsible for the partial water-insolubility of the β-glucan, because they provide surfaces that are capable of forming junction zones. A proposed structure for the cassava mucilage is modeled on the water extractable arabingalactan and is composed of a backbone of (1→3)-linked-β-D-galactopyranosyl (Galp) residues with sidechains of terminal a-L-arabinofuranosyl.

^{13}CNMR Analysis

A characteristic region of the ^{13}CNMR spectrum obtained from crude and purified cassava mucilages are shown in Figures 3 A & B. The dominating, moderately separated resonances in the spectrum are from terminal arabinose (A) units. The resonances at 102.8, 82.6, 77.5, 70 and 68 ppm are from C-1, C-2, C-3, C-4 and C-5, (Figure 3A), respectively. Generally, resonances originating from galactose (G) are less well defined, due to the occurrence of 1,3- (termed 3G), 1,6- (termed 6G) and 1,3,6-(termed 3,6G) linked units. The resonances from the anomer carbons of galactose at about 104ppm are dominated by 3,6G. Galacturonic acid was not detected by ^{13}CNMR due to its low content and/or spectral conditions unsuited for quantitative determination of uronic acids.

The purified polysaccharide was obtained after weak acid hydrolysis, giving rise to a much simpler ^{13}C-NMR spectrum (Figure 3B). The dominant resonances are from 1,6-linked galactose (6G- type units), where typical resonances from arabinose are less or lacking. This indicates that the cassava mucilage contains an abundant linkage of 1,6-linked galactose. The strong signals at about 102 ppm as well as strong intensity signals at 76, 74 and 71.2 are assigned to C1, C-3, C5, C-2 and C6 of 3G units, respectively.

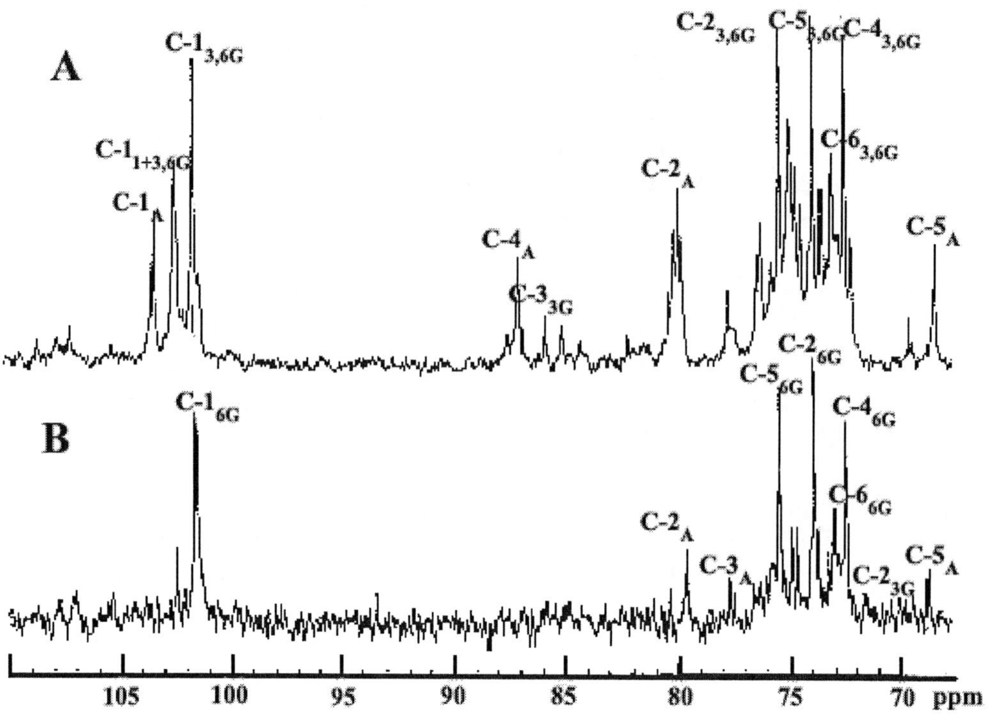

Source: Charles *et al.*, 2008.

Figure 3. ^{13}CNMR spectra of (A) crude and (B) purified cassava mucilage (Year 3). C refers to the different carbons. Lowercase A, 3,6G refer to arabinose, 1 3-linked,1,3,6-linked and 1,5-linked galactose, respectively.

Characterization of Cassava Mucilage Proteins by SDS Gel Electrophoresis

Analyses on the protein components of cassava polysaccharide extracts from 3 harvest cycles using SDS-PAGE suggested that molecular weights of the polysaccharides ranging from 15 - 20 kDa. The assumption made was that the 16.7 kDa polypeptide (similar to arabinogalactan; Du *et al.,* 1996) is the enzymatically active component of the mucilage. Analysis of the sds gel page revealed that protein concentration is higher from the third harvest cycle, suggesting that higher deposition of protein was associated with mucilage deposition in the aging root primary cell wall.

Arabinogalactan proteins (AGPs) are structurally complex macromolecules composed of a polypeptide, characterized by large branched glycans chains. AGPs are widely distributed in the plant kingdom, probably occurring in every cell of every plant from bryophytes to angiosperms. They are primarily in the extracellular matrix but also on the plasma membrane and the cell wall. These proteoglycans are typically 60 to 300 kD, and their carbohydrate moiety (usually more than 90%) is typically rich in arabinose and galactose, organized primarily as a type II arabino-3, 6-galactan. Their core polypeptide component (usually less than 10%) is typically rich in Hyp, Ala, Ser and Thr and the mature protein has 145 amino acid residues (16.7 kDa).

The tuber roots of *M. esculenta* contained several types of water extractable polysaccharides, predominantly water soluble mucilage. Starch is the major reserve polysaccharide of *M. esculenta* roots, (Charles *et al.*, 2006) and its content in the roots are influenced by the stage of maturation of the roots and the biosynthesis of the mucilage compound (Charles *et al.*, 2008). Whereas the starch is a complex polysaccharide, the mucilage is purely hemicellulosic of mainly arabinogalactans containing a (β-1, 3)-linked glucan backbone having side chain branching with non-reducing terminal residues mostly of (1, 6)-linked galactose and some arabinose and glucosyl moieties. Many plant polysaccharides with immunological activities have been reported, most of which are arabinogalactans.

Biological Properties of Sweet Cassava Arabinogalactan-Type Polysaccharides

Phase I Enzymes Expression

Feeding different dosages (0.2 or 0.5 mg/ml) of cassava polysaccharide to rats significantly influences CYP1A1 mRNA expression, when compared with the control. Similarly, CYP1A2 mRNA expression in the rat liver tissues is similarly influenced by the dosage treatment of cassava polysaccharide in the rat's diet. Results demonstrated that while there was no significant difference between the control and 0.1 ~ 0.2 mg/ml cassava polysaccharide dosage, increasing the dosage to 0.5 mg/ml cassava polysaccharide resulted in significant differences that can explain the inhibitory properties of cassava polysaccharide on phase I enzymes. These results suggest that cassava polysaccharide had CYP-inhibitory properties, *in vivo,* in the rat, as well as the ability to preferentially inhibit CYP-mediated catalytic activities *in vitro*.

Phase II Enzymes Expression

The analyses of mRNA RT-PCR products for the induction of phase II enzymes (GSTYa1 and UGT1A6) is illustrated in Figure 4. Feeding the test animals 0.2 or 0.5 mg/ml cassava polysaccharide resulted in significant differences in the expression of GSTYa1 and UGT1A6 mRNA, whereas, β-actin mRNA, used as a control, demonstrated a similar lack of diversity in its expression among the test rats. However, differential expressions of UGT1A6 were observed in Figures 4 A & B, as a result of incremental increase of the cassava polysaccharide. Figure 4A showed that the induction of the detoxifying enzyme UGT1A6 mRNA was influenced by the presence of cassava polysaccharide in the diet. UGT1A6 mRNA expressions were higher than the controls; however, there was no significant difference in the expression of UGT1A6 mRNA of the three groups of rats fed low dosage of SCP in the diet. It is apparent that cassava polysaccharide could significantly induce GSTYa1 mRNA expression in liver of rats in a dose dependent manner (Figure 4C). The hepatoprotective functions of cassava polysaccharide in rat liver were further investigated by quantification of the mRNA product of Phase II enzymes using the real-time PCR technique.

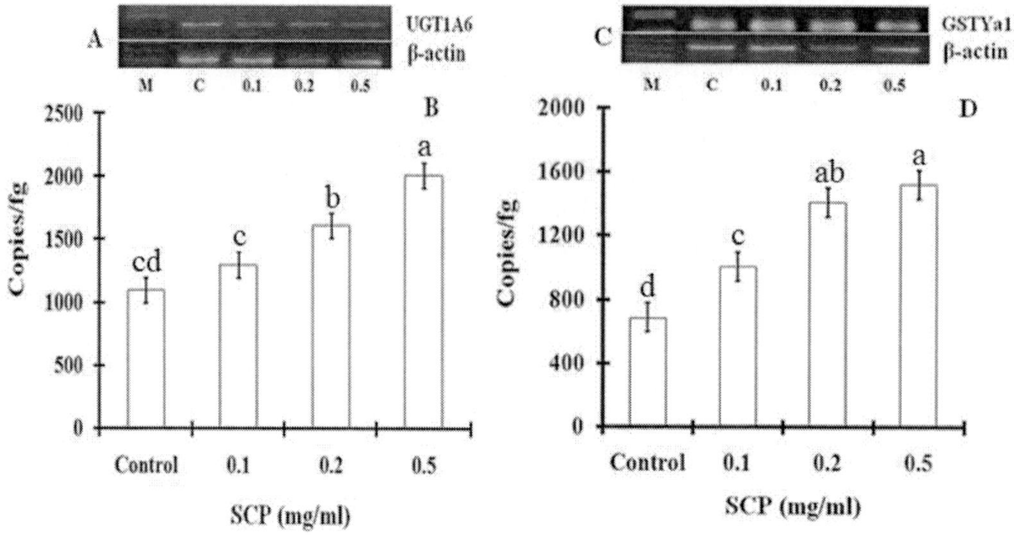

Source: Charles *et al.*, 2009.

Figure 4. Relative Competitive RT-PCR for (A, C) UGT1A6 and GSTY1a against control β- actin and (B, D) real-time PCR quantitative analyses of UGT1A6 and GSTY1a mRNA expression in rat liver tissue.

Analysis of Phase I and II Enzymes by Real-Time PCR

Analyses on differential expression of the phase I (CYP1A1 and CYP1A2) enzymes in rats' liver were conducted using real-time PCR. C_T values of CYP1A1 gene expression in liver were 20.892x, 23.235x and 26.150x and 26.687x, and 18.321x, 21.451x, 22.214x, and 27.012x of CYP1A2, of control, 0.1, 0.2, and 0.5 mg/ml cassava polysaccharide respectively. CYP1A1 and CYP1A2 gene expression in rat liver were higher in the control and linearly decreased with dosage of cassava polysaccharide. In similar experiments, the analysis on the expression of UGT1A6 and GSTYa1 genes in rat liver tissues are illustrated in Figures 4 B & D, and from the data obtained, the expression of the phase II enzymes were significantly higher than control and expression was determined at 1.186x, 1.237x, and 1.251x that of control, according to the doseages administered at 0.1, 0.2, and 0.5 mg/ml, respectively. Moreover, in earlier RT-PCR experiments, a similar trend was observed in the liver tissue samples tested for the detoxifying enzymes UGT1A6 and GSTYa1. In Figure 4D, GSTYa1 expression was significantly higher than control and expression was determined at 1.464x, 1.402x, and 10.8x that of control and increased linearly with the doseages of polysaccharide extracts indicated.

In many studies, Phase II enzyme genes are reported to be induced by external oxidant stress or by reactions from other products with antioxidative abilities. These antioxidative elements enhance the expression of specific Phase II enzymes and detoxifying ability. Other studies also pointed out that rats fed with extraction of *Myagropsis myagriodes*, *Sargassum henslowianum* and *Sargassum siliquastrum* can be used to treat carbon tetrachloride chemically-injured rat liver, while decreasing hepatitis and protecting rat liver. Additionally, in Morita (2002), rats fed with *Undaria pinnatifida*, *Hizikia fusiformis* and *Laminria japonica* revealed the benefit of administering dioxin, a component of the feed, which improves excretion and decreases dioxin absorption. This study reported that sweet cassava polysaccharides can help rats to excrete these poisons and decrease their accumulation in the body. Additionally, giving some insight on the inductive activity of SCP on liver detoxifying enzymes in rats, the expressed biological activity of the cassava polysaccharides inferred to antioxidative substances that were able to induce the phase II detoxifying enzymes analyzed in this study.

UGT, which catalyzes glucuronization, may play central roles in the conjugation and ultimately excretion and elimination of many drugs and xenobiotics. The higher levels of UGT in response to cassava polysaccharide dosage indicated the hepaprotective potential of cassava polysaccharide in inducing this phase II enzyme. A large number of carcinogens are metabolized by the cytochrome P450 (CYP) enzymes to chemically reactive species that covalently bind to the DNA and promote carcinogenesis. In addition, a decrease or inhibition of CYP1a1 and CYP1a2 mRNA expression is interpreted as the possible mechanism of chemoprevention involving either inhibition of the CYPs required for metabolizing carcinogens, such as the CYP1 family, or the induction of phase II enzymes such as GSTYa1 and UGT1A6.

It is likely that the cassava polysaccharide extracts, identified as polysaccharide extracts from FTIR, NMR spectroscopy, and HPAEC analyses (Charles *et al.*, 2008) were responsible for the expression of phase II detoxifying enzymes in rats' liver that could lead to improvements in excretion and prevention of poison buildup in the rats. Further viability or proliferation analyses of the cassava polysaccharide were carried to further determine and confirm the protective potential of the polysaccharide extract.

SOD Activity and SOD1 Induction of Sweet Cassava Polysaccharides

Recent research findings (Padmavathi *et al.*, 2005) has indicated anticarcinogenesis activity is attributed to the concomitant induction of phase II enzymes such as GST and DT-diaphorase and antioxidant enzymes such as SOD, catalase, GPX, and GR in the mouse liver. SODs were selected for this research project since they are ubiquitous metalloproteins that play a major protective role in living organisms by catalyzing the dismutation of superoxide into hydrogen peroxide and molecular oxygen. There are different types of SODs characterized by the redox-active metals at the catalytic site. In mammalian systems, Cu/Zn-SOD is located primarily in the cytosol and nucleus whereas Mn-SOD exists mainly in mitochondria.

Cassava polysaccharide extracts exhibited increment of percentage pyrogallol autoxidation decrease which implies that SOD scavenging activity increased with extract concentration. However, control samples (DMSO-treated cell cytosols) showed relatively higher autoxidation activity compared to the sample extracts. Superoxide radicals are normally formed during the first step of cellular oxidation reaction, and then they magnify their effect by producing other kinds of cell-damaging free radicals and oxidizing agents. The highest superoxide anion scavenging activity was observed in HepG2 cell lines. Analysis of the messenger RNA levels of the superoxide enzyme (SOD1) indicated that SOD1 enzyme was induced and that their scavenging activities were dosage dependent; however, transcriptions of the enzyme was in higher concentration in HepG2 cell lines (2.134x, 2.345x, and 3.452x) than C6 (0.876x, 1.056x, and 1.347x) when compared to the control (Beta-actin) (Charles *et al.,* 2009). Superoxide radical anions are formed when oxygen is reduced by the transfer of a single electron to its outer shell. This takes place not only in living cells but also in early products of protein glycation (Maillard reaction), such as Schiff base and Amadori compounds. These are possible key structural components involved in the generation of a superoxide anion. Therefore, inductions of superoxide anions scavenging activities of cassava polysaccharide in HepG2 and C6 cell lines increases its value not only as a functional food item but as a natural candidate for inducing antioxidant enzymes.

In Vitro Measurement of Cassava Polysaccharides on Cell Proliferation

The antiproliferative activities of different dosages of the polysaccharide extract were tested on the cell lines (3T3-L1, rat glioma C6 cell lines, and HepG2 human hepatoma) and were analyzed using the MTT ((3-(4,5-Dimethylthiazol-2-yl)-2,5-diphenyltetrazolium bromide) assay. C6 and HepG2 cell proliferation was inhibited in a dose-dependent manner, whereas, 3T3-L1 cell lines proliferation activity increased only at low concentrations of the extract. Other studies has revealed similar trends, in that although polysaccharides stimulated normal cell viability (94 % viability), it did not exhibit significant antiproliferative ability on 3T3-L1 cell line less than 200ppm. However, the extracts had significant antiproliferating activities on HepG2 and C6 tumor cells (~45.4 % inhibition) which linearly increased in a dose-dependent manner.

In other related experiments, antiproliferative activities of the different extract fractions from sweet cassava tubers on the growth of the human hepatoma HepG2 cell line, glioma C6 and Chang liver cell lines *in vitro* were studied (Figure 5). Cell proliferation was analyzed at 48 h after cells had been cultured with an extract fraction of 0, 0.5, 0.625, 1.25, 2.5 and 5 μg/mL of dry matter/mL in the media using the MTT assay. Cell proliferation was inhibited in a dose-dependent manner after exposure to the different extract fractions. The results showed that the polysaccharide extracts displayed cytotoxic effects against both HepG2 and C6 cell lines with lowered antiproliferation towards the control, Chang liver cells.

The antiproliferative activities of each fraction were expressed as the median EC_{50}, with a lower EC_{50} value indicating a higher antiproliferative activity. Under the experimental conditions of this test, enzyme activity and the number of formazan formed were proportional to the number of cells. Reduction in the number of cells by a particular agent (cytotoxicity) can generally be explained by cell killing and/or inhibition of cell proliferation. The EC_{50} value (drug concentration causing 50% inhibition of the normal and tumor cells) was used as

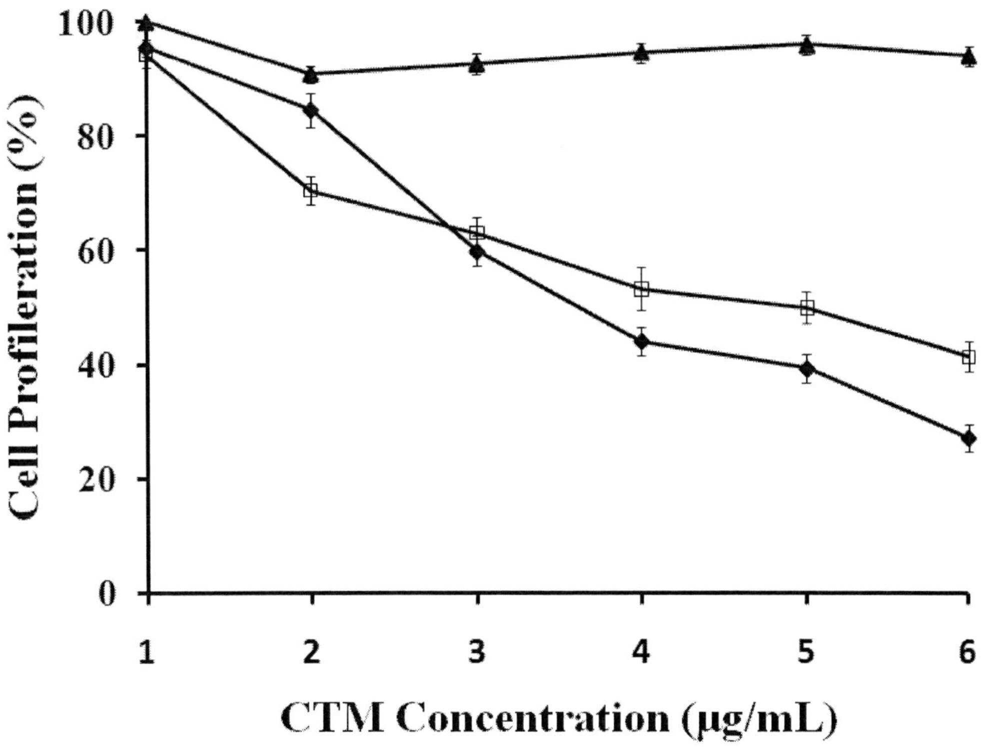

Figure 5. The cytotoxic effect of DMSO extract of CTM on (•) C6, (□) HepG2 and (▲) Chang Liver cell lines.

a parameter for cytotoxicity (Smit *et al.*, 1995). Any plant extract which has the EC_{50} value below 20 µg ml^{-1} can be accepted as a potent cytotoxic extract (Wall *et al.*, 1987). The EC_{50} values of CTM towards HepG2 and C6 were 2.67 and 2.06 µg ml^{-1}, respectively, and indicated potential toxicity potential of the cassava root extract. The CTM extracts exhibited a linear reduction in cytotoxic behavior towards Chang liver cell line (Figure 5) as the dosage concentration increased.

Peroxidation Inhibition of Sweet Cassava Polysaccharides

A comparative study was carried out to investigate the influence of cassava polysaccharide extracts on preventing lipid peroxidation. The standards used included butylated hydroxyanisole (BHA), ethylenediaminetetraacetic acid (EDTA), and Vitamin C. The synthetic metal chelater, EDTA, demonstrated the highest inhibition among the samples tested and was as good as that of the chelating capacity of Fe^{2+} in the FTC method. Peroxidation inhibition linearly increased with concentration of cassava polysaccharide. Results indicated that cassava polysaccharide antioxidative action against lipid peroxidation was just as efficient as the synthetic antioxidant BHA. Statistical analysis showed no significant differences among the standards and cassava polysaccharide dosages, except at 250 ppm. The sweet cassava is reportedly a source of antioxidant substances (Charles *et al.*,

2009), and in recent years, plant polysaccharides have been demonstrated to play an important role as free-radical scavengers and antioxidants in the prevention of oxidative damage in living organisms.

Hydrogen radical scavenging activity of cassava polysaccharide was determined as the inhibition of the hydroxyl radical generated in the H_2O_2-cell treated environment. Hydroxyl scavenging activity was dose-dependent with highest activity observed in the cassava polysaccharide extract at higher concentrations of 5 μg/ ml, which was slightly lower than that of both α-tocopherol and Butylated hydroxytoluene (BHT) in both cell lines used. The extracts of cassava polysaccharide also displayed slightly higher activity in scavenging hydroxyl radical, at low concentrations, than that of commercial antioxidants. The hydroxyl radical plays a major role in lipid oxidation initiation and possesses the strongest cell-damaging action among free radicals. The measurement of H_2O_2 scavenging activity is one of the useful methods determining the ability of antioxidants to decrease the level of prooxidants such as H_2O_2. The scavenging effect of cassava polysaccharide on hydrogen peroxide linearly increased with cassava polysaccharide concentration indicating the extracts as a potential source of antioxidants and corroborating the hepaprotective properties of cassava polysaccharide as exhibited in HepG2 cells.

Chemical Composition and Structural Properties of Seaweed Sulfated-Polysaccharides

Chemical composition analyses revealed crude seaweed polysaccharides (CSP) contained at least 54% total sugar, 7.2% crude protein, with a total extractable yield of 16%. CSP contained moderate amounts of ash (9.61-12.75%). Ash was attributed to counterions associated with sulfate ester groups and uronic acids of the polysaccharides. Sulfated polysaccharides mainly are composed of rhamnose, with variable contents of glucose and fucose, trace amounts of xylose, glactose and mannose.

Infrared spectra yielded useful information on the sulfate and monosaccharide content in sulfated polysacharides. From the FTIR spectra (Figure 6), the absorption bands of the purified polysaccharide are at 1250 and 817 for the sulfate group. The absorption peak at 817 indicates that the sulfate group is attached at the C-6 position of galactose. The IR spectrum of the sulfated polysaccharides from *M. nitidum* shows bands corresponding to a sulfate ester and uronic acids. Spectral peaks at 1256-1264 cm^{-1} is attributed to S-O stretching, and 849-853 cm^{-1} duc to C-O-S stretching. A sulfuric acid group peak is also observed at 1130 cm^{-1} and the signal at 770-789 cm^{-1} might correspond to the bending vibration of C-O-S of sulfate in equatorial position.

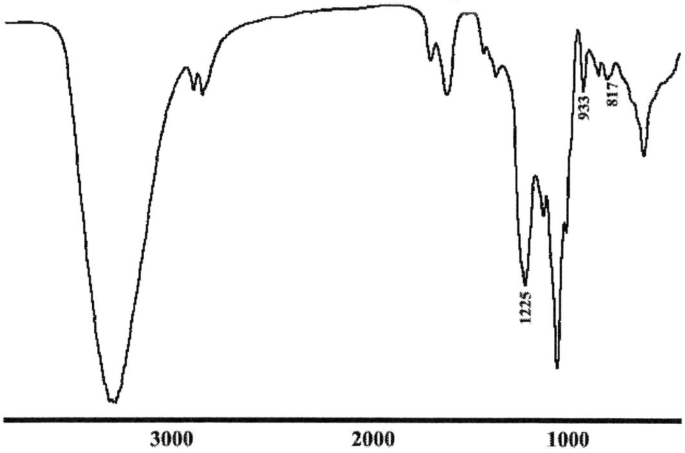

Source: Charles *et al.*, 2007.

Figure 6. An FTIR spectra of water extracts sulfated polysaccharide of *Monostroma nitidum*.

BIOLOGICAL PROPERTIES OF *MONOSTROMA NITIDUM* SULFATED POLYSACCHARIDES

Expression of Phase I and II Enzymes

The expression level of phase I (CYP1A1 and CYP1A2) and phase II (GST, and UGT) enzyme-coded genes were measured in liver microsomes of 30 Sprague Dawley rats fed sea weed (*Monostroma nitidum*). Quantitative and qualitative analysis of the detoxifying enzymes were investigated using reverse transcription polymerase chain reaction (RT-PCR) and real time polymerase chain reaction (real-time PCR) techniques. The antioxidative properties of seaweed were screened and investigated for its hepatoprotective activity in rat. There was no significant induction of GSTYa1, GSTYa2, and CYP1A2. However, an *M. nitidum* diet was found to significantly increase UGT1A1 and UGT1A6 mRNA levels and to decrease CYP1A1 mRNA levels in rat liver. Structural studies confirmed the presence of sulfated polysaccharides in the seaweed samples. The results demonstrate the potential of seaweed as a natural source of sulfated polysaccharide substances with potential use in chemoprevention medicine.

The expression of mRNA of CYP1A1, CYP1A2, and β-actin in liver tissues mRNA of rats fed *M. nitidum,* are illustrated in Figures 7 A & B. From Figure 7A, CYP1A1 mRNA expression in liver of rats is noticeably less than the control group and expression levels increased with dosage of *M. nitidum*. The expression of CYP1A1 increased with *M. nitidum* diet with decreases in CYP1A1 mRNA. Figure 7B shows that feeding 5 or 10 mg/ml *M. nitidum,* respectively, to rats did not significantly or clearly affect CYP1A2 mRNA expression, when compared with the control.

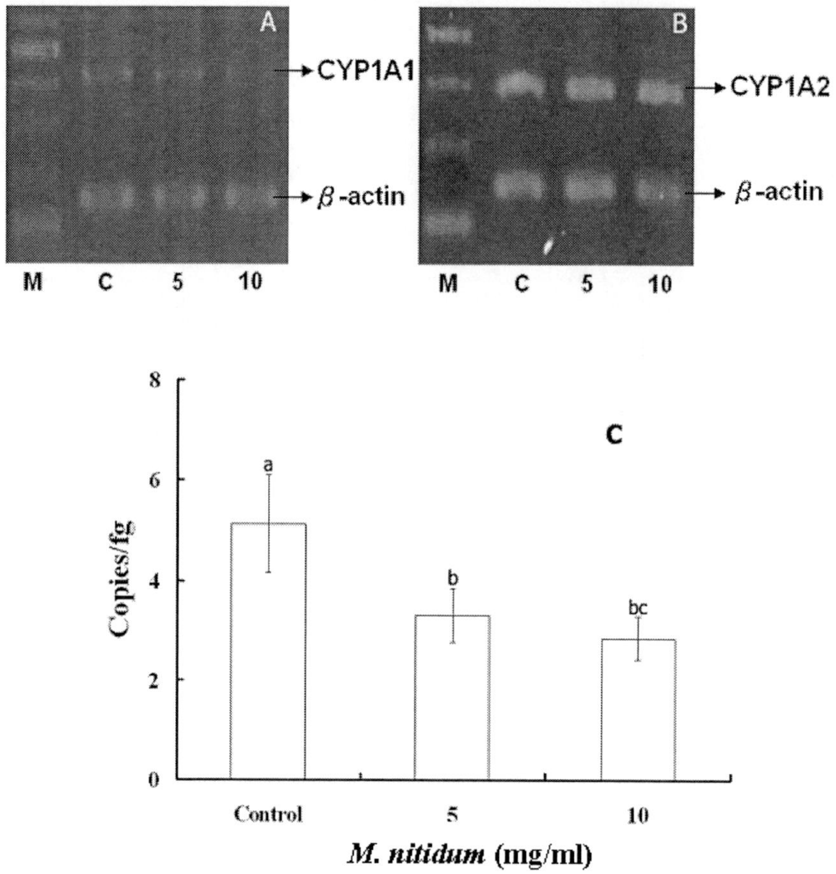

Source: Charles *et al.*, 2007.

Figure 7. Relative Competitive RT-PCR for (A) CYP1A1, (B) CYP1A2, against control β-actin and (C) real-time PCR quantitative analyses of CYP1A1 mRNA expression in rat liver tissue.

The analyses of of phase II enzymes (GSTYal, GSTYa2, UGT1A1, and UGT1A6) indicated that feeding the test animals 5 or 10 mg/ml *M. nitidum,* respectively, did not result in any apparent differences in the expression of GSTYa1 and GSTYa2 mRNA. However, differential expressions of UGT1A1 and UGT1A6 were observed as a result of incremental increase of the seaweed diet (Charles *et al.*, 2007). The induction of the detoxifying enzymes UGT1A1, UGT1A6 mRNA was apparently influenced by the presence of *M. nitidum* in rats' diet. UGT1A1 and UGT1A6 mRNA expressions were higher than the controls in all experiments.

Quantitation of Phase II Enzymes Expression Levels by Real-Time PCR

Analysis on differential expression of liver detoxifying enzymes in rats was conducted using real-time PCR. Figure 7C illustrates the expression of CYP1A1 gene using real-time

PCR technique. C_T values of CYP1A1 gene expression in liver were 26.505x, 27.145x and 27.364x of control, 5 or 10 mg/ml *M. nitidum,* respectively. The original copy was 0.643 and 0.545 times expression of control. CYP1A1 gene expression in liver in rats was less in control and linearly increased with amounts of *M. nitidum.*

Based on data on the expression of UGT1A1 and UGT1A6 genes in the three groups of rats analyzed, UGT1A1gene expression in test rats' liver tissues was calculated 1.318x and 1.320x expression that of the control. The results indicate that the UGT1A1 gene expression was much higher than the control. However, feeding 5 or 10 mg/ml *M. nitidum,* respectively, to rats, UGT1A1 gene expression was similar. Similarly in earlier RT-PCR experiments, a similar trend was observed among the liver tissues tested for the detoxifying enzyme UGT1A1. In other experiments, the UGT1A6 gene in liver was 1.220x and 1.420x the expression in control rat liver samples. The UGT1A6 gene of liver in rats fed with 5 or 10 mg/ml *M. nitidum,* respectively, was significantly higher than the control and increased linearly with increments of the seaweed in the animals' diet.

Additionally giving some insight on the inductive activity of seaweed on liver detoxifying enzymes in rats, the data reported in this study on the biological activity of the *M. nitidum* infers to antioxidative substances that were able to induce the phase II detoxifying enzymes (UGT family) that were investigated.

UGTs, which catalyze glucuronization, may play central roles in the conjugation and ultimately excretion and elimination of many drugs and xenobiotics. The higher levels of UGT in response to seaweed dosage indicate the protective potential of seaweed in inducing these protective phase II enzymes. A large number of carcinogens are metabolized by the cytochrome P450 (CYP) enzymes to chemically reactive species that covalently bind to the DNA and promote carcinogenesis. In addition, a decrease or inhibition of CYP1a1 mRNA expression is interpreted as the possible mechanism of chemoprevention involving either inhibition of the CYPs required for metabolizing carcinogens, such as the CYP1 family or the induction of phase II enzymes such as the UGT1A1 and UGT1A6.

It is likely that the extracts, identified as sulfated-polysaccharide extracts from FTIR and HPLC analyses (Charles *et al.*, 2007), could be responsible for the ability of *M. nitidum* to induce the phase II detoxifying enzymes in rats' liver and also to improve excretion and prevent poison buildup in rats. Sulfated polysaccharides of seaweeds are chemically differed from those of land plants, and may have special biological activity. Further viability or proliferation analyses of the *M. nitidum* SP1 were carried to further determine and confirm the protective potential of these seaweed species.

Influence of *M. Nitidum* Sulfated Polysaccharides on Cell Proliferation

In MTT studies, the antiproliferative activities of different dosages of the seaweed extract using cell systems; 3T3-L1 and rat glioma C6 cell lines treated using *M. nitidum* sulfate polysaccharide, was carried out. C6 cell proliferation was inhibited in a dose-dependent manner, whereas, normal 3T3-L1 cell lines proliferation activity was not affected, but increased growth was observed at low concentrations of the extracts. Although sulfated polysaccharides stimulated normal cell viability (94% viability) (Figure 8), it did not have significant antiproliferative ability to 3T3 cell of mice less than 0.2 mg/ml. However, the extracts had significant antiproliferating activities on C6 tumor cells (52.5% inhibition) which linearly increased in a dose-dependent manner.

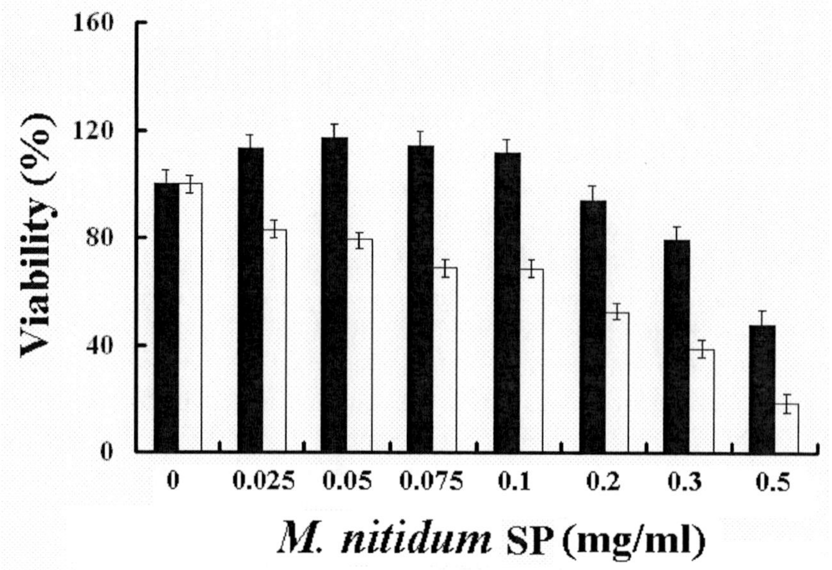

Source: Charles *et al.*, 2007.

Figure 8. Measurement of cell proliferation action of SP1 in (■) 3T3 and (□) C6 cell lines.

Peroxidation Inhibition of *M. Nitidum* Sulfated Polysaccharides

The properties of the sulfated polysaccharide extracts and their ability to behave as a detoxifying agent in rats were investigated using the lipid peroxidation inhibition system. The comparative study was carried out to investigate the influence of seaweed extracts on preventing lipid peroxidation. The standards and reference materials included BHA, EDTA, and Vit. C (Charles *et al.*, 2007). The synthetic metal chelater, EDTA, demonstrated the highest inhibition among the samples tested and was as good as that of the chelating capacity of Fe^{2+} in the FTC method. Peroxidation inhibition increased with concentration of sulfated polysaccharides. Results indicated that the antioxidative action against lipid peroxidation was just as efficient and effective as when the synthetic antioxidant, BHA was used. Statistical analysis showed no significant differences among the test materials and our extract dosages. This information confirms the high antioxidative potential of sulfated polysaccharides, as well as confirms the reports on the seaweed as a rich source of antioxidant substances. In recent years, algal polysaccharides have been demonstrated to play an important role as free-radical scavengers and antioxidants for the prevention of oxidative damage in living organisms.

CONCLUSION

Mucopolysaccharide extracts (cassava arabinogalactan-type polysaccharides) from cassava tuber mucilage was isolated from the roots of sweet cassava and studied for its characteristic physical, molecular, and structural properties. Sugar composition and molecular properties of purified samples were analyzed using high-performance anion exchange (HPAEC) and size exclusion (HPSEC- MALLS-RI) chromatographic techniques. CTM is mainly composed of the major neutral sugars sucrose, fructose, glucose, galactose, and arabinose, with a molecular average weight of 3.6×10^5, number average molecular mass (Mn) of 2.4×10^6, and a polydispersity index (Mw/Mn) of 1.5. On structural properties of crude and purified samples, fourier transformation infra red (FT-IR) revealed high protein content in the crude fractions, whereas purified fractions were mainly carbohydrate. Nuclear magnetic resonance spectroscopy using the ^{13}CNMR technique showed that the mucilage consists of 1, 3-linked glucan chains with 1, 6-linked galactan side chains attached to position 6. The mucilage is purely hemicellulosic and structurally an arabinogalactan, which was confirmed by SDS gel electrophoresis of crude cassava extracts, contains a (1, 3)-linked β-glucan backbone having side chain branching with non-reducing terminal residues mostly of (1, 6)-linked galactose and some arabinose and glucosyl moieties. There was significant induction of GSTYa1 and inhibition of CYP1A2. Moreover, a cassava polysaccharides diet was found to significantly increase UGT1A6 mRNA levels and to decrease CYP1A1 mRNA levels in chemically-injured rat liver. Cassava polysaccharides ethanol extracts exhibited hydroxyl radical and superoxide scavenging activities in a dose-dependent manner. *In vitro* and intracellular antioxidative enzyme activity assays demonstrated and confirmed the potential of cassava root extracts as a natural source of mucopolysaccharide substances with potential use in chemoprevention medicine.

Chemical composition analysis of seaweed sulfated polysaccharides revealed high content of rhamnose, with glucose, fucose and trace amounts of xylose, galactose and mannose. Infrared spectra confirmed sulfate groups linked to galactose, with signals indicating the presence of uronic acids in its structure. UDPGT, GSTYa1, and GSTYa2 mRNA levels were induced by *M. nitidum* sulfated polysaccharides in an increasing dose-dependent manner. Where RT-PCR analyses demonstrated no significant induction of GSTYa1, GSTYa2, and CYP1A2; quantitative analyses, using real-time PCR techniques, revealed that *M. nitidum* was found to significantly increase UGT1A1 and UGT1A6 mRNA levels and to decrease CYP1A1 mRNA levels in rat liver tissues.

Chemical properties and structural properties of seaweed polysaccharides were similar in sugar composition to cassava polysaccharides, which form the basis for the similar patterns in their biological properties in similar cellular and analytical systems.

REFERENCES

[1] Amin, El. S. (1956). Mucilages of Hibiscus esculentus and Corchorus olitorius. *Journal of American Chemical Society,* 78, 828-832.

[2] Charles, A. L., Chang, Y. H., Ko, W. C., Sriroth, K., & Huang T. C. (2004). Some physical and chemical properties of starch isolates of cassava genotypes. *Starch/Starke,* 56, 413-418.

[3] Charles, A. L., Sriroth, K., & Huang, T.C. (2005a). Proximate composition, mineral contents, hydrogen cyanide and phytic acid of 5 cassava genotypes. *Food Chemistry,* 92, 615–620.

[4] Charles, A. L., Chang, Y. H., Ko, W. C., Sriroth, K., & Huang, T. C. (2005b). Influence of amylopectin structure and amylose content on thegelling properties of 5 cultivars of cassava starches. *Journal of Food Agriculture and Chemistry*, 53, 2717-2725.

[5] Charles, A. L., Huang, T. C., Lai, P. Y., Chen, C. C., Lee, P. P., & Chang, Y. H. (2006). Study of wheat flour-cassava starch composite mix and the function of cassava mucilage in Chinese noodles. *Food Hydrocolloids.* 21, 368-378.

[6] Charles, Albert Linton; Chang, Cheng-Kai; Wu, Mei-Li; Huang, Tzou-Chi, 2007. Studies on the expression of liver detoxifying enzymes in rats fed seaweed (Monostroma nitidum). *Food and Chemical Toxicology,* 45, 2390-2396.

[7] Charles, A.L., Huang, T.C., & Chang, Y.H. 2008. Structural analysis and characterization of a mucopolysaccharide isolated from roots of Cassava (Manihot esculenta Crantz L.). *Food hydrocolloids,* 22, 184-191.

[8] Charles, A.L., & Huang, T.C. 2009. Sweet cassava polysaccharide extracts protects against CCl4 liver injury in Wistar rats. *Food hydrocolloids,* 23, 1494-1500. Cock, J. H., & Reyes, J. A. (1985). Cassava: research, production and utilization. UNDP, CIAT.

[9] Du, H., Simpson, R. J., Clarke, A. E., & Bacic, A. (1996). Molecular characterization of a stigma-specific gene encoding an arabinogalactan-protein (AGP) from Nicotiana alata. *The Plant Journal,* 9, 313-323.

[10] Duvivier, P., Hsieh, P.-C., Lai, P.-Y., & Charles A. L. 2008. Evaluation of drying methods on antioxidant activity, total phenolic and Total carotenoid contents of Sweet Potato (Ipomoea batatas (L.) Lam.) var. Tainong 73. *Journal of International Cooperation (JIC)* 3, 73-86.

[11] Jiang, G., & Ramsden, L. (1999). Characterisation and yield of the arabinogalactan-protein mucilage of taro corms. *Journal of the Science of Food and Agriculture,* 79, 671-674.

[12] Lin, H., & Huang, A. S. (1993). Chemical composition and some physical properties of a water-soluble gum in taro (Colocasia esculenta). *Food Chemistry*, 48, 403-409.

[13] Mao W. J., Zang, X.X., & Zhang, H.J. 2006. Sulfated polysaccharides from marine green algae Ulva conglobata and their anticoagulant activity. *Journal of Applied Phycollogy*, 18, 349–354.

[14] Morita, K., & Nakano, T., 2002. Seaweed accelerates the excretion of dioxin stored in rats. *Journal of Agriculture and Food Chemistry,* 50, 910-917.

[15] Osborn, H. M. I., Lochney, F., Mosely, L., & Read, D. (1999). Analysis of polysaccharides and monosaccharides in the root mucilage of maize (Zea mays) by gas chromatography. *Journal of Chromatography,* 831, 267-276.

[16] Osseni, R.A., Rat, P., Bogdan, A., Warnet, J.M., & Touitou, Y. 2000. Evidence of pro-oxidant and antioxidant action of melatonin on human liver cell line HepG2. *Life Science*, 68, 387-399.

[17] Samuelson, A. B., Paulsen, B. S., Wold, J. K., Knutsen, S H., & Yamada, H. (1998). Characterization of a biologically active arabinogalactan from the leaves of Plantago major L. *Carbohydrate Polymers,* 35, 145-153.

[18] Santisopasri, V., Kurotjanawong, K., Chotineeranat, S., Piyachomkwan, K., Sriroth, K., & Oates, C. G. (2000). Impact of water stress on yield and quality of cassava starch. Elsevier, 1-15.

[19] Smit, H.F., Woedenbag, H.J., Singh, R.H., Meulenbeld, G.J., Labadie, R.P., Zwaving, J.H. 1995. Ayurvedic herbal drugs with possible cytostatic activity. *Journal of Ethnopharmarcology,* 47, 75-84.

[20] Uhumwangho, M.U., Okor, R, S., & Ayomanor, M. (2005). Influence of mucilage viscosity on the globule structure and stability of certain starch emulsions. *Online Journal of Health Allied Sciences,* 13, 1-6.

[21] Western, T. L., Skinner, D. J., & Haughn, G. W. (2000). Differentiation of mucilage secretory cells of the arabidopsis seed coat. *Plant Physiology,* 122, 345-355.

[22] Wolf, C.R., 2001. Chemoprevention: Increased potential to bear fruit. *Proceedings of the National Academy Sciences,* 98, 2941-2943.

In: Food Hydrocolloids: Characteristics, Properties... ISBN: 978-1-60876-222-4
Editor: C. S. Hollingworth, pp. 205-224

Chapter 7

CHEMISTRY, STRUCTURE & ARCHITECTURE, SYNERGY & BIODIVERSITY OF GALACTOMANNANS

Ganesharanee Ravindran, Lara Matia-Merino and Ghulam M.U. Paracha
Institute of Food, Nutrition and Human Health, Massey University, Private Bag 11 222, Palmerston North 4442, New Zealand

1.1. ABSTRACT

Present knowledge on the chemistry, structure and architecture, and the functionality and health benefits of four commonly used seed galactomannans, namely, guar gum, locust bean gum, tara gum and fenugreek gum is reviewed to provide directions for research needed to improve the utilization of these gums as functional ingredients in the food industry.

1.2. INTRODUCTION

Galactomannans (GM) are a family of water soluble/dispersible, non-starch, non-cellulosic, high molecular weight, hetero-polysaccharides that are plant-derived and widely distributed in nature. They are generally classified as hydrocolloids or gums. Hydrocolloids that are commonly used in foods are many and diverse in its constituent members. They embrace many polysaccharides that are extracted from plants and plant exudates. These include galactomannans (guar gum, locust bean gum and tara gum), pectin, β-glucan, gum exudates (arabic, ghatti, karaya and tragacanth); sea weeds (agar, algin and carrageenan); microbial sources such as, xanthan, and, to a much lesser extent, more recently approved gellan gum. They also include physically or chemically or enzymatically modified biopolymers of starch and cellulose. These hydrocolloids are most often used as functional ingredients to improve or control the textural, stability and other physical attributes of processed food products. Each hydrocolloid has its own special properties that are reflective of its unique molecular structure. A common feature of these hydrocolloids is that they are all hydrophilic, high molecular weight biopolymers. It is the molecular inter-chain associations and entanglements of these molecules that enable thickening, water holding, viscosifying, gelling, stabilizing emulsions and dispersions, and other desired textural properties in foods. They conserve the appearance of foods by virtue of these characteristics. Nutritionally, being plant-based, environmentally friendly and naturally occurring biopolymers, gums are considered 'GRAS' (generally recognised as safe).

For many years, galactomannans have been used as hydrocolloids in different food applications, in particular, as structurizers and texturizers through hydration, water holding, thickening or viscosity building, aggregation and gelation behavior. In recent years, however, they have attracted considerable scientific and commercial interest not due to their versatility as functional ingredients to control texture, but also for their possible health benefits as an important source of dietary fiber. More recently, they have been used in foods such as breads and pasta to manipulate their potential glycemic response. The supplementation of soluble fiber through bread has been found to be more effective than powder form in lowering blood glucose levels (Glore & Van Treek, 1994). In addition to food and pet food industries, they have applications in paper, textile, cosmetic and pharmaceutical industries mainly due to their ability to manipulate the flow behavior of aqueous systems. Vendruscolo *et al.*, (2005) have reported the galactomannan to be a potential excipient for oral controlled drug release.

Amongst the galactomannans identified from the plant world, the two that are approved as food additives, are guar gum (GG) and locust bean gum (LBG). Another recent source of water soluble galactomannan that is gaining research attention is fenugreek gum (FG), extracted from the seeds of the legume, fenugreek (*Trigonella foenum-graecum* L.), known for its medicinal benefits. Tara (*Cesalpinia spinosa*) galactomannan, is yet another approved additive gum, but industrially it is less known and also minimally researched compared to GG and LBG. The commercial importance of the several other sources of galactomannans is reported in the literature (Srivastava & Kapoor, 2005), but these are yet to be explored.

The primary aim of this chapter is to critically review and compare the current literature on the chemistry, molecular structure, functions (technological and physiological), synergy or incompatibility (interactions) of the industrially emerging, new (FG) versus traditional galactomannans (GG and LBG) with each other or with other food components (starch, proteins, lipids, minerals, acids, etc.), with focus on actual and potential commercial applications. Tara gum is also a relatively new commodity industrially and therefore for food applications, compared to GG and LBG.

1.3. Sources of Galactomannans

Galactomannans are plant-based reserve polysaccharides sourced predominantly from the members of the family leguminosae (Srivastava & Kapoor, 2005), with microbial sources (yeast and fungi) a close second. Functional industrial gums are usually obtained from the endosperm of mature seeds, where they are present as thickened and enlarged deposits of cell wall storage components and energy reserves which are mobilized during germination. In seeds, aside from being a substrate reserve, galactomannan acts as an osmoprotectant. It retains water and plays an important physiological function by protecting the embryo from desiccation which would lead to protein denaturation and consequently dysfunction of enzymes essential for seed germination. The galactomannans are generally hard and compact in the native state, but when soaked, they imbibe significant amounts of water and become soft and mucilaginous.

Guar, obtained from the seeds of *Cyamopsis tetragonaloba* (L.) TAUB, is currently the foremost seed gum industrially used worldwide. The specific galactomannan component of guar gum is called guaran. The plant is cultivated mainly in India and Pakistan. Locust bean gum (also called carob gum) is obtained from the seeds of carob tree (*Ceratonia siliqua* L.) is also used widely in foods. Mediterranean and Middle East countries such as Italy, Spain and Cyprus are the major producers of this gum. Tara gum, also called Peruvian carob is obtained from the seeds of *Caesalpinia spinosa* Kuntze, a spiny evergreen shrub grown in South American countries (Bolivia, Northern Chile, Peru , and also Colombia, Chile, Ecuador and Venezuela), with Peru being the major producer for tanning purposes.

In fenugreek, the endosperm is composed almost entirely of galactomannan which accounts for about 30% of the dry weight of the seed. After full hydration of the seed, the endosperm contains about 70% of the imbibed water and the water buffering action of the endosperm has been attributed entirely to the properties of the galactomannan (Edwards et al., 1992). Currently, isolated fenugreek polysaccharide (galactomannan) products are commercially available under the trade names of FenuLife®, FenuPure® and Fenu fiber®.

Galactomannans as food hydrocolloids are usually processed from seed sources in a relatively pure form. The seeds are industrially processed by kibbling, milling and sifting operations to isolate the endosperms, which are ground or pulverised to a flour of desired fineness. The flour thus obtained is called crude gum or commercial grade gum and is reported to contain about 80- 85% pure galactomannan on a weight basis and make an important contribution to the hydrocolloid industry. The remainder of the galactomannan constitutes of 5- 6 % protein, 1-2% ash and 1-6% cellulose which are inherently associated with the polysaccharide. The commercial products are specified according to viscosity, protein content, acid insoluble residue (an indication of the residual hull content) and particle size distribution. Further purification involves dissolving in water and fractional precipitation with polar solvents, such as ethanol or isopropanol to remove impurities and much finer purification employs forming complexes with Cu^{++} or Ba^{++} salts and dialysis, and membrane filtration.

1.4. Composition, Structure and Architecture of Galactomannans

Mannans that contain more than 5% galactose are generally classified as galactomannans, but mannans without galactosyl substitution are rather rare (Aspinall, 1959). The functional polysaccharide have the fundamental molecular structure composed of a linear, poly (1 → 4) -β- linked backbone of D-mannopyranose units (similar to poly (1 → 4)-β -D-glucopyranose backbone in cellulose) with a statistical distribution of single α- D-galactopyranose side groups grafted on the main backbone *via* (1 → 6) -α-glycosidic linkages at varying degrees of substitution depending on the source.

Galactomannans from various plant sources exhibit diverse molecular structures. They can be distinguished from each other by their overall mannose:galactose ratios, but one of the most important characteristics of the fine molecular structure of galactomannan polysaccharide family is the degree of galactose substitution (DS_{Gal}), which confers hydrophilicity to the molecule and without which the isolated polysaccharide has no impact on the commercial functionality. The polydiversity in structure is also due to the patterning of these galactose groups (symmetric, random, and block), chain length or degree of polymerization (DP) and other steric disturbances due to inter- and intra- molecular associations, which are related directly to the mechanism of biosynthesis. These structural variations in turn have a major impact on the nature of the gum. On an ascending order, the molar ratios of galactose to mannose of FG, GG, TG and LBG are1:1, 1:2, 1:3 and 1:4, respectively.

The common chemical structure of an average theoretical building block of the different galactomannans and that of fenugreek gum is shown in Figures 1 and 2, respectively. Of the four gums, the galactomannan of LBG has the fewest branch units and its structure is more irregular compared to the other three. Fenugreek gum is unique in having a saturated ratio of galactose/mannose and, is the richest in galactose and the largest in size. In both GG and FG, the galactosyl monomer units are fairly evenly placed along the backbone. Ramesh *et al.* (2001) have employed integrated approaches of 2D NMR correlations to support the substitutional pattern in FG.

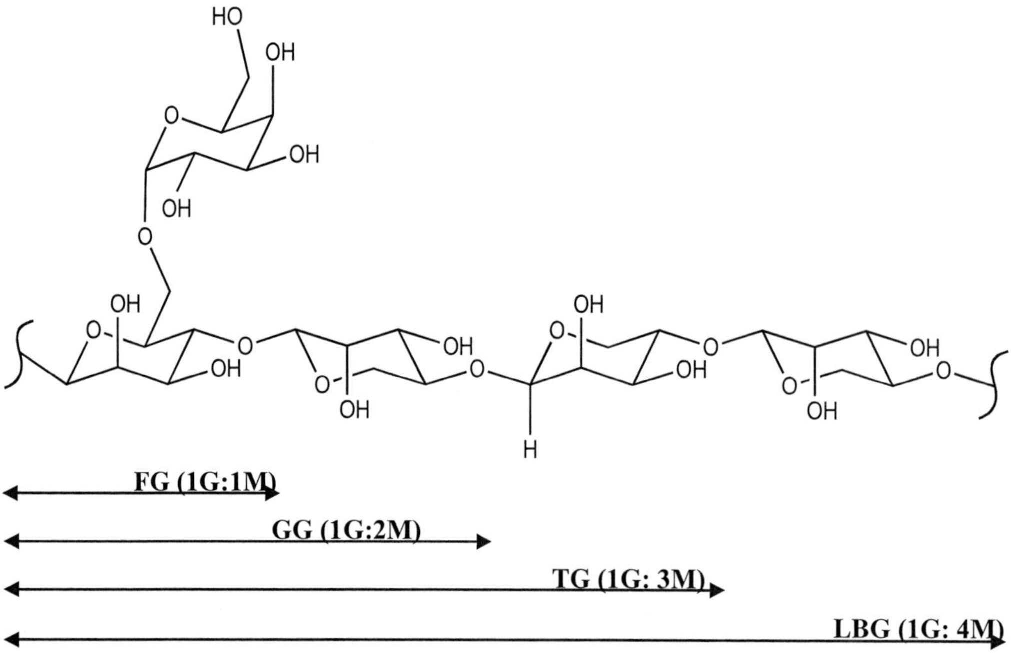

Figure 1. Representative repeating segments of fenugreek, guar, tara and locust bean galactomannans.

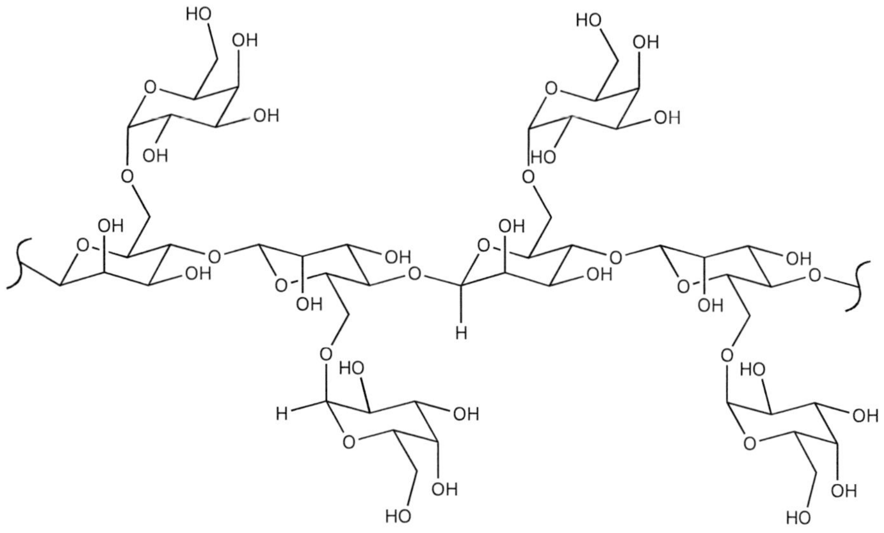

Figure 2. Repeating segment of fenugreek galactomannan.

While these ratios are predominantly regular in the fine structure of these galactomannans, variations in terms of galactose distribution as well as in the mannose backbone are reported with 5-11 unsubstituted mannosyl units in a row for GG and many more for LBG in addition to block condensation of galactosyl units (Daas *et al.*, 2002; McCleary, 1985). Tara gum and LBG are reported to have both random and blockwise distribution of galatose side groups, while the guar gum possesses a block-wise distribution and FG, a very regular distribution. Such variations in the mannose:galactose ratio in galactomannan depend on the source of seeds, cultivation and harvesting conditions and, methods of isolation and purification. Song *et al.* (1989) reported a galactose-mannose ratio of 0.93 and found FG to be physically more regular, exhibiting markedly less loss of crystallinity upon drying when compared to locust, guar or tara gums.

Edwards *et al.* (1992) investigated the mannose/galactose ratio during galactomannan formation in three types of legume seeds, namely, guar, fenugreek and senna, and related the differences in the DS_{Gal} to a cause-and-effect relationship between the relative bioactivities of mannosyl- and galactosyltransferases, the deposition of GM during biosynthesis; and involvement of α - galactosidase in galactose removal during late GM deposition. In other words, D_{Gal} have been attributed to the presence of differing ratios of galactosyl to mannosyl transferases between species and also post-deposition modifications (Edwards *et al.*, 1992). In addition, metabolic degradation, environmental variations and methods of extraction and purification also contribute to the proportional variations of galactose to mannose in seed galactomannans (Reid & Meier, 1970; Gidley & Reid, 2006).

The average molecular weight of these galactomannans varies significantly, typically ranging from 0.2 to 3 million. In aqueous solutions, galactomannans predominantly exist in a semi-flexible random-coil conformation and a ribbon like flat structure in the solid state (Petkowicz, *et al.*, 1998). The galactomannans of GG, LBG and FG are non-ionic, *i.e.*, exhibit a neutral charge. However, the hydroxyl groups can be derivatised to provide non-ionic, anionic (e.g., carboxylate, sulphate groups), cationic (e.g., amino groups) and amphoteric derivatives. All of these structural features contribute to their differences in physical and functional properties, such as solubility; synergy or incompatibility with each other or with other ingredients; thickening, gelling, emulsifying, encapsulating and similar properties.

Galactomannans can also be modified physically or chemically in order to extend functionality or to meet a specific challenge. Each of the galactose side units have four –OH groups, the unsubstituted mannose unit, three –OH groups and the substituted mannose, two secondary –OH groups. These hydroxyl groups show practically the same reactivity and enable synergistic interactions with other components.

1.5. Functional Interactions with Food Components

All gums must be hydrated to function in a food system; they are known as hydrocolloids, with 'hydro' meaning water and 'colloid' referring to the dispersion of small particles in another medium. Hydrated gums occupy a large hydrodynamic volume in aqueous medium, can form structure with water, thereby allowing other ingredients in a food system to be dispersed and suspended.

Within the large family of hydrocolloids, large differences in functional properties exist between various food biopolymers depending on the detailed chemical structure and sensitivity to solution conditions (temperature, pH, ionic strength, co-solutes or specific ions). The functionality of an individual polymer in foods is also affected by its interactions with other food components – proteins, starch and non-starch polysaccharides, lipids, sugars and salts.

The functionality of the galactomannans is mainly due to their ability to change the rheological aspects of aqueous systems. They can create a wide range of rheologies, depending on the choice of single gum or the synergistic blend of more than one gum. All galactomannans are efficient thickening agents, if dissolved in water, and are able to interact with other hydrocolloids such as, agar-agar, carrageenans and xanthan gum to form and/or reinforce three-dimensional stable structures. The thickening power of the galactomannans depends upon the size or the length and association of the macromolecules and thus on their apparent molecular weights. These thickening and gelling agents are widely used in food products, mainly to make them appealing and attractive to the consumer and to improve their shelf life by binding water, to control texture, to inhibit ice and crystal formation, to prevent creaming or settling, to improve the freeze-thaw behaviour, to encapsulate and control release flavors, to prevent syneresis, to prevent the retrogradation of starch products, to form films, to maintain turbidity in soft drinks and juices, and as soluble dietary fibers. This is reflected in their versatile applications as functional ingredients in many food products.

1.5.1. Galactomannan and Water

1.5.1.1. Solubility

Pure mannan without galactose substitution is completely insoluble in water, similar to cellulose, due to chain association. Substitution of the mannan chain by more than 12% wt. of galactose makes the galactomannan water soluble (Reid & Edwards, 1995). Increasing galactose content increases the solubility of the polymer by allowing it to become extended. If the galactose content is reduced to less than 12%, the final product becomes insoluble even in hot water, highlighting the crucial role of the hydrophilic nature of the galactosyl side groups in determining the ease with which a galactomannan could be dissolved.

A possible explanation lies in their structure, especially the galactose/mannose ratios. Of the three gums, FG has the largest amount of galactose side groups and LBG, the lowest. Higher the galactose substitution, lesser is the inter-chain association due to steric hindrance and consequently, the amount of crystallization. In other words, large amounts of galactose side units prevent strong cohesion of the main backbone, so that no extensive crystalline regions can be formed. Water can easily penetrate between the molecules to hydrate or dissolve the accessible gum. Thus FG and GG swell and disperse almost completely in both cold and hot water, whereas LBG needs to be heated for complete solubility. Tara gum is 70% soluble in cold water, but mild heating is required to disrupt aggregation and full dissolution.

In LBG, due to low substitution of galactose, tiny crystalline regions alternating with much bigger amorphous regions are formed which makes its hydration and solubility properties different. This also enables LBG to intra- or inter-associate with other gums under defined conditions to improve textural characteristics. The minute crystalline regions can be better dissociated by hot water, with concentrations of up to 5% w/v being possible.

1.5.1.2. Rheology

Much of the interest in galactomannans is due to their unique property to provide aqueous solutions of very high viscosities at low concentrations. The rheological behaviour of galactomannans is characteristic of random coil polysaccharides. The viscosity of galactomannans is shear–dependent at high concentrations, while Newtonian behaviour has been observed at low concentrations. Solutions at 1% of FG, LBG and GG show obvious shear-thinning (*i.e.*, reduction in viscosity with increasing shear rate) after a "Newtonian plateau" region of constant viscosity at low shear rates (Brummer *et al.*, 2003). These authors have also reported weaker rheological properties of FG compared to LBG and GG. Additionally, TG has been reported to have a similar rheological behaviour to LBG (Sittikijyothin *et al.*, 2005). For each galactomannan, a range of rheologies can be obtained depending on sucrose concentration and thermal history (Richardson & Norton, 1998). In general, the viscosity of galactomanans tends to remain constant over a broad pH 4–9 although degradation can occur at extreme acidic and alkaline conditions, especially above pH 11 and at high temperatures (Srivastava & Kapoor, 2005).

Dynamic measurements on galactomannans, at relatively high concentrations, show a liquid-like behaviour at low frequencies and a solid-like behaviour at higher frequencies. The 'cross-over frequency' ($G'=G''$) is typically moved to lower frequency values when the concentration is increased. The change on the mechanical spectra of these galactomannans shows the transition from dilute to semi-dilute behavior with increasing gum concentration (Cui *et al.*, 2007).

Variations in the bulk rheology of polysaccharides tend to reflect differences at molecular level. Intrinsic viscosity and radius of gyration decrease in the following order LBG > GG > FG, with the molecular weights (Mw) increasing in the same order (Brummer *et al.*, 2003). This is consistent with the fact that the introduction of galactosyl residues on the mannan backbone induces a lowering in chain dimension, thus the intrinsic viscosity is reduced (Petkowicz *et al.*, 1999) .

The transition from a dilute solution of individual coils to an entangled network, as polymer concentration is increased, occurs at lower degree of space occupancy (or coil overlap parameter, $c[\eta] \approx 2.5$) for GG, LBG and FG compared to other random coil polysaccharides ($c[\eta] \approx 4$). The log plots of 'zero-shear' specific viscosity versus $c[\eta]$ also show a steeper slope in the semi-dilute region (when neighbouring coils interpenetrate to form an entangled network) for these galactomannans. This deviation from the 'normal' concentration–dependence behaviour has been attributed to association ('hyperentanglement') of unsubstituted mannan sequences in the galactomannan chains. This original concept has recently being refined based on viscosity studies on FG (Doyle *et al.*, 2009) where a perpendicular association to the direction envisaged initially has been hypothesized. This is, in the semi-dilute region, galactomannan hyper-entanglements are formed but not necessarily involving unsubstituted regions of mannan backbone as previously believed.

1.5.1.3. Gelation

Galactomannans, in general, are non-gelling due to their inability to form a three-dimensional network that is essential for gel formation. Because GG and FG have their galactosyl units fairly evenly placed along the chain, there are few locations on the chains that are galactose-free and suitable for formation of junction zones. However, LBG with its long exposed smooth chain segments can synergistically interact with the helical sections of other hydrocolloids such as xanthan, to form junction zones and rigid gels. LBG can also form cohesive gels with κ-carrageenan, but by a different mechanism involving mainly a random aggregation with the surface of carrageenan crystallites and to a much lesser extent by intermolecular binding (Cairns, 1987; Morris *et al.*, 1998). TG also exhibits synergistic interactions with xanthan and carrageenan in a similar manner to LBG and GG. The trend appears to be related to the galactose substitution. The higher the galactose content, the more pronounced are the synergistic interactions and less is the ability to form gels on its own. In line with this trend, FG is a non-gelling galactomannan which exhibit little synergistic interaction with other gums, but shows resistance against freeze-thaw cycles like GG. On the contrary, LBG and other galactomannans of low galactose content form 'cryo-gels' on freezing and thawing and also under prolonged storage at low temperatures (Dea *et al.*, 1977). The presence of other co-solutes like sucrose can also modify the gelling behaviour of such galactomannans (Richardson & Norton, 1998)

Guar gum can form a gel at ambient alkaline conditions in the presence of borax. The borate ion reacts with the hydroxyl groups at the 2- and 3- positions of mannose and those at the 3- and 4- positions of the galactose side stubs (Kesavan & Prudhomme, 1992).

1.5.2. Galactomannan and Polysaccharides

As stated above, it is generally accepted that unsubstituted mannan regions in galactomannan chains can bind to bacterial polysaccharides, such as xanthan, and to aggregated double helices of algal polysaccharides in the agar/carrageenan series (Morris, 1995). The synergy between LBG and these polysaccharides results in gels that are elastic, very cohesive, and relatively free from syneresis. It is also known that GG does not gel with xanthan and FG has shown little interaction with other gums (Cui *et al.*, 2007). In general, galactomannan gums with lower degrees of substitution have stronger interactive properties than higher-galactose gums. Yet differences have been observed between galactomannans with identical degrees of substitution. It seems that synergistic interactions are favored by a higher frequency of longer stretches of contiguous, unsubstituted mannose residues (Dea *et al.*, 1986).

Studies of the interaction between mannan-based polysaccharides and bacterial cellulose have shown that a wide range of composite microstructures can be generated, reflecting a diversity of interaction mechanisms (Whitney *et al.*, 1998) . For example, solutions of cellulose and GG can form interpenetrating networks containing cellulose fibres and an aggregated GG network.

Galactomannans also interact with mucilages such as yellow mustard mucilage, YMM (Cui *et al.*, 2006). Synergistic interactions between this mucilage and galactomannans resulting in an increase in viscosity and gel formation are also strongly dependent on the molar ratio of galactose to mannose (LBG > GG > FG), similar to the interactions of galactomannans with xanthan (Bresolin *et al.*, 1999; Casas & Garcia-Ochoa, 1999).

1.5.2.1. Interactions with Starch

Galactomannans have been used as starch modifiers to improve the texture and stability of starch-based products. Such products depend on starch as the primary component for viscosity; but galactomannans occupy the aqueous phase of gelatinized starch and have the potential to synergistically improve the overall pasting, textural and viscoelastic properties. There are many reports on the effects of GG and LBG on the gelatinization, retrogradation and rheological behaviour of starch (Alloncle & Doublier, 1991; Funami *et al.*, 2005; Kim *et al.*, 2005). However, the interactions of starch and FG galactomannan are less studied.

Only recently, FG has been reported to possibly reduce the amount of amylose leaching, granule swelling, and accelerate retrogradation in soup systems containing corn starch and FG (Ravindran & Matia-Merino, 2009). Moreover, the functions of FG on the gelatinization and retrogradation behaviours of corn starch depend on the coil overlap parameter $c[\eta]$ or weight-average molecular weight *Mw* of FG (Funami *et al.*, 2008).

Previous studies on fenugreek–supplemented wheat flour have shown that the incorporation of FG results in less cohesive and elastic doughs with higher degree of stability (Hooda & Jood, 2003). A synergistic effect has been reported also in doughs (Brennan *et al.*, 2006) where the inclusion of FG can significantly increase the pasting properties of wheat flour or starch mixes. Furthermore, in soup systems formulated with corn starch and FG, significant improvements in viscoelasticity have been reported (Ravindran & Matia-Merino, 2009). Figure 3 shows an example of the synergistic effect between corn starch and FG in these systems. When starch and FG are added together, the enhancement on the viscosity is much greater than when the two ingredients are added individually.

In general, the thickening ability of galactomannans and/or possible molecular interactions between the galactomannans and amylose and/or amylopectin components can explain the effects on the gelatinization, retrogradation behaviour and final viscoelasticity of starch-based products (Alloncle & Doublier, 1991; Funami *et al.*, 2005). Other factors which can contribute to the final effect is that galactomannans reduce the amount of water available for starch and also can induce phase separation due to thermodynamic incompatibility. An incipient microphase separation or segregation of components could lead to an increase in the effective concentration of both polymers. Regardless of the mechanism involved, a synergistic effect has been detected.

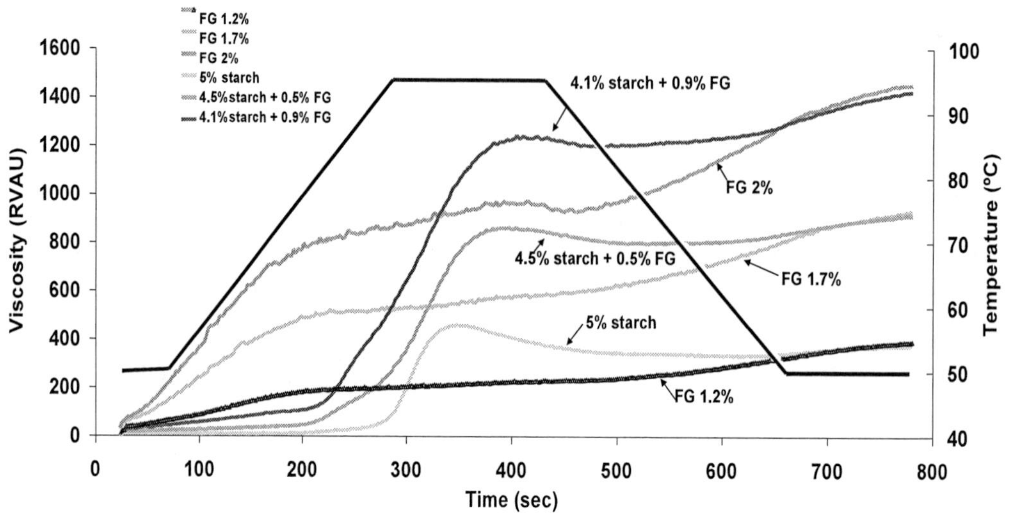

Figure 3. Pasting curves of FG soup formulations with and without starch.

Phase separation is concentration-dependent of both species (starch and galactomannan), and is related *inter alia* to the extent of starch granule disintegration, amylopectin content, molecular weight of each component and charge of the polysaccharide added (Annable, *et al.*, 1994; Closs, *et al.*, 1999). Phase separation has also been detected in sago starch and galactomannans such as GG and LBG (Ahmad & Williams, 2001) and between amylopectin or waxy maize starch and purified GG (Closs *et al.*, 1999).

1.5.3. Galactomannans and Lipids

Hydrocolloid gums do not generally exhibit significant surface activity in comparison to proteins due to their pronounced hydrophilic nature. It is generally accepted that hydrocolloids are not true emulsifiers, since they do not actively adsorb onto liquid surfaces and the stability that they impart is mainly via their rheological effects on the continuous aqueous phase, *i.e.*, increasing viscosity and decreasing mobility. However, galactomannans such as GG and FG are reported to exhibit sufficient hydrophobic character to adsorb strongly at the oil-water interface, allowing certain functional groups to re-orient at the oil-water interface and anchor into the oil (Garti & Reichman, 1993; Dickinson *et al.*, 2002), thereby having the capacity for making and stabilising oil-in-water emulsions.

The interfacial activity of galactomannans was initially ascribed to the presence of residual surface-active impurities in the crude gum (Gaonkar, 1991; Anderson, 1986). The role of residual proteins in crude galactomannans is not yet fully understood. Studies with several gum purification steps to obtain almost protein-free polysaccharides have demonstrated that no significant improvement in emulsion stability is found upon using a purified gum over the "protein-rich" gums (Garti *et al.*, 1997; Garti & Leser, 2001). These authors have classified such galactomannan gums LBG, GG and FG as a new category of natural-occurring hydrocolloid-emulsifiers of importance for emulsion technologists.

Recent studies on model emulsions comparing the stabilizing effect of GG and LBG with other polysaccharides are available on oil-in-water beverage emulsions (Mikkonen *et al.*, 2009) as well as on mayonnaise-like systems (Hayati *et al.*, 2009).

Compared to LBG and GG, FG seems to have extremely good emulsification properties, with most oil droplets at 0.3% gum concentration being in the range of 2–5 μm (Garti *et al.*, 1997). The adsorption load for FG (18 $mg.m^{-2}$) is characteristic of amphiphilic biopolymers and is achieved at much lower gum concentrations compared to LBG and GG which show lower surface loads (2.8 and 4 $mg.m^{-2}$, respectively). Compared to other hydrocolloids such as propylene glycol alginate, gellan, gum arabic and micro-crystalline cellulose, FG has been shown to have the best potential as a food emulsifier (Huang *et al.*, 2001).

1.5.4. Galactomannans and Proteins

Various proteins have been conjugated with galactomannans through Maillard-type glycosylation reactions and the emulsifying properties and surface activity of the complexes have been studied; complexes from bovine serum albumin–galactomannan conjugates (Kim *et al.*, 2003) to more recent studies on tofu whey–galactomannans conjugates (Matemu *et al.*, 2009). Whey protein isolate–fenugreek conjugates have also been tested on the stability of double emulsions (o/w/o) (Benichou *et al.*, 2007). However, these conjugates were found to be unstable due to fenugreek gum being non-ionic and less interactive with the charged residues in the protein molecule compared to xanthan gum.

The phase behaviour of mixed proteins and galactomanans is often studied in solution as well as in emulsions. Segregative phase separation has been observed in various systems such as sodium caseinate–GG aqueous solutions (Neirynck *et al.*, 2007) with references to possible synergistic interactions between caseins and galactomannans (Perissutti *et al.*, 2002) Formation of cloud-like flocs has been detected in mixtures of gelatine with LBG or GG (Alves *et al.*, 2001; Harrington & Morris, 2009) attributed to association of unsubstituted 'smooth' regions of the mannan backbone in response to segregative interactions with gelatine. However, FG seems to behave in a different manner which may be attributed to the lack of smooth regions.

1.6. Commercial Applications in Foods

Guar gum and LBG are the most extensively used galactomannans due to its availability, low cost and functional versatility. They provide multiple applications as functional ingredients in many foods to improve textural, structural, flavour and shelf-life characteristics, and as dietary fibers (Table 1). They are also used in pet foods, pharmaceuticals and cosmetic products to obtain the same attributes. They are generally used in the range of 0.05-0.50% in order to obtain the desired 'mouth feel' characteristics. In extruded products, however, the galactomannans could be incorporated into cereal-legume mixtures to much higher levels; 5-15% to develop breakfast cereals or snack products of acceptable quality resulting in significant addition of fiber and low glycemic index (Gamlath & Ravindran, 2009; Ravindran et al., 2010). Higher levels in many other products affect the palatability due to excessive thickening caused by high viscosity. However, levels of addition could be improved if the galactomannans are enzymatically depolymerised and a partially hydrolysed guar gum (Benefiber®) is currently available. Such gum products could be administered into most foods without compromising sensory appeal.

Fenugreek gum, GG, TG and LBG are compatible with most other food gums, with the former two known for its synergistic increase in viscosity and the latter for gelation purposes. LBG and TG are rarely used alone, rather used in combination with agar, carrageenans, carboxy methyl cellulose and xanthans, and when used in right concentrations and proportions, a synergistic interaction takes place, for example, as in ice cream to result in a smooth, uniform product resistant to melting and handling. About 85% of the LBG is used in dairy and frozen dessert products, where it is used at levels of 0.05 – 0.25% in combination with κ-carrageenan and xanthan to take advantage of the synergistic gel-forming properties. LBG is also popular as a chocolate substitute, known as carob chocolate.

LBG and GG are often added to ice cream mixes to slow down ice crystal growth, therefore helping to maintain a smooth texture during constant and fluctuating temperatures (Bolliger *et al.*, 2000; Patmore *et al.*, 2003). Again both galactose content and distribution dictates the effectiveness of a galactomannan against recrystallization. FG has also been reported to be a very efficient stabilizer in ice creams (Balyan *et al.*, 2001). Another useful functionality that is of technological significance is the fact that GG and FG, in contrast to TG and LBG, are tolerant to repeated 'freeze-thaw cycles' and prevent large ice crystal formation during freezing.

Addition of GG and FG to the bread has been reported to lower the glycemic index. Inclusion of these gums does not compromise the elasticity or extensibility or the water retention properties of the gluten in wheat flour dough. Tara gum is the least exploited functional food ingredient amongst the galactomannans, but with a galactose/mannose ratio in the middle of GG and LBG, it could be expected to exhibit intermediate functionality and can be used as an alternative to these galactomannans.

1.7. Physiological Effects and Health Aspects

A review of the literature reveals no reports of clinically significant harmful effects by these gums. As non-starch polysaccharides, galactomannans contribute significantly to the intake of dietary fiber, a necessary supplement to any diet. Fiber is considered to be beneficial for lessening the chance of developing diverticulitis, hemorrhoids and colo-rectal cancer principally by minimising the colonic transit time of fecal mass and improving regularity. There are other health benefits associated with ingestion of fibers, especially soluble fibers such as FG. By creating a high level of viscosity in the gut, they limit the access to nutrients and delay the break down of foods and the digestion of nutrients, eventually resulting in slower absorption of nutrients. Such a delay in gastric emptying and diminished post-prandial glucose release, suppress appetite, improve satiety, limit overconsumption of calories and aid

Table 1. Commercial applications of galactomannans in foods

Food product	Galactomannan	Function and application
Dairy products	LBG	To improve the smoothness and uniformity of the body, resistance to melting and handling, to prevent ice crystal growth and whey separation in ice cream.
	LBG, GG	To impart a smooth texture, extensibility and reduce syneresis in processed cheeses, cheese spreads
	LBG	To control syneresis in a variety of gels
	GG & LBG	To add fiber to yoghurt
	GG	To retain moisture and enhance flavour
Diabetic products	FG, GG, LBG	To act as dietary fiber; to reduce calories; to lower the glycemic response in breads cereal bars and other bakery products, snack products, milk formulations
Chocolate products	LBG	As a fat substitute
Convenience and snack food	FG, GG, LBG	To lower the glycemic index (GI); to improve overall mouth feel characteristics
Canned foods	GG	To thicken and stabilise
Extruded products	FG, GG, LBG	To add fiber ; to lower GI; to improve texture
Bakery products	GG, FG	To lower the glycaemic response; improve moisture retention, dough elasticity and bread structure;
Frozen and dessert products	GG, LBG	To form a variety of gels; to tolerate freeze-thaw cycles and control syneresis; as a fat substitute in sorbet.
Fruit products	GG	To thicken and stabilise pie fillings; to prevent syneresis in jellies, low-sugar jams and conserves; to maintain turbidity in juices and drinks
Meat products	LBG, GG	To provide binding, thickening and lubrication to comminuted meat products, canned meat
Soups, sauces and salad dressings	GG	To prevent creaming or settling; to obtain good mouth feel
Starch-based products	GG	To obtain a synergistic increase in viscosity and to accelerate retrogradation
Pet-foods	All	To provide economical thickening
Powdered products	All	To disperse and stabilise dessert pudding powders, coffee whiteners
Low-calorie products	GG, FG	To reduce the energy level of high calorie products

in weight management and overall health. Fiber-rich foods could also increase secretion of the anorexigenic and insulinotropic hormone, glucagon-like peptide-1 (GLP-1) to improve glucose tolerance and reduce weight gain (Reimer & Russell, 2008). Another favorable effectof the fibers is as a result of their fermentation by the colonic microflora to produce beneficial short chain fatty acids. These acids contribute to colonic and cardiac health through well characterised biochemical pathways (Wong *et al.*, 2006).

Soluble fibers are also experimentally documented for their hypolipidemic potential. In a number of animal and clinical trials, they have been shown to effectively reduce cholesterol levels, in particular, the low-density lipoprotein-cholesterol and triglycerides, highlighting theirs potential to reduce risk of heart attack. Soluble and insoluble fibers appear to be protective against high C-reactive protein levels in liver which are implicated to be the latest markers for the onset of diabetes, cardiac and other inflammatory diseases like arthritis (Kumar & Maliakel, 2008). The functional, physiological, and multiple health effects of fenugreek have recently been reviewed (Srinivasan, 2006).

Beneficial physiological effects of galactomannans (FG, GG and LBG) in reducing cholesterol and post-prandial glucose levels are relatively well established. However, of the various sources of these gums, fenugreek is the most recognised in terms of its long history of therapeutic use in traditional medicine for a wide array of conditions, including, diabetes, diarrhea, gastric stimulation, bronchitis, sore throat, anorexia, to increase lactation, constipation, to reduce inflammation and heal wounds, ulcers and eczema (Srinivasan, 2006). It is most commonly used as an expectorant in veterinary practice.

Potential anticancer benefits of FG are currently being explored, but inhibitory effects of fenugreek seed powder on colon carcinogenesis have been reported and attributed to suppressed activities of carcinogen-liberating enzymes, β-glucuronidase, and mucinase, which breaks down the protective mucin and exposes the intestinal epithelial cells to carcinogens (Devasena, & Menon, 2003).

Clinical studies in support of many of these claims, however, are limited except for diabetes where the potential benefit of FG in modulating carbohydrate and ameliorating altered lipid metabolism during diabetes are well recorded. Studies have been carried out in animals as well as humans to identify the bioactives in fenugreek responsible for its hypoglycaemic and hypolipidimic activities (Srinivasan, 2006). Through these studies, it has been determined that the hypoglycaemic effect is largely due to the galactomannan fraction of the seed which constitutes 50% (20-25% soluble fiber) and alters the rate of starch digestion. In addition to galactomannans, saponins are also implicated in the hypocholesterolemic effect of fenugreek seeds through increased biliary cholesterol excretion. However, fenugreek as seeds, are not accepted world-wide due to its pungent odour and bitter taste, which also limits its wider application in the food industry without significant processing. This bitterness is due mainly to the oil, steroidal saponins and alkaloids (Srinivasan, 2006).

Currently standardized fenugreek polysaccharide products containing greater than 80% soluble galactomannan, such as FenuLife® and FenuPure® are commercially available in deodorized, debittered forms. However, several questions remain about their effectiveness as functional foods.

1. How effective is the isolated gum in comparison with the whole fenugreek seed flour in terms of its acclaimed health benefits?

2. Is it merely another fiber? How does it compare with the other members of the galactomannan family? Do they perform equally in similar amounts or dose-related?
3. What is the mechanism behind structure-property-process-functional (technological and physiological) relationships that govern sensory and nutritional or health supporting attributes of consumer relevance?

Studies examining these questions are limited. A study that tested GG, LBG and FG for their stability during digestion found FG to be more resistant to enzyme degradation than the other two gums (Mathern, 2003) and concluded that FG, due to its sustained water binding capacity throughout the digestive tract, should be able to exert better health effects than the other gums. Clinical studies with different doses of FG also report a significant, linear dose-dependent effect in lowering blood sugar in normal healthy individuals (Mathern, 2003).

Evans *et al.* (1992), also found FG to be more effective than GG and LBG in a study on their effects on cholesterol metabolism in rats and attributed this to differences in their chemical composition and structure. All three galactomannans lowered the concentrations of cholesterol in the liver and plasma, decreased the rate of hepatic synthesis of cholesterol, increased cecal volatile fatty acids, and increased the amount of water in the feces. However, the effects were more pronounced with FG, which was also found to result in the highest viscosity of the digesta. Similar observations are also reported by Venkatesan et al. (2007) who examined the effect of a cocktail of fenugreek seed powder, GG and wheat bran (Fibernat™) on cholesterol homeostasis and lipid metabolism in rats. This cocktail was found to reduce liver cholesterol, atherogenic lipoproteins, low-density lipoprotein and very low density lipoprotein and oxidative stress.

1.8. Conclusions and Directions for Future Research

The incidence of 'metabolic syndrome or pre-diabetic syndrome' when insulin resistance gives rise to impairment of glucose metabolism leading to a larger issue such as type-2 diabetes is dramatically increasing globally. This syndrome further aggravates the risk of cardiovascular morbidity and mortality. Foods containing functional ingredients form the primary foundation for lipid and blood sugar control, often referred to as glycemic control. Eating foods that promote glycaemic balance or improve lipid parameters is the key to a healthy diet, not just for those with insulin resistance, but for the general population. Type-2 diabetes is a manageable disease with appropriate lifestyle changes. Thus, a low-risk, inexpensive, functional ingredient such as galactomannans that not only improves the structural/textural properties of the food product but also modulate glucose homeostasis and potentially improve lipid parameters aimed at normalizing the metabolic milieu, would be of immense practical interest in food systems.

In terms of physiological effects, fenugreek fiber provides the best choice out of the four galactomannans since it contains both types of fiber fractions in high percentages. The history of the usage is such that it is a better dietary ingredient to be able to resonate between functional food and medicine. In addition, from the point of view of functionality, FG offers the best emulsification properties.

It is clearly evident that most of the technological as well as the health supporting functions of galactomannans are a direct result of their molecular structures. It is specifically reflected in their solubility and inter- and intra-molecular associations that are relevant to the extent and pattern of galactose substitution. A strategic area of research would be the development of bioactive, biocompatible galactomannan polymers by modifying a single gum or blending with other gums or anchoring on to other functional polymers tailored to deliver specific technological and therapeutic functions.

Overall, incorporation of galactomannan gums as functional ingredients in food products remains an under-exploited area of research. Although the results from the limited studies are promising, conclusions drawn from them are still preliminary and the underlying mechanisms remain unclear. Demonstrating the efficacy of these gums needs more consistent dose-related preparatory procedures and measurement of outcomes through well-designed, long term food based intervention studies. More rigorous scientific research is warranted to prove the promise that these gums hold as technological and health supporting functional ingredients.

REFERENCES

[1] Ahmad, F. B. & Williams, P. A. (2001). Effect of galactomannans on the thermal and rheological properties of sago starch. *Journal of Agricultural and Food Chemistry, 49,* 1578-1586.

[2] Alloncle, M., & Doublier, J. L. (1991). Viscoelastic properties of maize starch hydrocolloid pastes and gels. *Food Hydrocolloids*, 5, 455–467.

[3] Alves, M. M., Garnier, C., Lefebvre, J., & Goncalves, M. P. (2001). Microstructure and flow behaviour of liquid water-gelatin-locust bean gum systems. *Food Hydrocolloids, 15,* 117-125.

[4] Annable, P., Fitton, M. G., Harris, B., Phillips, G. O., & Williams, P. A. (1994). Phase behaviour and rheology of mixed polymer systems containing starch. *Food Hydrocolloids*, 8, 351–359.

[5] Aspinall, G.O. (1959). Structural chemistry of the hemicelluloses. *Advances in Carbohydrate Chemistry*, 14, 429-468.

[6] Balyan, D. K., Tyagi, S. M., Singh, D., & Tanwar, V. K. (2001). Effect of extraction parameters on the properties of fenugreek mucilage and its use in ice cream as stabilizer. *Journal of Food Science and Technology-Mysore, 38,* 171-174.

[7] Benichou, A., Aserin, A., & Garti, N. (2007). O/W/O Double emulsions stabilized with WPI-polysaccharide conjugates. *Colloids and Surfaces A-Physicochemical and Engineering Aspects, 297,* 211-220.

[8] Bolliger, S., Wildmoser, H., Goff, H. D., & Tharp, B. W. (2000). Relationships between ice cream mix viscoelasticity and ice crystal growth in ice cream. *International Dairy Journal, 10,* 791-797.

[9] Brennan, C. S., Suter, M., Matia-Merino, L., Luethi, T., Ravindran, G., & Goh, K. (2006). Gel and pasting behaviour of fenugreek–wheat starch and fenugreek–wheat flour combinations. Starch-Starke, 58, 527–535.

[10] Bresolin, T. M. B., Milas, M., Rinaudo, M., Reicher, F., & Ganter, J. L. M. S. (1999). Role of galactomannan composition on the binary gel formation with xanthan. *International Journal of Biological Macromolecules, 26,* 225-231.

[11] Brummer, Y., Cui, W., & Wang, Q. (2003). Extraction, purification and physicochemical characterization of fenugreek gum. *Food Hydrocolloids, 17,* 229-236.

[12] Cairns, P. et al., (1987)X-ray fibre-diffraction studies of synergistic, binary polysaccharide gels, *Carbohydrate Research*, 160, 411-423.

[13] Casas, J. A. & Garcia-Ochoa, F. (1999). Viscosity of solutions of xanthan/locust bean gum mixtures. *Journal of the Science of Food and Agriculture, 79,* 25-31.

[14] Closs, C. B., Conde-Petit, B., Roberts, I. D., Tolstoguzov, V. B., & Escher, F. (1999). Phase separation and rheology of aqueous starch/galactomannan systems. *Carbohydrate Polymers*, 39, 67–77.

[15] Cui, S. W., Eskin, M. A. N., Wu, Y., & Ding, S. D. (2006). Synergisms between yellow mustard mucilage and galactomannans and applications in food products - a mini review. *Advances in Colloid and Interface Science, 128,* 249-256.

[16] Cui, S.W., Ikeda, S. and Eskin, M.N.A. (2007). Seed polysaccharide gums. In Biliaderis, C. & Izydorczyk, M. (Ed.) Functional food carbohydrates (pp.127-138). New York: CRC press, Taylor & Francis group.

[17] Dea, I. C. M., Clark, A. H., & Mccleary, B. V. (1986). Effect of galactose-substitution-patterns on the interaction properties of galactomannans. *Carbohydrate Research, 147,* 275-294.

[18] Dea, I. C. M., Morris, E. R., Rees, D. A., Welsh, E. J., Barnes, H. A., & Price, J. (1977). Associations of like and unlike polysaccharides - mechanism and specificity in galactomannans, interacting bacterial polysaccharides, and related systems. *Carbohydrate Research, 57,* 249-272.

[19] Daas,P., Schols, H., and de Jongh, H. (2002) Toward the recognition of structure-function relationships in galactomannans. *Journal of Agriculture and Food Chemistry.* 50, 4282-4289.

[20] Devasena, T. & Menon, V.P.(2003). Fenugreek affects the activity of beta-glucuronidase and mucinase in the colon. *Phytotherapy Research*, 17, 1088-1091.

[21] Dickinson, E. (2002). Hydrocolloids at interfaces and the influence on the properties of dispersed systems. *Food Hydrocolloids*, 17, 25-39.

[22] Doyle, J. P., Lyons, G., & Morris, E. R. (2009). New proposals on "hyperentanglement" of galactomannans: solution viscosity of fenugreek gum under neutral and alkaline conditions. *Food Hydrocolloids, 23,* 1501-1510.

[23] Edwards, M., Scott, C., Gidley, M.J., & Reid, T.S.G. (1992) Control of mannose/galactose ratio during galactomannan formation in developing legume seeds. *Planta*, 187, 67-74.

[24] Evans AJ, Hood RL, Oakenfull DG and Sidhu GS. (1992). Relationship between structure and function of dietary fibre: a comparative study of three galactomannans on cholesterol metabolism in the rat. *British Journal of Nutrition.* 1: 217-229.

[25] Funami, T., Kataoka, Y., Omoto, T., Goto, Y., Asai, I., & Nishinari, K. (2005). Food hydrocolloids control the gelatinization and retrogradation behavior of starch. 2b. Functions of guar gums with different molecular weights on the retrogradation behavior of corn starch. Food Hydrocolloids, 19, 25–36.

[26] Funami, T., Kataoka, Y., Noda, S., Hiroe, M., Ishihara, S., Asai, I., Takahashi, R., & Nishinari, K. (2008). Functions of fenugreek gum with various molecular weights on the gelatinization and retrogradation behaviors of corn starch. 1: Characterizations of fenugreek gum and investigations of corn starch/fenugreek gum composite system at a relatively high starch concentration; 15 w/v%. Food Hydrocolloids, 22, 763–776.

[27] Gamlath, S., & Ravindran, G. (2009). Extruded products with fenugreek (*Trigonella foeum-graecium*), chickpea and rice: Physical properties, sensory acceptability and glycaemic index. *Journal of Food Engineering*, 90, 44-52.

[28] Garti, N. & Leser, M. E. (2001). Emulsification properties of hydrocolloids. *Polymers for Advanced Technologies, 12,* 123-135.

[29] Garti, N., Madar, Z., Aserin, A., & Sternheim, B. (1997). Fenugreek galactomannans as food emulsifiers. *Food Science and Technology-Lebensmittel-Wissenschaft & Technologie, 30,* 305-311.

[30] Glore, S. R., Van Treeck, V.D., Knehans, A. W. & Guild, M. (1994).Soluble fiber and serum lipids: a literature review. J-Am-Diet-Assoc. 94(4): 425-436.

[31] Gidley, M.J. & Reid, J.S.G. (2006). Galactomannans and other cell wall storage polysaccharides in seeds. In Stephen et al., (Ed.) food polysaccharides and their applications, 2nd ed., (pp. 181-209). New York: CRC press, Taylor & Francis group.

[32] Harrington, J. C. & Morris, E. R. (2009). An unusual manifestation of phase separation in mixtures of disordered gelatin with konjac glucomannan. *Food Hydrocolloids, 23,* 460-467.

[33] Hayati, I. N., Man, Y. B. C., Tan, C. P., & Aini, I. N. (2009). Droplet characterization and stability of soybean oil/palm kernel olein o/w emulsions with the presence of selected polysaccharides. *Food Hydrocolloids, 23,* 233-243.

[34] Hooda, S. & Jood, S. (2003). Physicochemical, rheological, and organoleptic characteristics of wheat-fenugreek supplemented blends. *Nahrung-Food, 47,* 265-268.

[35] Huang, X., Kakuda, Y., & Cui, W. (2001). Hydrocolloids in emulsions: particle size distribution and interfacial activity. *Food Hydrocolloids, 15,* 533-542.

[36] Kesavan, S. & Prudhomme, R.K. (1992). Rheology of guar and HPG- cross linked by borate, Macromolecules, 25, 2026-2032.

[37] Kim, H. J., Choi, S. J., Shin, W. S., & Moon, T. W. (2003). Emulsifying properties of bovine serum albumin-galactomannan conjugates. *Journal of Agricultural and Food Chemistry, 51,* 1049-1056.

[38] Kim, C., Lee, S. P., & Yoo, B. (2006). Dynamic rheology of rice starch–galactomannan mixtures in the aging process. *Starch-Starke*, 58, 35–43.

[39] Kumar, K. I.M. & Maliakel, B.P. (2008). Fenugreek dietary fibre a novel class of functional food ingredient. *Supplement to AgroFood industry health, Mar/Apr* 19(2), 18-21.

[40] Matemu, A. O., Kayahara, H., Murasawa, H., & Nakamura, S. (2009). Importance of size and charge of carbohydrate chains in the preparation of functional glycoproteins with excellent emulsifying properties from tofu whey. *Food Chemistry, 114,* 1328-1334.

[41] Mathern, J. (2003). Overview of galactomannan. *AGROFood industry hi-tech* 57-59.

[42] McCleary, B. (1985). The fine structures of carob and guar galactomannans, *Carbohydrate Research*, 139, 237-260.

[43] Mikkonen, K. S., Tenkanen, M., Cooke, P., Xu, C. L., Rita, H., Willfor, S., Holmbom, B., Hicks, K. B., & Yadav, M. P. (2009). Mannans as stabilizers of oil-in-water beverage emulsions. *Lwt-Food Science and Technology, 42,* 849-855.

[44] Morris, V.J. (1998) Gelation of polysaccharides, In: Hill, S.E., Ledware, D.A., and Mitchell, J.R., (Eds), Functional Properties of Food Macromolecules (pp143), Aspen Publishers, Gaithersburg, MD.

[45] Neirynck, N., Van lent, K., Dewettinck, K., & Van der Meeren, P. (2007). Influence of pH and biopolymer ratio on sodium caseinate-guar gum interactions in aqueous solutions and in o/w emulsions. *Food Hydrocolloids, 21,* 862-869.

[46] Patmore, J. V., Goff, H. D., & Fernandes, S. (2003). Cryo-Gelation of galactomannans in ice cream model systems. *Food Hydrocolloids, 17,* 161-169.

[47] Perissutti, G. E., Bresolin, T. M. B., & Ganter, J. L. M. S. (2002). Interaction between the galactomannan from mimosa scabrella and milk proteins. *Food Hydrocolloids, 16,* 403-417.

[48] Petkowicz, C.L.O., Reicher, F., and Mazeau, K., (1998). Conformational analysis of galactomannans: from oligomeric segments to polymeric chains. *Carbohydrate polymers*, 37, 25-39.

[49] Petkowicz, C. L. O., Milas, M., Mazeau, K., Bresolin, T., Reicher, F., Ganter, J. L. M. S., & Rinaudo, M. (1999). Conformation of galactomannan: experimental and modelling approaches. *Food Hydrocolloids, 13,* 263-266.

[50] Ramesh, H.P., Yamaki, K., Ono, H., & Tsushida, T (2001). Two-dimensional NMR spectroscopic studies of fenugreek (*Trigonella foenum-graecum L.*) galactomannan without chemical fragmentation. *Carbohydrate polymers*, 45, 69-77.

[51] Ravindran, G., & Matia-Merino, L. (2009). Starch-fenugreek (*Trigonella foenum-graecum L.*) polysaccharide interactions in pure and soup systems. *Food Hydrocolloids*, 23, 1047-1053.

[52] Ravindran, G., Nalle, C., Carr, A & Hardacre, A. (2010). Comparative effects of three galactomannans on the functionality of extruded cereal-legume blends. *Food Chemistry.* (In Press)

[53] Reid, J.S.G., & Meier, H. (1970). Formation of reserve galactomannan in the seed of *Trigonella foenum-graecum*, Phytochemistry, 9, 513-520.

[54] Reid, J.S., & Edwards, M.E. (1995) Galactomannans and other cell wall storage polysaccharides in seeds. In A. M. Stephen (Ed.), Food polysaccharides and applications (pp. 155–186). New York: Marcel Dekker.

[55] Reimer, R.A & Russell, J.C. (2008). Glucose tolerance, lipids and GLP-1 secretion in JCR: LA-cp rats fed a high protein fiber diet. *Obesity*, 16, 40-46

[56] Richardson, P. H. & Norton, I. T. (1998). Gelation behavior of concentrated locust bean gum solutions. *Macromolecules, 31,* 1575-1583.

[57] Sittikijyothin, W., Torres, D., & Goncalves, M. P. (2005). Modelling the rheological behaviour of galactomannan aqueous solutions. *Carbohydrate Polymers, 59,* 339-350.

[58] Song, B.K., Winter, W.T., & Taravel, F.R. (1989). Crystallography of highly substituted galactomannans: fenugreek and lucerne gums. *Macromolecules*, 22, 2641-2644.

[59] Song, J., Kwon, J., Chol, J., Kim, Y., & Shin, M. (2006). Pasting properties of non-waxy rice starch–hydrocolloid mixtures. *Starch*, 58, 223–230.

[60] Srinivasan, K. (2006). Fenugreek (Trigonella foenum-graecum): a review of health beneficial physiological effects. *Food Reviews International*, 22, 203–224.

[61] Srivastava, M. & Kapoor, V. P. (2005). Seed Gallactomannans: An Overview. *Chemistry & Biodiversity, 2,* 295-317.

[62] Vendruscolo, C.W., Andreazza, I.F., Ganter, J.L.M.S. & Bresolin, T.M.B. (2005). Xanthan and galactomannan (from M.scabrella) matrix tablets for oral controlled delivery of theophylline. *International journal of Pharmaceutics*. 271, 197-205.

[63] Venkatesan, N, Devaraj, S.N. & Devaraj, H. (2007). A fibre cocktail of fenugreek, guar gum and wheat bran reduces oxidative modification of LDL induced by an atherogenic diet in rats. Molecular and Cellular Biochemistry, 294, 145-153.

[64] Wielinga, W.C. & Maehall, A.G. (2000). Galactomannans. In Phillips, G.O. & Williams, P.A., (Ed.) Handbook of hydroclloids (chap. 8). New York: Woodhead Publishing Limited, CRC Press.

[65] Wong, J.M.W, de Souza, R.R.D., Kendall, C.W.C, Azadeh, E. & Jenkins, D.J.A. (2006). Colonic health: fermentation and short chain fatty acids. *Journal of Clinical Gastroenterology*, 40, 235-243.

[66] Whitney, S. E. C., Brigham, J. E., Darke, A. H., Reid, J. S. G., & Gidley, M. J. (1998). Structural aspects of the interaction of mannan-based polysaccharides with bacterial cellulose. *Carbohydrate Research, 307,* 299-309.

In: Food Hydrocolloids: Characteristics, Properties… ISBN: 978-1-60876-222-4
Editor: C. S. Hollingworth, pp. 225-241

Chapter 8

SUCROSE PECTIN INTERACTION FROM SOLUTION TO GELS

Donatella Bulone, Daniela Giacomazza, Mauro Manno, Vincenzo Martorana and Pier Luigi San Biagio
National Research Council-Institute of Biophysics
Via U. La Malfa, 153 I90146 Palermo, Italy

ABSTRACT

Pectin is a very important polysaccharide in food technology, due to its ability to thicken or gel in aqueous solution under specific conditions. One of the most common uses of pectin is in making jellies or jams, whichs requires the addition of a large amount of sucrose to acidic solutions. The role of sucrose in promoting pectin gelation has been ascribed to the strengthening of hydrophobic interactions consistently with its well-known stabilizing effect on protein structure. On the other hand, more specific effects on dimension and stiffness of pectin chains have been suggested by computational work. Here we present measurements of rheology, static and dynamic light scattering on samples of pectin containing different amounts of sucrose, ranging from few percent up to close to the solubility limit of sucrose. This corresponds to spanning from low viscous liquid samples to strong gels. Results show that below the threshold value of 55% w/w, sucrose acts as a critical parameter for the sol/gel transition by increasing the strength of excluded volume and hydrodynamic interactions between polymer chains. A further increase of sucrose above the critical concentration yields gels with a higher viscoelastic component corresponding to a higher amount of frozen structural inhomogeneities.

INTRODUCTION

Understanding the chemical and physical principles of food making is a fascinating task from many different points of view. Indeed, besides the reward of enlightening the cause-effect relationships implied in many tricks of the art of cooking, it is also an exciting challenge to deal with systems that are among the most complex to be studied. A new interesting perspective in studying such systems is offered by a modern approach based on concepts from soft matter and colloid physics [1,2].

Here we discuss the case of jam or jelly making by using pectin, a natural polysaccharide obtained from citrus peel or apple pomace. Homemade jam or jelly recipes have one to dissolve the pectin powder in hot water with a large amount of sucrose and added lemon juice, and let the solution cool to room or lower temperature until a firm clear gel is formed. With the aim of tracing the gelation mechanism of this system on a chemical physics ground, it is worthwhile considering what is known about the ingredients involved. Pectin is mostly made of a linear chain of α-(1-4)-linked D-galacturonic acid with some of the carboxyl groups esterified with methanol. In some regions of the regular polygalacturonate backbone, the presence of 1-2 linked L-rhamnose residues increases the structural flexibility and provides sites for the attachment of some neutral sugar chains [3]. Depending on the proportion of methyl ester groups, pectin is classified as high methoxyl (HM) having 50% or higher esterification degree, or low methoxyl (LM). Both types of pectin form gels at concentrations of a few grams per liter, but under different conditions: the gelation of LM pectin requires the presence of calcium ions that bind the carboxyl groups of different chains according to the so-called "egg-box" model [4]; HM pectin, which is the type most used in jam or jelly making, forms gels at acid pH (lower than 3.5) only in the presence of a large amount of sugar or similar cosolutes [5-7] which are known to reduce water activity [7-9]. HM pectin has other applications in food industry as a thickening agent or stabilizer.

Sucrose is one of the most common additives in food sweetening or preservation. It is well known that sucrose, like other simple sugars, stabilizes globular proteins against thermal denaturation [10]. The mechanism of this effect has been explained in terms of "preferential hydration" [11], that is a non-homogeneous distribution of the components of a mixed solvent around the macromolecular surface. The preferential hydration results from the balance between the excluded volume effect, due to the different sizes of the two solvent components, and the free energy change associated with the transfer of the macromolecule from pure to mixed solvent [11]. According to the preferential hydration model, sucrose falls in the category of preferentially excluded cosolutes. In fact, the model can explain much experimental data on sucrose effects on protein functionality, but it fails in some cases where there is a tendency to favor the exposure of hydrophobic surfaces [12,13] that could be attributed to a direct interaction between sucrose and protein. Sucrose's effects on protein gel formation are too complex to be explained by a single mechanism. Protein gels are formed only upon protein unfolding; an increase of the gelation temperature at increasing sucrose concentration is commonly observed [12-17] due to an increase in thermal stability. What remains puzzling is the non-monotonic behavior of the gelation rate and gel strength on varying sucrose concentration. The competition between the reinforcement of molecular interactions and the viscosity increase that could hinder molecular encounters, has been proposed as a possible explanation [13, 15-17].

Even more controversial are the opinions about the effect of sucrose and other sugars on the gelation of biopolymers. Similar to the case of proteins gels, a different behavior is observed at relatively low or high sugar concentration. Starting from very low values and increasing sugar concentration up to about 40%, an increase in the gel strength and melting temperature is commonly observed [18-20]. Further addition of sugar causes an abrupt decrease in the elastic modulus in several biopolymeric systems except for case of gelatin [20]. Finally, a fast growth of both elastic and loss modulus is observed on entering the higher concentration region where a rubbery-glass transition occurs [21]. The gel structure at high sugar content is also matter of controversy, since the gel network has been alternately described as made of aggregated particles [22] or flexible and extended chains, lightly cross-linked [23]. In this scenario the sucrose (and more generally sugars) may play different roles through modulation of the hydrophobic interaction, direct binding to polymeric chains, or bulk viscosity increase. It has been recently shown that the gelation behaviour of a high sugar/polymer system can also dramatically depend on the type of sugar added [24], thus showing that no general rules can be easily drawn.

The gelation kinetics of HM pectin with 60% sucrose at different temperatures has been studied by rheology experiments [25]. A non-monotonic dependence of the gelation rate on temperature, with a maximum rate at about 30 ^{0}C, was observed and attributed to the different temperature dependence of hydrogen bonding and hydrophobic interactions, both responsible for gelation. Rheological studies at different temperatures on gelled samples with high sugar content have shown that the system has a mechanical response with frequency increase that is typical of a rubbery-glass transition [26-27].

Here we present data on pectin at 0.2 % concentration in sucrose-aqueous solution and gel, spanning a range of sucrose concentration from a few percent up to well above the minimum value (55% w/w) required for observing gel formation at room temperature. We do not present data on gelation kinetics, but simply change the sucrose concentration and study the system, both in sol and gel state, by Light Scattering and Rheology measurements. Light Scattering results on the liquid side had been the subject of our previous work [28] and are here briefly summarized with the intent of providing a more complete description of the system. Data on gelled samples from Dynamic Light Scattering are analyzed in terms of Mode Coupling Theory (MCT), originally developed [29-31] to describe the dynamical properties of supercooled liquids on approaching the glass transition. A brief account of MCT applications to the description of gelling colloidal systems is given in the "Theory" section.

Materials and Methods

Pectin Samples

Slow-set high methoxyl pectin (esterification degree 64.5%) was a kind gift from Hercules Inc. (Wilmington, DE). Water was Millipore Super Q filtered with 0.22 ím filters. All other chemical compounds used were analytical grade from Merk Co. A stock solution at a pectin concentration of 2%w/w was preliminarily prepared by dissolving the powder in pure water at 100 °C for few minutes in a high-speed mixer. The solution was then cooled at room temperature overnight, stored at 4 °C, and used up to 2 or 3 days. Samples at a pectin concentration of 0.2% and different sucrose concentrations were prepared by heating to 100 °C while stirring the potassium citrate buffer (pH 3.4) with the appropriate amount of water and sucrose. Stock pectin solution was then added and boiled for few minutes. The final buffer concentration was 12 mM. Finally, a few microliters of 50% acid citric solution was added to adjust the pH value to 3.1, and samples were filtered at 80 °C through 0.45 μm Corning filters directly into quartz cuvettes. A duplicate of each sample, but without pectin, was prepared by the same procedure to be used for the solvent scattering measurement.

Samples at sucrose concentrations of 55% or higher were prepared and divided in several aliquots in glass tubes. One aliquot was put in a quartz cuvette. All the aliquots were put in a thermostated bath at 20 ^{0}C. The falling balls method was used to check the samples over time until they were found to be capable of sustaining the ball weight (70 mg).

Static and Dynamic Light Scattering Experiments

Samples that do not form a gel, were allowed to equilibrate at 20 °C for 1 h in the thermostated cell compartment of a Brookhaven Instruments BI200-SM goniometer. The temperature was controlled to within 0.1 °C using a thermostated recirculating bath. The light scattered intensity and time autocorrelation function were measured by using a Brookhaven BI-9000 correlator and a 100 mW Ar laser (Melles Griot) tuned at 514.5 nm or a 35 mW He-Ne laser (Melles Griot) tuned at 632.8 nm. Measurements were taken at different scattering vectors $q = 4\pi n\lambda_0^{-1}\sin(\theta/2)$ where n is the refraction index of the solution, λ_0 is the wavelength of the incident light, and θ is the scattering angle. Values of the refractive index at different sucrose concentrations were measured by an Abbe refractometer and found almost equal to those of the corresponding sucrose-water solutions [32].

A particular procedure was followed for measurements on gelled samples. Indeed, gels are known to exhibit a nonergodic behavior [33], i.e. the intensity autocorrelation function obtained by time-averaging is different from that of the ensemble average. This effect arises from the presence of frozen structures that are unable to relax over the experimental time scale. The ensemble-average correlation function was obtained as a collection of more than 100 time-averaged measurements each lasting 10 minutes and taken at different positions by using a motor-driven cell holder. The total was normalized for the total number of detected photons and the number of individual measurements.

Rheological Experiments

Rheological measurements under low amplitude oscillatory shear were performed on a controlled stress AR-1000 rheometer (TA Instruments, UK) using a standard-size double concentric aluminium cylinder (rotor outer radius 21.96 mm, rotor inner radius 20.38 mm, stator outer radius 20.00 mm, cylinder immersed height 59.50 mm, gap 500 mm). The filtered hot solution was loaded into the cylinder set at 20 ^{0}C. The cylinder–cylinder upper gap was coated with silicone oil to prevent the evaporation.

Viscosity measurements on liquid samples were performed by applying a shear stress ramp ranging from 0 to 80 Pa and recording the corresponding values of shear strain rate. Viscosity was obtained as the slope of the shear stress-shear strain rate graphs. Three measurements were done for each sample.

Gelling solutions were loaded and left to form the gel. The viscoelastic spectra were measured in a frequency range of 0.02-30 Hz at a strain of $4x10^{-3}$, well within the linear viscoelastic region.

THEORY

The Mode Coupling Theory describes the dynamics of a system close to its glass transition limit, through the inclusion in the equations of motion of non-linear delayed interactions (or coupling) between density fluctuations occurring over different lengthscale [29-31]. On varying the parameters that control the static properties, the strength of the dynamic coupling increases with a consequent slowing down of the system dynamics, until a critical value is reached. At this point a dynamic arrest occurs without any change in the structural properties. Close to the critical point, φ_c, the dynamics is governed by the separation parameter:

$$\sigma = k(\varphi - \varphi_c)/\varphi_c \tag{1}$$

Two structural relaxation times with diverging time scales are observed: one, called α-relaxation, is relative to the macroscopic scale and becomes completely arrested at the glass transition point; the other one, called β-relaxation, is relative to the rattling motion of particles confined in cages formed by the neighbor molecules and persists on traversing the critical point. Over the range of the β-relaxation time scale, intermediate between those corresponding to microscopic motion and long-time α-relaxation, the correlation function of density fluctuations is factorizable and takes the general form:

$$f(q,\tau) = f_c(q) + h(q)c_\sigma g_\pm(\tau/\tau_\beta) \tag{2}$$

where the "±" referes to a positive or negative σ value, $f_c(q)$ is the nonergodicity parameter, that is the contribution of the frozen structure, $h(q)c_\sigma$ is the amplitude of the fluctuating part, with $c_\sigma = c_0 |\sigma|^{1/2}$ (c_0 is a system-dependent constant). The temporal dependence is all contained in $g_\pm(\tau/\tau_\beta)$ with $\tau_\beta = t_0 |\sigma|^{-\delta}$, $\delta = 1/(2a)$ where a is a critical exponent. Two limiting forms of $g_\pm$ are:

for

$$\tau \ll \tau_\beta \qquad g_\pm = (\tau/\tau_\beta)^{-a} \tag{3}$$

for

$$\tau_\beta \ll \tau \ll \tau_\alpha \qquad g_- = -B(\tau/\tau_\beta)^b \tag{4}$$

The α-relaxation, occurring only on the liquid side, is described by :

$$f(q,\tau) = f_c(q)G(q,\tau/\tau_\alpha) \tag{5}$$

with $\tau_\alpha = t_0\,|\sigma|^{-\gamma}$, $\gamma = 1/(2a)+1/(2b)$. No universal form for G(q,τ) is predicted, even if a stretched exponential is often observed [31].

The validity of MCT in describing the dynamics at the glass transition in a colloidal suspension of hard spheres was first proved by van Megen and Pusey [34] by Dynamic Light Scattering experiments. Non-aqueous solutions of poly-methylmethacrylate particles stabilized against aggregation were concentrated and used to obtain metastable states capable of undergoing a glass transition at a critical concentration value. Further experiments on the same system inside the glass phase [35] confirmed that predictions of MCT correctly capture the dynamical features of the caging effect described by the β-relaxation. Some years later, Bergenholtz and Fuchs [36] pointed out that gelation of dilute solutions of colloidal particles interacting through strong, short range attraction, can be described in terms of the MCT picture. The main parameter is, in this case, the ratio between the particle size and the range of the attractive interaction, as the particle entrapment is due to the very high energy cost of liberating a caged particle.

RESULTS

On the Sol Side

By Static Light Scattering we measured the form factor of 0.2 % (w/w) HM pectin at T=20 ^{0}C in aqueous solution containing different amounts of sucrose (0-50%). A detailed data analysis has been reported in Ref. 28. Here we recall that the form factor, shown in Figure 1, is well described by the empirical expression found to be valid in the case of fractal clusters [37-38]:

$$P(q) = \left(1 + a\left(qR_g\right)^2\right)^{-d_f/2} \qquad a = \frac{2}{3d_f} \tag{6}$$

where d_f is the fractal dimension and R_g is the gyration radius in dilute solution. The expression also applies to the semidilute regime with R_g being a quantity proportional, with a factor of order one [39], to the static screening length. The latter represents the monomer-monomer correlation length beyond which excluded-volume interactions are screened by the presence of monomers of other chains [39-40]. We found d_f = 1.55 ±0.1. The dependence of R_g on sucrose concentration is shown in the inset of Figure 1. The small, steady, increase of R_g is at odds with the expected reduction of solvent –exposed surface that sucrose should induce. Rather, the decrease of the solvent dielectric constant could be responsible for the increased chain stiffness through reinforcement of electrostatic interactions.

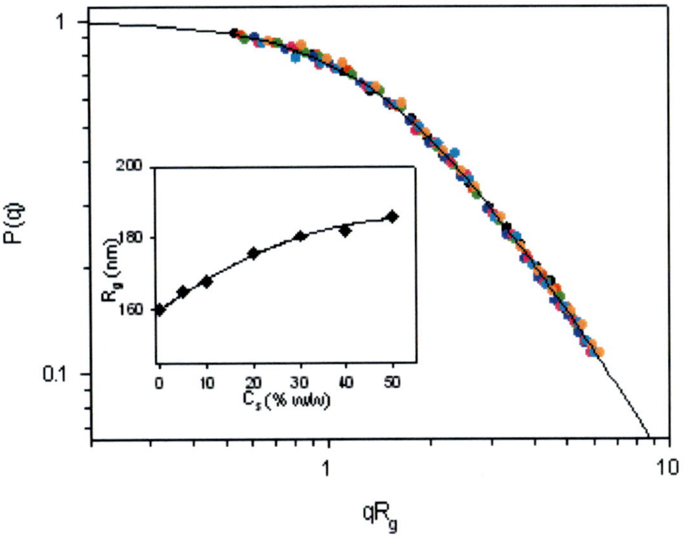

Figure 1. Form factor, P(q), of 0.2% HM-pectin vs qRg at different sucrose concentrations: 0% (black); 5% (red); 10% (green); 20% (blue); 30% (pink); 40% (cyan); 50% (orange). The continuous line is the fit to eq 1 in the text with d_f=1.55. The inset shows the Rg values at different sucrose concentrations.

Changes in the sample dynamic properties were explored by Dynamic Light Scattering experiments. Electric field autocorrelation functions at different sucrose concentrations are plotted all together in Figure 2. On increasing sucrose concentration from 0 to 50%, the main decay time increases by more than one order of magnitude, while the contribution of a faster decay becomes more visible. A simple exponential decay fit to data was found to be very poor even at 0% sucrose, where the fast initial decay is not clearly visible. As we'll see in the following, the fast decay becomes even more visible in gelled samples where it can be clearly ascribed to the presence of small clusters of sucrose molecules [41]. However, due to the small amplitude of the fast mode and the shortage of experimental points in this time scale, we decided to consider the slow mode only and fit data to a stretched exponential function:

$$g^{(1)}(\tau) = A \exp\left[-\left(\frac{t}{\tau}\right)^{-\beta}\right] \qquad (7)$$

This expression is often used to describe the heterogeneous dynamics of disordered materials such polymeric melts or supercooled liquids close to the glass transition [31]. This dynamics is the result of a distribution of structural relaxation times, whose width is characterized by the value of β **$(0 < \beta < 1)$**, with smaller values corresponding to wider distributions. We found a gradual decrease of β from 0.9 to 0.8 on changing sucrose concentration from 0 to 50%, indicating an increase of the dynamic heterogeneity.

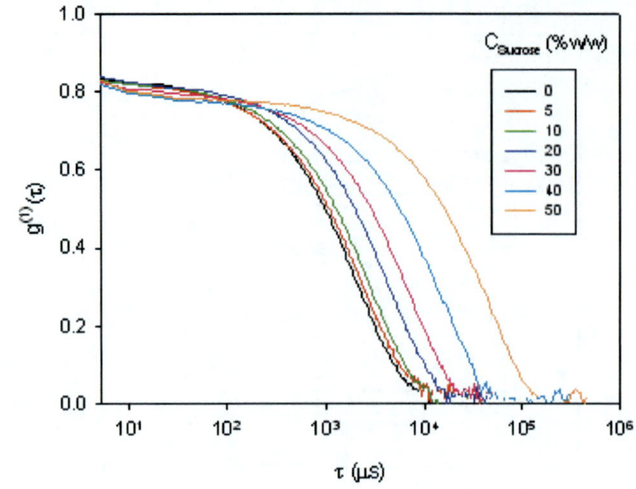

Figure 2. Electric field auto-correlation functions g $^{(1)}$ (τ) at 90° scattering angle for solutions of HM-pectin with different sucrose content. Color code as in Figure 1.

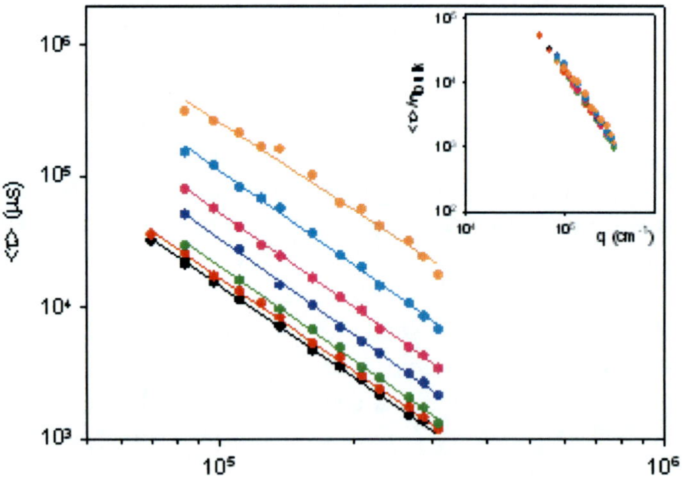

Figure 3. Mean relaxation time, τ, vs q of HM-pectin in aqueous solution with different sucrose content on a log-log scale. In the inset τ-values were normalized for the solution bulk viscosity. Color code as in Figure 1.

The mean relaxation time, <τ> , is calculated by:

$$\langle \tau \rangle = \frac{\tau}{\beta} \Gamma\left(\beta^{-1}\right) \tag{8}$$

where *Γ* is the gamma function. Figure 3 shows on a log-log scale the mean relaxation time vs. q at different sucrose concentrations. A similar q^n dependence at any sucrose concentration is found. The exponent n= 2.3 indicates a dynamics with a main contribution from Brownian diffusion (n=2) together with vibrational motions (n=3) [40]. The inset of Figure 3 shows the relaxation times relative to each solution, normalized by the respective bulk viscosity values. The latter were taken from the literature [32]. The collapsing of data to a single curve indicates that the slowing down of pectin dynamics is entirely due to the bulk viscosity increase.

The viscosity values of HM-pectin solutions were determined by rheological measurements. Figure 4 shows the resulting viscosity vs. the sucrose concentration, as compared with the solvent viscosity at the same sucrose concentration. The comparison shows that the diverging behavior of the sample's viscosity cannot be totally ascribed to the solvent viscosity, since the pectin contribution is seen to increase with sucrose concentration. Finally, we note that no sign of shear thinning, as observed by other authors [42], is visible under our conditions probably due to the lower value of pectin concentration used.

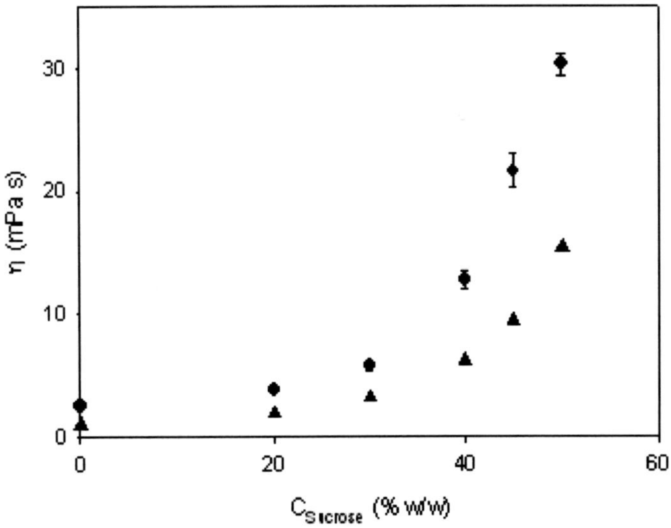

Figure 4. Viscosity of aqueous solutions of HM-pectin with different sucrose content vs sucrose concentration (circles). The corresponding viscosity value of aqueous sucrose solution without pectin is also shown for comparison (triangles).

From Sol to Gel

In studying the sol-gel transition in systems that form physical gels, the kinetic aspects deserve to be carefully considered, especially when the tuning effect of a given parameter has to be evaluated. Indeed, depending on the technique employed [43], it may be arduous to exactly define the sol-gel transition point or the attainment of a steady state. The latter may be practically never gained in several physical gels as a further phase (aging) of minor rearrangements is often observed. At 0.2% pectin concentration and 60 % sucrose, a gel forms in a few hours, while some days are necessary at 58 % sucrose. For the purpose of comparing samples at equivalent final stages of gelation kinetics, we used the classical method of the falling balls for determining when a self-supporting network was formed. Samples at different sucrose concentrations were prepared and kept in a thermostated bath at 20 ^{0}C to be periodically checked until they were found capable of sustaining the weight of a small steel ball. Although this empirical method probably overestimates the time necessary for the formation of a stable structure, it provides a convenient tool when dealing with very long kinetics. Table 1 reports the so estimated times at different sucrose concentrations.

Table 1. Waiting time (t_w)for the formation of a self-supporting structure

$C_{Sucrose}$ (%w/w)	t_w (hr)
55	320
56	150
57	72
58	35
60	6
62.5	1.5
65	0.5

Gelled Samples

Static light scattering data from gelled samples with different amounts of sucrose were collected by ensemble averaging as a function of the q-vector. Results are shown on a log-log plot in Figure 5. No differences were observed in the q-dependence of the different samples. The fractal dimension, given by the slope of log (I) vs. log(q), is almost constant with a value ranging between 1.5 and 1.6, practically the same as that obtained in liquid samples.

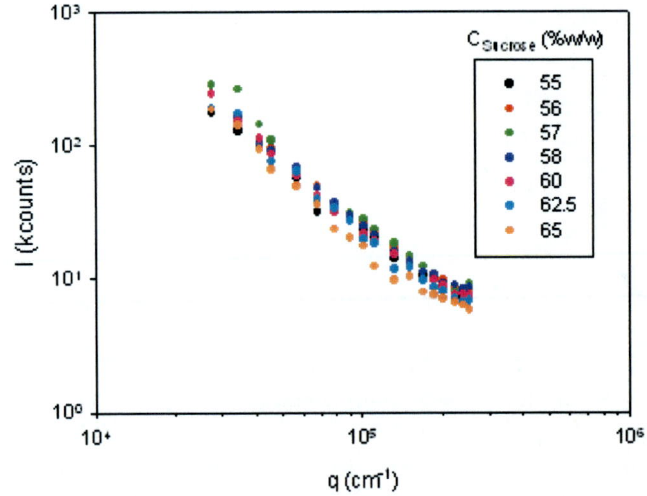

Figure 5. Static light scattered intensity vs. q for gels of HM-pectin with different sucrose content: 55% (black); 56% (red), 57%(green), 58%(blue), 60%(pink); 62.5%(cyan), 65% (orange). The intensity values were obtained by averaging over different sample regions.

Ensemble-average correlation functions measured by DLS on gels with different sucrose content are shown in Figure 6. Only the results for three samples are reported to prevent overcrowding of the figure. The high value of the non-decaying component indicates that the major fraction of the scattered light comes from species that do no relax over the experimental time scale. Two decays are also visible: the first one, occurring on a time scale of a few tens of microseconds, is well described by an exponential; the second one takes a much longer time and can be represented by a stretched exponential decay. To ascertain the kind of motion associated to the first decay, we looked at its q-dependence in the gel at 60% sucrose. A linear relationship between Γ and q^2 was observed as shown in Figure 7. This assures that the motion corresponds to a free diffusion of small objects inside the gel matrix. The latter could be sucrose clusters, whose presence has been observed in aqueous solution even at very low sucrose concentration [41]. This hypothesis was tested by measurements on a sample with 60% sucrose prepared following the same protocol, but without pectin. Similar Γ values were obtained as shown in Figure 7. The slope of Γ vs. q^2 gives a diffusion coefficient of 8.1×10^{-7} cm^2/s. To estimate the hydrodynamic radius of the diffusing sucrose clusters, the value of the local viscosity needs to be known. Using the bulk viscosity gives an unphysical value of 0.4 Å , whereas that of pure water gives 2.6 nm.

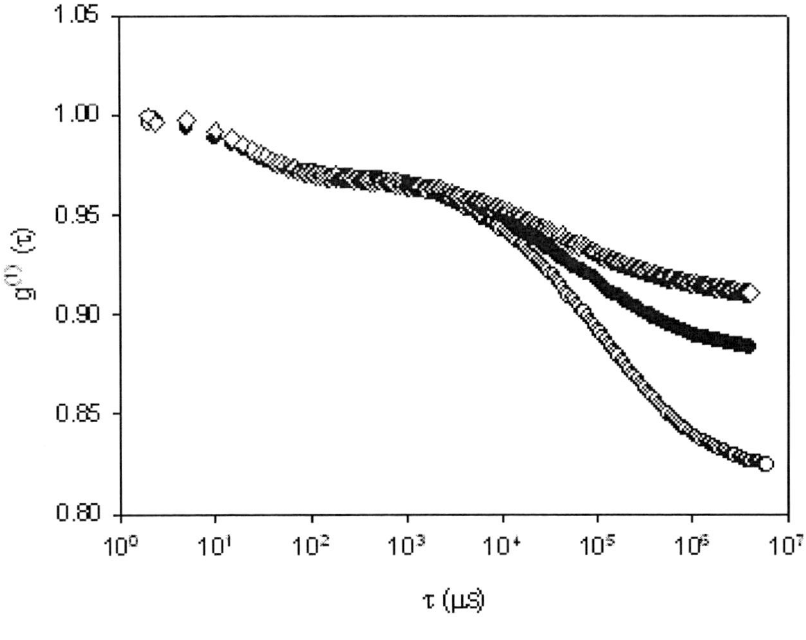

Figure 6. Ensemble average electric field auto-correlation function measured at 90^0 scattering angle on gels of HM-pectin with sucrose content of: 55% (empty circles), 58% (full circles), 60% (empty diamonds).

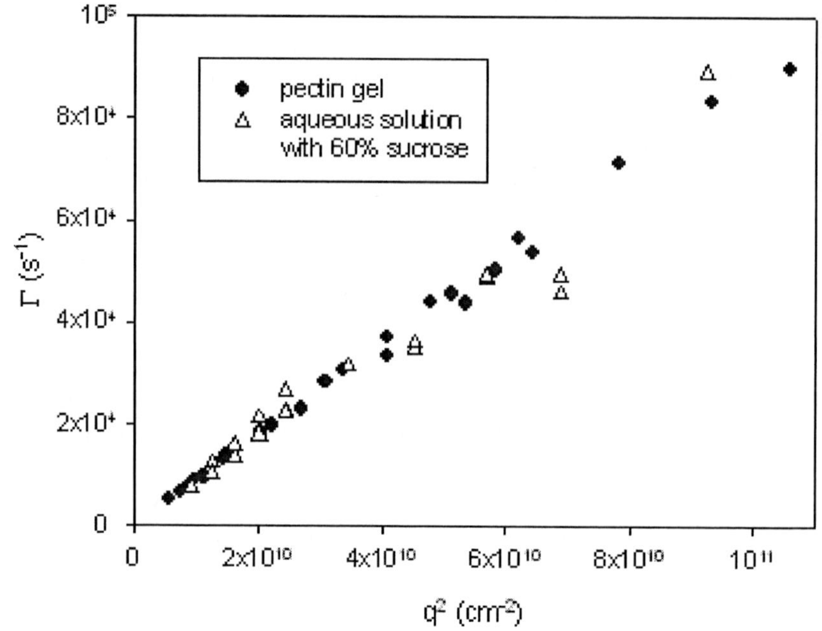

Figure 7. q-dependence of the relaxation time of the initial fast decay time for a gel of HM-pectin with 60% sucrose (black circles), and an aqueous solution with 60% sucrose (white triangles).

Following the suggestion that a glass transition could be entered on increasing sucrose concentration, as indicated by the results on liquid samples, we attempted a data analysis in terms of MCT. Data relative to the correlation function tail were fit to Eq. 2 with $g_{\pm}$ given by Eq. 3. In a first run all fit parameters were left free. The small spread in the values of the exponent a suggested that we repeat the fitting procedure by keeping it constant and equal to the fit average value of 0.345. MCT predicts $0 < a < 0.5$, and a value of 0.305 was found in the case of hard colloidal spheres [33]. Figure 8 shows the correlation functions scaled according to the results of the fitting procedure. The quantity shown in the figure is $[f(\tau)-f_C]/k$ with the $k = [h(q)*c_{\sigma}]*(\tau_{\beta})^{a}$. In this way, each correlation function was normalized by its own parameters (the baseline offset f_C, the amplitude of the β-relaxation $[h(q)*c_{\sigma}]$, and the scaling time τ_{β}) so that a unique time dependence obeying the MCT prediction should be observed at large τ. The continuous line in the figure represents the master function $g_{\pm} = (\tau)^{-a}$ for the β-relaxation with $\tau_{\beta} = 1$. Upon scaling, the long-time tails of data related to different sucrose concentrations are made coincident over two orders of magnitude of time. This result confirms that, at least from a phenomenological point of view, MCT is applicable in our case, so indicating that the dynamic behavior is the same for each sample. For a quantitative comparison, the theoretical values of $f_c(q)$ and $h(q)$ need to be known, as calculated starting from an appropriate interaction potential. In its absence we can look at the dependence of the fit parameters on sucrose concentration. One prediction that can be easily tested is the dependence of τ_{β} on $/\sigma/^{1/2a}$. In panel A of Figure 9, we plot $(1/\tau_{\beta})^{2a}$ vs. sucrose concentration to test the hypothesis that sucrose concentration plays the role of a critical parameter. The hypothesis appears reasonably valid. In panel B of the same figure the dependence of f_c and $[h(q)*c_{\sigma}]$ are shown. The f_c behavior reflects the increasing degree of the freezing of the system with increasing sucrose concentration. No conclusions can be drawn from the decrease of the motional amplitude *[h(q)*c$_\sigma$]* with increasing sucrose concentration since the variation in *h(q)* with changes in potential interaction at varying sucrose concentration is not known.

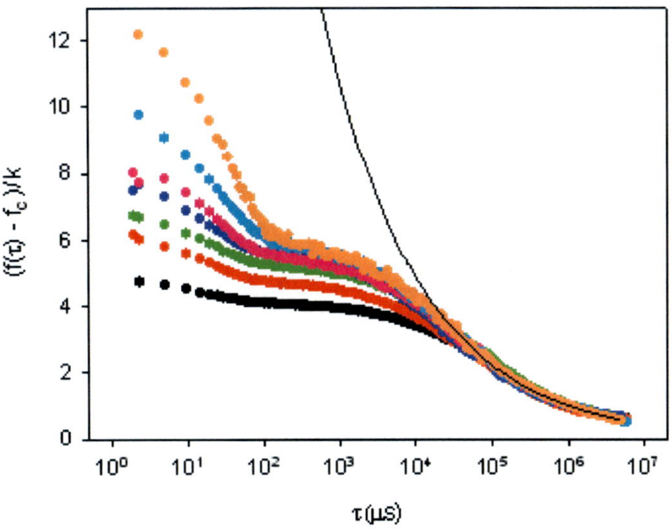

Figure 8. Ensemble-average electric field autocorrelation functions scaled according to Eq. 2. The critical parameters $f_c(q)$ and $[h(q)*c_\sigma]$ are given in Figure 9. The solid line represents the master function $g_\pm(\tau)$ with τ_β=1 for the β process of Mode Coupling theory.

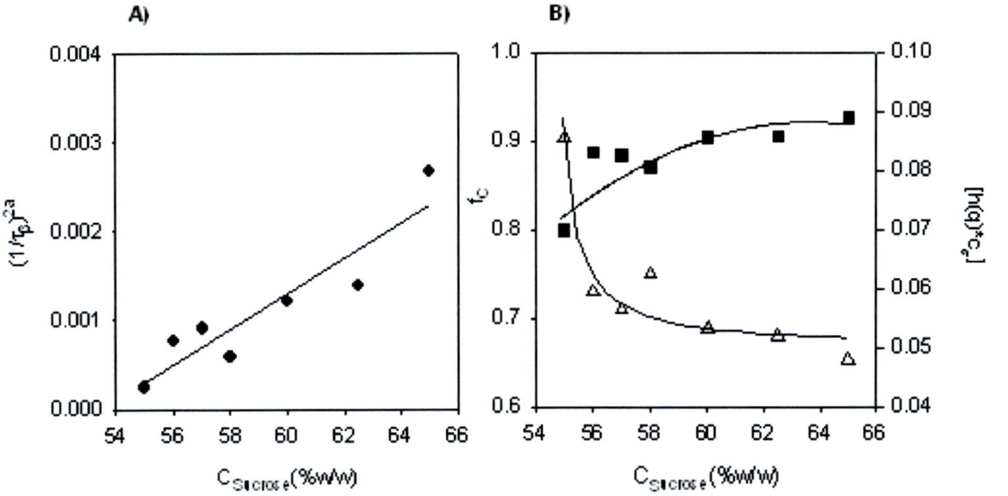

Figure 9. Fit parameters to Eq. 2 of the correlation functions of gelled samples vs. sucrose concentration. Panel A: $(1/\tau_\beta)^{2a}$; panel B: f_c (full square); $[h(q)\ c_\sigma]$ (empty triangles);

Mechanical properties of the gelled samples were characterized by measuring the viscoelastic spectra. Figure 10 shows the dependence on frequency of the storage (G'(ω)) and loss-modulus (G"(ω)). The solid-like character of any sample is manifested by the nearly flat behavior of G'(ω), whose magnitude increases with the sucrose concentration. The viscous modulus G"(ω) increases quite linearly with ω over the entire frequency range and crosses G'(ω) at a frequency value that increases with increasing sucrose concentration. Except for the magnitude, no differences are seen in G'(ω) and G"(ω) behavior at varying sucrose concentration, thus suggesting that a master curve would be easily obtained. This was accomplished by firstly scaling G'(ω) and G"(ω) of each data set by the respective plateau modulus (A_G), and subsequently determining the scaling factors (A_ω) for the frequency. The master curve shown in Figure 11 is very similar to that observed for solidlike samples of weakly attractive colloidal particles above a critical value of particle concentration or interaction energy [44]. The relationship between the scaling factors is shown in the inset of Figure11. The scaling of the viscoelastic data evidences that the increased solid-like character of gels with different sucrose contents is due to a higher extent of frozen structure formed by the same mechanism.

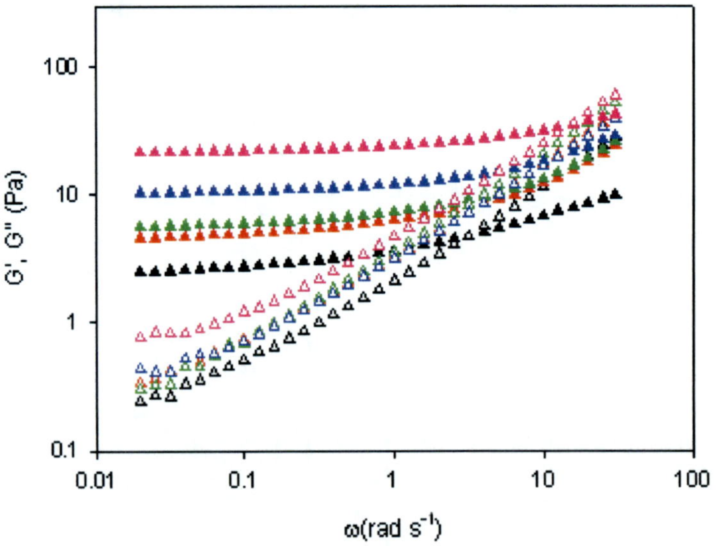

Figure 10. Mechanical spectra (G'(ω) (full symbols), G"(ω) (empty symbols)) measured on HM-pectin gels with different sucrose content: 56% (black); 58% (red), 60% (green); 62.5% (blue), 65% (pink).

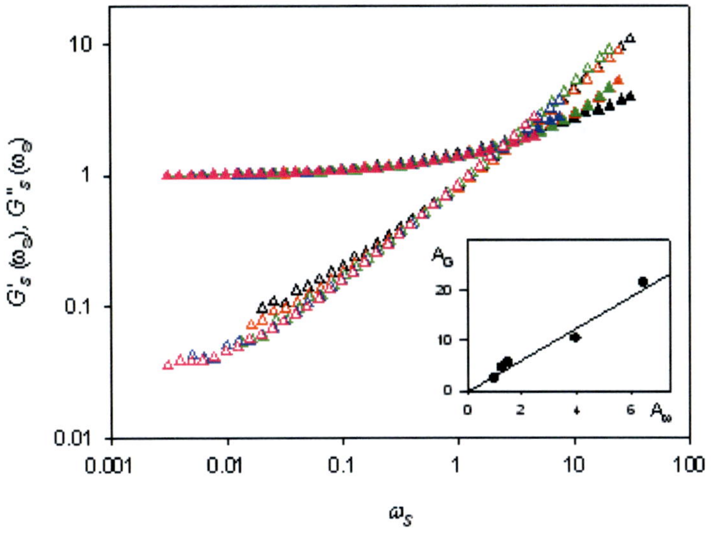

Figure 11. Master curve showing the scaled moduli $G'_S(\omega)$ and $G''_S(\omega)$ vs. the scaled frequency ω_S. Color code as in Figure 10. The inset shows the relationship between the scaling factors.

Conclusion

Tracing the mechanism of pectin gelation in the presence of sucrose to changes in the specific interactions between pectin, sucrose and water is made difficult by the lack of a clear understanding of the sugars effects on the gelation of biomolecules, as discussed in the Introduction. Here we have presented results on aqueous solutions of HM-pectin at varying sucrose concentrations, so as to span from liquid to gelled samples.

Pectin in sol samples appears as a system of branched clusters with strong hydrodynamic interactions. An increase in sucrose content induces a moderate increase in the screening length and a strong slowing down of the dynamics, proportional to the divergence in the bulk viscosity. The residue dynamics in gelled samples are seen to agree well with predictions of Mode Coupling Theory for the glass transition in supercooled liquids. This may appear quite surprising considering the complexity of the system under study, made of a natural polymer in sucrose-water solution. This finding further supports the idea that different phenomena characterized by an abrupt dynamic arrest can be described in a unified view [45]. In particular, it has been proposed that a jamming phase diagram, whose axes are density, interaction energy and applied stress, is capable of describing a wide variety of phenomena occurring in attractive colloidal systems such as the colloidal glass transition, gelation and aggregation [46]. This approach also provides useful indications of the relevant parameters for tailoring system properties. If the MCT model is used to describe the behavior of our system, sucrose concentration appears to play the role of a critical parameter for pectin gelation. Further work is necessary to determine whether the viscosity increase or the reinforcement of short range interactions, or both of these, are responsible for this effect.

Acknowledgements

We would like to thank Prof. J. Newman for his comments and a critical reading of the manuscript.

We also thank Dr. R. Noto for discussions and suggestions. Technical help from Mrs. F. Giambertone, G. Lapis and M. Lapis is also acknowledged.

References

[1] Mezzenga, R.; Schurtenberger, P.; Burbidge, A.; Michel, M. *Natures Materials* 2005, 4, 732-740.

[2] van der Sman, R. G. M. ; van der Goot, A. J. *Soft Matter* 2009, 5. 601-510.

[3] Willats, W.G.T.; Knox, J.P.; Mikkelsen, J.D. *Trends Food Sci. Tech.* 2006, 17, 97-104.

[4] Grant, G. T.; Smith, P. J. C.; Thom, D. *FEBS Lett.* 1973, 32, 195-198.

[5] Morris, V. J. In Functional Properties of Food Macromolecules Mitchell, J. R.; Ledward, D. A.; Eds.; (series name and number) Elsevier Applied Science Publishers: London, (UK) 1986; Chapter 3, p 121.

[6] Rees, D. A. *Adv. Carbohydr. Chem. Biochem.* 1969, 24, 267-332.

[7] Oakenfull, D.; Scott, A. J. *Food Sci.* 1984, 49, 1093-1098.

[8] Somero, G. N. *Am. J. Physiol.* 1986, 251, R197-R213.

[9] Timasheff, S. N. In Water and Life; Somero, G. N.; Osmond, C. B.; Bolis, C. L.; Eds.; Springer-Verlag: Berlin, 1990; Chapter 6, p 70-.

[10] Arakawa, T.; Timasheff, S.N. *Biophys. J.* 1985, 47, 411.

[11] Lee, J.C.; Timasheff, S.N. *J. Biol. Chem.* 1981, 256, 7193-7201.

[12] Semenova, M.G.; Antipova. A.S.; Belyakova, L.E. *Curr. Opin. Colloid Interface Sci.* 2002, 7, 438-444.

[13] Baierand, S.K.; McClements, D. J. *Comp. Rev. Food Sci. Food Safety.* 2005, 4, 43-54.

[14] Baier, S.; McClements, D.J. *J. Agric. Food Chem.* 2001, 49, 2600-2608

[15] Dierckx, S.;, Huyghebaert, A. *Food Hydrocolloids.* 2002, 16, 489-497.

[16] Kulmyrzaev, A.; Bryant, C.; McClements, D.J. *J. Agric. Food Chem.* 2000, 48, 1593-1597.

[17] Kulmyrzaev, A.; Cancelliere, C.; McClements, D.J. *J. Sci, Food Agric.* 2000, 80, 1314-1318.

[18] Evageliou, V.; Kasapis, S.; Hember, M. W. N. *Polymer.* 1998, 39,3909–3917.

[19] Nishinari, K.; Watase, M.;Miyoshi, E.;Takaya, T.; Oakenfull, D. *Food Technology.* 1995, 49, 90–96.

[20] Kasapis, S.; Al-Marhoobi, I.M.; Deszczynski, M.; Mitchell, J.R.; Abeyesekera, R. *Biomacromolecules.* 2003, 4, 1142-1149.

[21] Kasapis, S. *Food Hydrocolloids.* 2001, 15, 631–641.

[22] Nickerson, M.T.; Paulson, A.T.; Speers, R.A. *Food Hydrocolloids.* 2004, 18, 783–794.

[23] Al-Amri, I.S.; Al-Adavi, K.M.; Al-Marhoobi, I.M.; Kasapis, S. *Carbohydrate Polymers.* 2005, 61, 379–382.

[24] Loret, C.; Ribelles, P.; Lundin, L. *Food Hydrocolloids.* 2009, 23, 823–832.

[25] da Silva, J.A.L.; Gonçalves, M.P.; Rao, M.A. *Int. J. Biol. Macromol.*1995, 17, 25-32.

[26] Al-Ruqaie, I.M.; Kasapis, S.; Richardson, R.K. *Polymer.* 1997, 38, 5685-5694.(pectina)
[27] Kasapis, S.; Al-Alawi, A.; Guizani, N.; Khan A.J.; Mitchell, J.R. *Carbohydrate Polymers.* 2000, 329, 399-407 (pectina).
[28] Bulone, D.; Martorana, V.; Xiao, C.; San Biagio, P.L. *Macromolecules.* 2002, 35, 8147-8151.
[29] Bengtzelius, U.; Götze, W.; Sjölander, A. *J. Phys.* C 1984, 17, 5915-...
[30] Leutheusser, E. *Phys. Rev.* A 1984, 29, 241-...
[31] Götze, W.; Sjögren, L. *Rep. Prog. Phys.* 1992, 55, 241-376.
[32] Handbook of Chemistry and Physics; Weast, R. C., Astle, M. J., Beyer, W. H., Eds.; CRC Press: Boca Raton, FL, 1984; p D-265.
[33] Pusey, P.; van Megen, W. *Physica A* 1989, 157, 705-741.
[34] van Megen, W.; Pusey, P.N. *Phys. Rev. A* 1991, 43, 5429-5441.
[35] van Megen, W.; Underwood, S. M. *Phys. Rev. E* 1994, 49, 4206-4220.
[36] Bergenholtz, J.; Fuchs, M. *Phys. Rev. E* 1999, 59, 5706-5715.
[37] Kallala, M.; Sanchez, C.; Cabane, B. *Phys. Rev.* A 1993, 48, 3692-3704.
[38] Trappe, V.; Bauer, J.; Weissmüller, M.; Burchard, W. *Macromolecules.*1997, 30, 2365-2372.
[39] Brown, W.; Nicolai, T. In Dynamic Light Scattering; Brown, W., Ed.; Clarendon Press: Oxford, 1993; Chapter 6, p 166.
[40] Schaefer, D. W.; Han, C. C. In Dynamic Light Scattering; Pecora, R., Ed.; Plenum Press: New York, 1985; Chapter 5, p 181.
[41] Kaufman, S.L.; Dorman, F. D. *Langmuir.* 2008, 24, 9979-9982.
[42] Richter, S. Macromol. *Chem. Phys.* 2007, 208, 1495–1502.
[43] Sato, Y.; Kawabuchi, S.; Irimoto, Y, Miyawaki, O. *Food Hydrocolloids.* 2004, 18, 527-534.
[44] Trappe, V.; Weitz, D.A. *Phys. Rev. Letters.* 2000, 85, 449-452.
[45] Liu, A.J.; Nagel, S.R. *Nature.* 1998, 396, 21-22.
[46] Trappe, V.; Prasad, V.; Cippelletti, L.; Segre, P.N.; Weitz, D.A. *Nature.* 2001, 411, 772-775.

In: Food Hydrocolloids: Characteristics, Properties... ISBN: 978-1-60876-222-4
Editor: C. S. Hollingworth, pp. 243-251

Chapter 9

UTILISATION OF GLUCOMANNANS FOR HEALTH

R. F. Tester and F. H. * *Al-Ghazzewi*
Glycologic Limited, c/o Glasgow Caledonian University,
Glasgow, G4 0BA, UK

ABSTRACT

Glucomannans are a source of soluble dietary fibre. In nature they are distributed in both softwood tissues such as roots or corms and the hemicellulose of hardwood plants where they serve as energy reserves and in some cases structural roles. As glucomannans contain components which may have a positive effect on health, for instance by selectively stimulating the gut-friendly bacteria and inhibiting pathogens, they can be utilised to serve as valuable 'functional foods'. Outside the gut, the impact of ingested glucomannans for broader health benefits include cholesterol lowering, reducing the risk of chronic constipation, diabetes management, cancer prevention, treating diverticular disease, stimulating the immune system and contributing towards weight reduction.

INTRODUCTION

Glucomannans are carbohydrate polymers that are considered to be source of soluble dietary fibre. They are distributed in nature in softwood plants (such as the roots or corms of konjac) or hemicellulose in hardwood plants (including the wood of conifers and dicotyledons) where they serve as energy reserves and in some cases structural roles. Glucomannans are differentiated from other soluble fibres by their extraordinarily high viscosity (McCarty 2002). The polymers comprise, in most cases, mannose residues with glucose as the second sugar. They also contain some acetylated residues and galactose side chains (Millane and Hendrixson 1994; Khanna 2003) where the degree of branching is about 8% through β-(1$\rightarrow$6)-glucosyl linkages. In konjac glucomannans, the sugars are arranged in

* Correspondence to: Dr Farage H. Al-Ghazzewi, Glycologic Ltd, c/o Glasgow Caledonian University, Cowcaddens Road, Glasgow G4 0BA, UK. E-mail: f.h.alghazzewi@glycologic.co.uk Tel:+44(0)1413313054 Fax:+44(0)1413318079

blocks of mannose residues interrupted with one or two glucose residues which are β-(1–4) linked with typically 1.6:1 mannose to glucose residues within the polysaccharides (Kato and Matsuda 1969; Sugiyama 1972; Shimahara *et al.* 1975; Maeda 1980; Ogawa 1991; Katsuraya *et al.* 2003). Acetyl groups are esterified to C3 at approximately every nineteenth residue (Cescutti *et al.* 2002).

Sources of glucomannans are presented in Table 1. From this Table, it is evident that glucomannans can be extracted from a broad range of botanical sources where there is variability in molecular weight and the mannose to glucose ratio. Glucomannan polysaccharides are fermented in the large intestine of man (Matsuura 1998; Nakajima *et al.* 2002) This fermentation releases short-chain fatty acids (Topping and Clifton 2001) which then may be absorbed into the blood stream and are reported to have beneficial effects against gut cancer (Wong *et al.* 2006). For more details regarding the fermentation of polysaccharides in the gut readers are referred to a recent review by Topping and Clifton (2001).

Table 1. Glucomannans from different sources

Source	Mannose:Glucose Ratio (MGR)	Degree of Polymerisation (DP)
Eastern white pine (Pinus strobes)	3.8:1	90
Higanbana (Lycoris radiata)	4.0:1	730
Konjac (Amorphophallus konjac)	1.6:1	>6,000
Lily (Lilium auratum)	2.7:1	220
Orchid (Tubera salep)	3.2:1	600
Ramie (Boehmeria nivea)	1.8:1	nd
Redwood (Sequois sempervirens)	4.2:1	60
Suisen (Narcissus tazetta)	1.5:1	nd

nd: not determined
Adapted from Khanna (2003)

As glucomannans contain components which may have a positive effect on human and animal health (Al-Ghazzewi *et al.* 2007; Elamir *et al.* 2008; Sutherland *et al.* 2008), for instance by selectively stimulating the gut-friendly bacteria and inhibiting the pathogens, they serve as valuable 'functional foods'. According to Jones (2002) functional foods are defined broadly as 'foods that provide more than simple nutrition; they supply additional physiological benefit to the consumer'. These functional foods are apparently important component of the diet as they promote gut transit of food provides bulk and satiety and provides a fermentation matrix in the colon.

Glucomannans (such as konjac) have broader nutritional benefits outside the gut itself which include cholesterol lowering (Arvill and Bodin, 1995; Gallaher *et al.*, 2002; Martino *et al.*, 2005), reducing the risk of chronic constipation (Passaretti *et al.*, 1991; Loening-Baucke *et al.*, 2004), diabetes management (Chen *et al.*, 2003), cancer prevention (Luo, 1992) and weight reduction (Walsh *et al.*, 1984). Mannans may prevent fat storage through inhibiting the intestinal absorption of dietary fat in high fat diets (Takao *et al.*, 2006).

POTENTIAL HEALTH BENEFITS

Conclusions from *in vitro* and *in vivo* studies have shown several potential health benefits of the glucomannans from various sources.

CHOLESTEROL

Glucomannans have been demonstrated to reduce significantly serum cholesterol and low density lipoprotein cholesterol in obese patients (Livieri *et al.* 1992; Walsh *et al.* 1984). Glucomannans have also been tested in children as dietary aids to reduce circulatory cholesterol. Decreased values for plasma total cholesterol and low density lipoprotein cholesterol have been observed in children (Martino *et al.* 2005). For other studies, short term ingestion (4 weeks) of 3.9g of glucomannan by healthy volunteers showed a decreased total cholesterol (10%), low density lipoprotein (7%), triglyceride (23%) and systolic blood pressure dropped by 2.5% (Arvill and Bodin 1995). Clinical trials showed that when compared with placebos, glucomannan from konjac reduced effectively plasma cholesterol (11%), low density lipoprotein (20%), total/high density lipoprotein - cholesterol ratio (15%) and fasting glucose (23%). When glucomannan was used in conjunction with chitosan, it decreased serum cholesterol in 21 overweight normocholesterolaemic subjects, possibly by increasing steroid excretion via the faeces (Gallaher *et al.* 2002).

DIABETES

Glucomannan may be emerging as an alternative therapeutic adjunct for type 2 diabetes mellitus sufferers (Vuksan *et al.* 2001). Studies by Huang and coworkers (1990) showed that glucomannans from konjac were useful for the prevention and treatment of hyperglycaemia. In general, a high-carbohydrate diet with fibre and in particular mannans improves the blood lipid profile and alleviates the fasting blood glucose levels of type 2 diabetics (Vuksan *et al.* 1999; 2000; Chen *et al.* 2003). Volunteers consuming glucomannan enriched biscuits as a breakfast showed a reduction in glycaemic increments after breakfast. It was concluded that the decreased insulin secretion and the reduction of insulin need can preserve the functional reserve of beta-cells (Melga *et al.* 1992).

OBESITY

There is evidence that glucomannans exert their beneficial effects by promoting satiety and weight loss (Keithley and Swanson 2005). This effect is probably due to the fibre absorbing water which subsequently forms a viscous gel – like mass promoting satiety and faecal energy loss. In twenty obese patients, consuming 1g of glucomannan with 250ml of water 1 hour before each of 3 meals daily over 8 weeks, there resulted in a significant average weight lose of 5.5 pounds (Walsh *et al.* 1984). In another trial, 2-3g/diet purified glucomannan decreased significantly excess weight for obese patients (Livieri *et al.* 1992).

Severely obese patients treated with glucomannans (~4g/diet in 3 doses) have been shown to exhibit a significant weight loss (Vita *et al.* 1992).

Constipation

Glucomannans as a soluble fibre source are often proposed as an ideal therapeutic tool for the management of constipation (Passaretti *et al.* 1991; Loening-Baucke *et al.* 2004). In the treatment of chronic constipation, glucomannans improve symptoms of constipation and are well-tolerated (Passaretti *et al.* 1991). The polymers may relieve constipation by decreasing faecal transit time (Marzio *et al.* 1989). Supplementation of glucomannans into a low-fibre diet promotes defaecation frequency in healthy adults, possibly by increasing the stool bulk (Chen *et al.* 2006). Clinical trials over 12 weeks have shown that in neurologically impaired children, glucomannans also increase stool frequency (Staiano *et al.* 2000). When glucomannans have been used in combination with lactulose for the treatment of constipation in pregnant women, a return to normal frequency of weekly number of defaecations and a parallel control of weight gain was achieved (Signorelli *et al.* 1996).

Cancer

A number of rodent studies have suggested that glucomannans may have the potential to decrease cancer risk (Mizutani and Mitsuoka 1982; 1983) and possibly even slow the aging process due to the regulation of IGF-I activity (McCarty 1998; 2002). As glucomannans are fermented to short chain fatty acids (SCFAs) by intestinal anaerobic bacteria (Matsuura 1998), they might have some utility in preventing colonic cancer. Franceschi and coworkers (2001) reported that the positive associations of glycaemic index and load with colorectal cancer suggest a detrimental role of refined carbohydrates in the etiology of the disease.

Immune System

Glucomannans have been shown to enhance lipid metabolism and immune functions in rats (Innami *et al.* 1999; Yamada *et al.* 2003). In other hosts such as fish and dogs, glucomannans also stimulated the immune system and improved infection resistance (Swanson *et al.* 2002; Torrecillas *et al.* 2007). Glucomannans and in particular those from konjac root have been found to have immunomodulatory function by suppressing the development of allergic rhinitis-like symptoms, prevent the increase of plasma immunoglobulin E, immunoglobulin G (Oomizu *et al.* 2006; Onishi *et al.* 2007) and spontaneously occurring dermatitis (Onishi *et al.* 2005).

DIVERTICULITIS

Although there are limited studies involving the use of glucomannans for treating diverticular disease, available work has revealed that about one-third to one-half of the subjects tested were found to benefit from glucomannan consumption (Papi *et al.* 1995; Petruzziello *et al.* 2006). Fibre – rich diets like glucomannans reduce intraluminal pressure and slow down the process of colonic diverticular disease (Latella *et al.* 2003; Papi *et al.* 2005; Frieri *et al.* 2006).

TOXICITY

In foods, glucomannans (from konjac roots) have been found to completely inactivate several food poisoning bacteria including *E.coli, Salmonella enteritidis, Vibrio parahemolyticus* and *Staphylococcus aureus*. A four-log reduction also occurred with spore forming bacteria such as *Bacillus subtilis*, *Bacillus cereus*, *Clostridium perfringens* and *Clostridium botulinum* type A and E (Kitamoto *et al.* 2003). In feed, supplementation with esterified glucomannan was effective in reducing aflatoxin B-1 induced hepatic injury in ducklings (Banlunara *et al.* 2005) and in mice (Madrigal-Santillan *et al.* 2007). Fermented unhydrolysed konjac glucomannan by colonic lactic acid bacteria *in vitro* produced an antioxidative capacity by preventing the initiation of ferrous ion-induced peroxidation, whereas the fermentation of konjac glucomannan oligosaccharides did so by increasing the radical scavenging ability and by eliminating lipid peroxide formation (Wang *et al.* 2008).

ADHERENCE TO PATHOGENS

Carbohydrates and in particular mannose appear to play a role in the inhibition of bacterial adhesion to epithelial cells (Schaeffer et al. 1980; Michaels *et al.* 1983; Wang *et al.* 2008). A study by Mangan and Snyder (1979) reported that D-mannose or its derivatives inhibit specifically adhesion of *E. coli* to epithelial cells. More recently, King and co-workers (2000) reported that mannose and N-acetyl-D-galactosamine inhibited adhesion of *E. coli* and *P. aerugenosa* to epithelial cells.

CONCLUSION

The impact of ingesting glucomannans on health is not only limited to its fermentation to short chain fatty acids which might have some utility in preventing and treating diseases including colonic cancer. They also prevent pathogens such as *E. coli* from binding to the gut mucosa or urinary tract. Although it needs further investigation, glucomannans could be used as a therapeutic tool for the treatment of a range of physiological disorders.

References

[1] Al-Ghazzewi, FH; Khanna, S; Tester, RF; Piggott, J. The potential use of konjac glucomannan hydrolysate as a prebiotic. *J Sci Food Agric*, 2007 87, 1758-66.

[2] Arvill, A; Bodin, L. Effect of short-term ingesion of konjac glucomannan on serum cholesterol in healthy men. *Am J Clin Nutr*, 1995 61, 585-9.

[3] Banlunara, W; Bintvihok, A; Kumagai, S; Immunohistochemical study of proliferating cell nuclear antigen (PCNA) in duckling liver fed with aflatoxin B-1 and esterified glucomannan. *Toxicon*, 2005 46, 954-7.

[4] Cescutti, P; Campa, C; Delben, F; Rizzo, R. Structure of the oligomers obtained by enzymatic hydrolysis of the glucomannan produced by the plant *Amorphophallus konjac*. *Carbohydr Res*, 2002 337, 2505-11.

[5] Chen, H-L; Sheu, W-H; Tai, T-S; Liaw, Y-P; Chen, Y-C. Konjac supplement alleviated hypercholesterolemia and hyperglycemia in type 2 diabetic subjects – a randomised double – blind trial. *J Am Coll Nutr*, 2003 22, 36-42.

[6] Chen, H-L; Cheng, H-C; Liu, Y-J; Liu, S-Y; Wu, W-T. Konjac acts as a natural laxative by increasing stool bulk and improving colonic ecology in healthy adults. *Nutr*, 2006 22, 1112-19.

[7] Elamir, AA; Tester, RF; Al-Ghazzewi, FH; Kaal, HY; Ghalbon, AA; Elmegrahai, NA; Piggott, JR. Effects of konjac glucomannan hydrolysates on the gut microflora of mice. *Nutr Food Sci*, 2008 38, 422-9.

[8] Franceschi, S; Maso, LD; Augustin, L; Negri, E; Parpinel, M; Boyle, P; Jenkins, DJA; La Vecchia, C. Dietary glycemic load and colorectal cancer risk. *Ann Oncol*, 2001 12, 173-8.

[9] Frieri, G; Pimpo, MT; Scarpignato, C. Management of colonic diverticular disease – focus on rifaximin. *Digestion*, 2006 73, 58-66.

[10] Gallaher, DD; Gallaher, CM; Mahrt, GJ; Carr, TP; Hollingshead, CH; Hesslink, R; Wise, J. A glucomannan and chitosan fibre supplement decreases plasma cholesterol and increases cholesterol excretion in overweight normocholesterolemic humans. *J Am Coll Nutr*, 2002 2, 428-33.

[11] Huang, CY; Zhang, MY; Peng, SS; Hong, JR; Wang, X; Jiang, HJ; *et al.* Effect of konjac food on blood glucose level in patients with diabetes. *Biomed Environ Sci*, 1990 3, 123-31.

[12] Innami, S; Shimizu, J; Kudoh, K.Dietary fibre and gastrointestinal functions. *Hydrocolloids*, PT 2- *Fundam Appl Food Biol Med*, 1999, 383-92.

[13] Jones, PJ. Clinical nutrition: 7. Functional foods – more than just nutrition. *Can Med Ass J*, 2002 166, 1555–63.

[14] Kato, K; Matsuda, K. Studies on chemical structure of konjac mannan. 1. Isolation and characterisation of oligosaccharides from partial acid hydrolysate of mannan. *Agric Biol Chem*, 1969 33, 1446–53.

[15] Katsuraya, K; Okuyama, K; Hatanaka, K; Oshima, R; Sato, T; Matsuzaki, K. Constitution of konjac glucomannan: chemical analysis and 13C NMR spectroscopy. *Carbohydr Polym*, 2003 53, 183–9.

[16] Keithley, J; Swanson, B. Glucomannan and obesity: a critical review. *Altern Ther Health Med*, 2005 11, 30-4.

[17] Khanna, S. The chemical, physical and nutritional properties of the plant polysaccharide konjac glucomannan. *PhD Thesis,* 2003, Glasgow Caledonian University, UK.

[18] King, SS; Young, DA; Nequin, LG; Carnevale, EM. Use of specific sugars to inhibit bacterial adherence to equine endometrium *in vitro*. *Am J Vet Res*, 2000 61, 446-9.

[19] Kitamoto, N; Kato, Y; Ohnaka, T; Yokota, M; Tanaka, T; Tsuji, K. Bactericidal effects of konjac fluid on several food poisoning bacteria. *J Food Prot*, 2003 66, 1822-31.

[20] Latella, G; Pimpo, MT; Sottili, S; Zippi, M; Viscido, A; Chiaramonte, M; *et al.* Rifaximin improves symptoms of acquired uncomplicated diverticular disease of the colon. *Int J Colorectal Dis*, 2003 18, 55-62.

[21] Livieri, C; Novazi, F; Lorini, R. The use of highly purified glucomannan based fibres in childhood obesity. *Pediatr Med Chir*, 1992 14, 195-8.

[22] Loening-Baucke, V; Miele, E; Staiano, A. Fiber (glucomannan) is beneficial in the treatment of childhood constipation. *Pediatrics*, 2004 113, E259-64.

[23] Luo, DY. Inhibitory effect of refined Amorphophallus konjac on MNNG-induced lung cancers in mice. *Zhonghua Zhong Liu Za Zhi*, 1992 14, 48-50.

[24] Madrigal-Santillan, E; Alvarez-Conzalez, I; Marquez-Marquez, R; Velazquez-Guadarrama, N; Madrigal-Bujaidar, E. Inhibitory effect of mannan on the toxicity produced in mice fed aflatoxin B-1 contaminated corn. *Arch Environ Contam Toxicol*, 2007 53, 466-72.

[25] Maeda, M; Shimahara, H; Sugiyama, N. Studies of mannan and related-compound. 5. detailed examination of the branched structure of konjac glucomannan. *Agric Biol Chem*, 1980 44, 245–52.

[26] Mangan, DF; Snyder, IS. Mannose-sensitive interaction of *E. coli* with human peripheral leukocytes *in vitro*. *Infect Immun*, 1979 26, 520-7.

[27] Martino, F; Martino, E; Morrone, F; Carnevali, E; Forcone, R; Niglio, T. Effect of dietary supplementation with glucomannan on plasma total cholesterol and low density lipoprotein cholesterol in hypercholesterolemic children. *Nutr Metabol Cardiovascular Dis*, 2005 15, 174-80.

[28] Marzio, L; Del Bianco, R; Donne, MD; Pieramico, O; Cuccurullo, F. Mouth – to – cecum transit time in patients affected by chronic constipation: effect of glucomannan. *Am J Gastroenterol*, 1989 84, 888-91.

[29] Matsuura, Y. Degradation of konjac glucomannan by enzymes in human faeces and formation of short-chain fatty acids by intestinal anaerobic bacteria. *J Nutr Sci Vitaminol*, 1998 44, 423–36.

[30] McCarty, MF. Dietary glycemic index may influence cancer risk by modulating IGF-I activity: a hypothesis. *J Medicinal Food*, 1998 1, 123-40.

[31] McCarty, MF. Glucomannan minimises the postprandial insulin surge: a potential adjuvant for hepatothermic therapy. *Medical Hypotheses*, 2002 58, 487-90.

[32] Melga, P; Giusto, M; Ciuchi, E; Giusti, R; Prando, R. Dietary fibre in the dietetic therapy of diabetes mellitus. Experimental data with purified glucomannans. *Riv Eur Sci Med Pharmacol*, 1992 14, 367-73.

[33] Michaels, EK; Chmiel, JS; Plotkin, BJ; Schaeffer, AJ. Effect of D-mannose and D-glucose on *E. coli* bacteriuria in rats. *Urol Res*, 1983 11, 97-102.

[34] Millane, RP; Hendrixson, TL. Crystal structures of mannan and glucomannans. *Carbohydr Polym*, 1994 25, 245–51.

[35] Mizutani, T; Mitsuoka, T. Effect of konjac mannan on spontaneous liver tumorigenesis and faecal flora in C3H/He male mice. *Cancer Lett*, 1982 17, 27-32.

[36] Mizutani, T; Mitsuoka, T. Effect of konjac mannan on 1, 2-dimethylhydrazine – induced intestinal carcinogenesis in Fischer 344 rats. *Cancer Lett*, 1983 19, 1-6.

[37] Nakajima, N; Ishihara, K; Matsuura, Y. Dietary-fiberdegrading enzymes from a human intestinal Clostridium and their application to oligosaccharide production from non-starchy polysaccharides using immobilized cells. *Appl Microbiol Biotechnol*, 2002 59, 182–9.

[38] Ogawa, K; Yui, T; Mizuno, T. X-ray diffraction study of glucomannans and their acetates. *Agric Biol Chem*, 1991 55, 2105–11.

[39] Oomizu, S; Onishi, N; Suzuki, H; Ueda, K; Mochizuki, M; Morimoto, K; *et al.* Oral administration of pulverised konjac glucomannan prevents the increase of plasma immunoglobulin E and immunoglobulin G levels induced by the injection of syngeneic keratinocyte extracts in BALB/c mice. *Clin Exp Allergy*, 2006 36, 102-10.

[40] Onishi, N; Kawamoto, S; Nishimura, M; Nakano, T; Aki, T; Shigeta, S; *et al.* A new immunomodulatory function of low-viscous konjac glucomannan with a small particle size: its oral intake suppresses spontaneously occurring dermatitis in NC/Nga mice. *Int Arch Allergy Immunol*, 2005 136, 258-65.

[41] Onishi, N; Kawamoto, S; Ueda, K; Yamanaka, Y; Katayama, A; Suzuki, H; *et al.* Dietary pulverised konjac glucomannan prevents the development of allergic rhinitis – like symptoms and IgE response in mice. *Biosci Biotechnol Biochem*, 2007 71, 2551-6.

[42] Papi, C; Ciaco, A; Koch, M; Capurso, L. Efficacy of rifaximin in the treatment of symptomatic diverticular disease of the colon. A multicentre double – blind placebo – controlled trial. *Aliment Pharmacol Ther*, 1995 9, 33-9.

[43] Papi, C; Koch, M; Capurso, L. Management of diverticular disease: Is there room for rifaximin. *Chemotherapy*, 2005 51, 110-14.

[44] Passaretti, S; Franzoni, M; Comin, U; Donzelli, R; Rocca, F; Colombo, E; Ferrara, A; Dinelli, M; Prada, A; Curzio, M. Action of glucomannans on complaints in patients affected with chronic constipation: a multicentric clinical evaluation. *Ital J Gastroenterol*, 1991 23, 421-5.

[45] Petruzziello, L; Iacopini, F; Bulajic, M; Shah, S; Costamagna, G. Review article: uncomplicated diverticular disease of the colon. *Aliment Pharmacol Ther*, 2006 23, 1379-91.

[46] Schaeffer, AJ; Amundsen, SK; Jones, JM. Effect of carbohydrates on adherence of *E. coli* to human urinary tract epithelial cells. *Infect Immun*, 1980 30, 531-7.

[47] Shimahara, H; Suzuki, H; Sugiyama, N; Nizakawa, K. Partial purification of beta-mannanases from konjac tubers and their substrate-specificity in relation to structure of konjac glucomannan. *Agric Biol Chem*, 1975 39, 301–12.

[48] Signorelli, P; Croce, P; Dede, A. A clinical study of the use of a combination of glucomannan with lactulose in the constipation of pregnancy. *Minerva Ginecol*, 1996 48, 577-82.

[49] Staiano, A; Simeone, D; Del Giudice, E; Miele, E; Tozzi, A; Toraldo, C. Effect of the dietary fibre glucomannan on chronic constipation in neurologically impaired children. *J Pediatr*, 2000 136, 41-5.

[50] Sugiyama, N; Shimahara, H; Andoh, T. Studies on mannan and related compounds. I.The purification of konjacmannan. *Bull Chem Soc Japan*, 1972 45, 561–3.

[51] Sutherland, A; Tester, R; Al-Ghazzewi, F; McCulloch, E; Connolly, M. Glucomannan hydrolysate (GMH) inhibition of *Candida albicans* growth in the presence of *Lactobacillus* and *Lactococcus* species. *Microbial Ecology in Health and Disease*, 2008 20, 127-34.

[52] Swanson, KS; Grieshop, CM; Flickinger, EA; Bauer, LL; Healy, HP; Dawson, KA; *et al.* Supplemental fructooligosaccharides and mannanoligosaccharides influence immune function, ileal and total tract nutrient digestibilities, microbial populations and concentrations of protein catabolites in the large bowel of dogs. *J Nutr*, 2002 132, 980-9.

[53] Takao, I; Fujii, S; Ishii, A; Han, LK; Kumao, T; Ozaki, K; Asakawa, A. Effects of mannooligosaccharides from coffee mannan on fat storage in mice fed a high fat diet. *J Health Sci*, 2006 52, 333-7.

[54] Topping, DL; Clifton, PM. Short-chain fatty acids and human colonic function: roles of resistant starch and nonstarch polysaccharides. *Physiol Rev*, 2001 81, 1031–64.

[55] Torrecillas, S; Makol, A; Caballero, MJ; Montero, D; Robaina, L; Real, F; *et al.* Immune stimulation and improved infection resistance in European sea bass (*Dicentrarchus labrax*) fed mannan oligosaccharides. *Fish Shellfish Immunol*, 2007 23, 969-81.

[56] Vita, PM; Restelli, A; Caspani, P; Klinger, R. Chronic use of glucomannan in the dietary treatment of severe obesity. *Minerva Med*, 1992 83, 135-9.

[57] Vuksan, V; Jenkins, DJ; Spadafora, P; Sievenpiper, JL; Owen, R; Vidgen, E; *et al.* Konjac – mannan (glucomannan) improves glycemia and other associated risk factors for coronary heart disease in type 2 diabetes. A randomised controlled metabolic trial. *Diabetes Care*, 1999 22, 913-19.

[58] Vuksan, V; Sievenpiper, JL; Owen, R; Swilley, JA; Spadafora, P; Jenkins, DJ; *et al.* Beneficial effects of viscous dietary fibre from konjac – mannan in subjects with the insulin resistance syndrome: results of a controlled metabolic trial. *Diabetes Care*, 2000 23, 9-14.

[59] Vuksan, V; Sievenpiper, JL; Xu, Z; Wang, EY; Jenkins, AL; Beljan-Zdravkovic, U; *et al.* Konjac-mannan and American ginsing: emerging alternative therapies for type 2 diabetes mellitus. *J Am Coll Nutr*, 2001 20, 370-80.

[60] Walsh, DE; Yaghoubian, V; Behforooz, A. Effect of glucomannan on obese patients: a clinical study. Int J Obesity, 1984 8, 289-93.

[61] Wang, C-H; Lai, P; Chen, M-E; Chen, H-L. Antioxidative capacity produced by Bifidobacterium – and Lactobacillus acidophilus – mediated fermentations of konjac glucomannan and glucomannan oligosaccharides. *J Sci Food Agric*, 2008 88, 1294-1300.

[62] Wang, F-Y; Arisawa, T; Tahara, T; Nagasaka, M; Fujita; Hirata, I; *et al.* The role of mannan – binding lectin (MBL) gene polymorphism in ulcerative colitis. *J Clin Biochem Nutr*, 2008 42, 54-8.

[63] Wong, JMW; de Souza, R; Kendall, CWC; Emam, A; Jenkins, DJA. Colonic health: Fermentation and short chain fatty acids. *J Clin Gastroenterology*, 2006 40, 235–43.

[64] Yamada, K; Tokunaga, Y; Ikeda, A; Ohkura, K; Kaku-Ohkura, S; Mamiya, S; *et al.* Effect of dietary fibre on lipid metabolism and immune function of aged Sprague-Dawley rats. *Biosci Biotechnol Biochem*, 2003 67, 429-33.

In: Food Hydrocolloids: Characteristics, Properties... ISBN: 978-1-60876-222-4
Editor: C. S. Hollingworth, pp. 253-271

Chapter 10

NOVEL APPLICATIONS OF POLYSACCHARIDES FROM LACTIC ACID BACTERIA: A FOCUS ON KEFIRAN

***Analia G Abraham**[*1,3]; **Micaela Medrano**[1]; **Judith A Piermaria**[1,3] **and Fernanda Mozzi**[2]*

[1]Centro de Investigación y Desarrollo en Criotecnología de Alimentos (CIDCA-CONICET), Facultad de Ciencias Exactas (UNLP), La Plata, Buenos Aires, Argentina.
[2]Centro de Referencia para Lactobacilos (CERELA-CONICET), Tucumán, Argentina.
[3] Área Bioquímica y Control de Alimentos, Facultad de Ciencias Exactas,Universidad Nacional de La Plata, La Plata, Argentina

ABSTRACT

The food industry is constantly searching for new food components to improve the texture and mouth-feel of food products; moreover, health-promoting capacity of food ingredients is also desirable.

In recent years, exopolysaccharide (EPS)-producing lactic acid bacteria (LAB) have gained attention due to their major role in functional foods and for improving the sensorial properties of fermented products. Certain LAB excrete a wide diversity of polysaccharides with respect to their monomer composition, molecular weight (MW), structure, and functionalities. These extracellular components or EPS can impart functional properties to food and additionally, may confer beneficial health-promoting properties. The role of EPS from LAB in the dairy food industry (i.e. fermented milks and cheeses) is well documented. Also, the addition of EPS-producing LAB to cereal products or their use as food additives have been successfully applied thus broadening their field of application.

Kefiran, a high-MW EPS produced by microorganisms present in the grains of the fermented milk kefir, has displayed novel applications in the food industry. Kefiran, which has a Newtonian behaviour in diluted solutions but pseudoplastic at high concentrations, improves the viscosity and viscoelastic properties of acid milk gels. This EPS is also able to form gels at low temperatures in addition to brittle and rigid films. However, glycerol addition confers it extremely high elongation values, allowing

* corresponding author: Analia G Abraham, E mail: aga@biol.unlp.edu.ar, analiaabraham@yahoo.com.ar

comparable flexibilities to those of synthetic material. Moreover, several health-promoting properties of kefiran such as immunomodulation or epithelium protection have been reported. Furthermore, kefiran has the ability to antagonize biological activity of *Bacillus cereus* on Caco-2 cells thus providing new perspectives for the role of this bacterial EPS in functional foods.

Exopolysaccharides from Lactic Acid Bacteria

Certain lactic acid bacteria (LAB) produce exopolysaccharides (EPS), which can be capsular polysaccharides (CPS) if they are tightly associated with the cell surface or slime-EPS if they are secreted into the environment (De Vuyst & Degeest, 1999). In spite of the widespread ability to synthesize EPS among bacteria, the physiological role of these polymers, particularly in LAB, remains to be clarified.

Classification and Diversity

According to their chemical composition, EPS can be divided into homopolysaccharides (HoPS), which contain only one type of monosaccharide, and heteropolysaccharides (HePS) that are formed of repeating units composed of two or more types of monosaccharides (De Vuyst & Degeest, 1999; Font de Valdez *et al.*, 2003; De Vuyst & Vaningelgem, 2004). This review will only deal with HePS from LAB as the LAB involved in the production of kefiran that synthesize this type of polymer.

HePS are composed of a backbone of repeating subunits, branched or unbranched, that consist of three to eight monosaccharides, derivatives of monosaccharides or substituted monosaccharides. These HePS usually contain D-galactose, D-glucose, and L-rhamnose although at different ratios. In a few cases fucose, ribose, acetylated amino sugars (such as N-acetylglucosamine and N-acetylgalactosamine), uronic acids and other non-carbohydrate compounds such as phosphate, acetate, and glycerol are also present (De Vuyst & Degeest, 1999; De Vuyst *et al.*, 2001). Industrially relevant LAB species of both mesophilic (i.e. *Lactobacillus casei, Lactobacillus paracasei, Lactobacillus rhamnosus, Lactobacillus sakei, Lactococcus lactis* subsp. *cremoris*) and thermophilic (i.e. *Lactobacillus acidophilus, Lactobacillus delbrueckii* subsp. *bulgaricus, Lactobacillus delbrueckii* subsp. *lactis, Lactobacillus helveticus, Streptococcus thermophilus, Streptococcus macedonicus*) group synthesize HePS (Mozzi *et al.*, 2006). A great diversity of HePS from LAB exists with regard to their composition and structure, molecular mass (MM), yield, and functionalities (De Vuyst & Degeest, 1999; van Kranenburg *et al.*, 1999; De Vuyst & Vaningelgem, 2004; Vaningelgem *et al.*, 2004; Mozzi *et al.*, 2006). The ability of LAB to synthesize CPS or to display ropiness is variable and strain-dependent; ropiness has been correlated with high-MM HePS (Mozzi *et al.*, 2006). Mozzi *et al.* (2006) studied the diversity of HePS from both thermophilic and mesophilic LAB strains and classified the EPS produced into nine different groups according to their monomer composition; the number of different monomers varying from one to four. Four of these HePS groups were similar to the six groups reported by Vaningelgem *et al.* (2004) for *S. thermophilus* strains. It is interesting to note that most HePS from LAB contain galactose in their chemical composition (van Kranenburg *et al.*, 1999;

Marshall *et al.*, 2001; Ruas-Madiedo *et al.*, 2002; Vaningelgem *et al.*, 2004; Mozzi *et al.*, 2006) with only a few exceptions (Robijn *et al.*, 1995; Harding *et al.*, 2003; Sánchez *et al.*, 2006).

More than fifty HePS structures from LAB have been determined using nuclear magnetic resonance (NMR) spectroscopy, 31 being unique structures (De Vuyst & Degeest, 1999; Laws & Marshall, 2001; Faber *et al.*, 2002; Landerjö *et al.*, 2002; Ruas-Madiedo *et al.*, 2002; Van Calsteren *et al.*, 2002; Lipinski *et al.*, 2003; Maeda *et al.*, 2004a; Harding *et al.*, 2005; Sánchez-Medina *et al.*, 2007). HePS may vary in charge, spatial arrangements, rigidity, and the ability to interact with proteins because of their different compositions and MM (Duboc & Mollet, 2001). All these characteristics contribute to the EPS properties such as solubility, rheology, and other physical characteristics of the polymers (Monchois *et al.*, 1999). An interesting property of the HePS produced by LAB is the high thickening power displayed by some of them at very low concentrations.

Biosynthesis and Genetic Organization of HePS

The biosynthesis of HePS involves several enzymes for the production of sugar nucleotide precursors: a glycosyl 1-phosphate transferase (named also *priming*-glycosyltransferase) that transfers the first sugar onto a phosphorylated carrier lipid, one or more glycosyltransferases that sequentially add new sugars to the growing repeating unit, and additional enzymes that participate in *eps* gene regulation, membrane translocation of the repeating units, and polymerization/chain length determination (Broadbent *et al.*, 2003). In contrast to the biosynthesis of HoPS, the precursor repeating-units, are formed intracellularly. The reactions mentioned commonly involve additional proteins whose functions are not unique for this process (Delcour *et al.*, 1999). A key intermediate linking the pathways of HePS production and those of sugar degradation seems to be glucose 6-phosphate, in which the flux of carbon bifurcates towards the formation of fructose 6-phosphate entering the glycolysis, for ATP formation and cell component synthesis, and toward the biosynthesis of sugar nucleotides for HePS formation. In many metabolic pathways leading to the synthesis of the required sugar nucleotides, glucose 1-phosphate is a key metabolite formed from glucose 6-phosphate (Ramos *et al.*, 2001). A correlation between HePS production, the activity of the sugar nucleotide biosynthetic enzymes, and the monomers present in the HePS repeating units has been demonstrated by several researchers (Grobben *et al.*, 1996; Degeest & De Vuyst, 2000; Degeest *et al.*, 2001; Mozzi *et al.*, 2001, 2003; Escalante *et al.*, 2002); however, differences among strains have been reported. In addition, glycosyltransferases are key enzymes in HePS biosynthesis as they catalyse the transfer of sugar moieties from activated donor molecules to specific acceptor molecules, thereby forming glycosidic linkages. The wide structural diversity of HePS from LAB suggests that these bacteria contain a large pool of specific glycosyltransferases for the donor and the acceptor sugar molecules (Stingele *et al.*, 1999). Different glycosyltransferases involved in the assembling of the HePS repeating units from LAB have been described (De Vuyst *et al.*, 2001; Jolly & Stingele, 2001; Jolly *et al.*, 2002; Lamothe *et al.*, 2002; Provencher *et al.*, 2003; Dabour & LaPointe, 2005). The mechanisms of polymerization and export of HePS remain poorly understood (Jolly & Stingele, 2001).

HePS genetic determinants can be located either on plasmid or chromosomal DNA. Genes involved in HePS biosynthesis (*eps* genes) are generally organized in clusters highly conserved with respect to their organization and sequence (Boels *et al.*, 2001). Genes or gene clusters involved in the biosynthesis of HePS from LAB have been reported for *S. thermophilus* (Stingele *et al.*, 1996; Low *et al.*, 1998; Bourgoin *et al.*, 1999; Almirón-Roig *et al.*, 2000), *L. helveticus* (Jolly *et al.*, 2002), *L. lactis* (van Kranenburg *et al.*, 1999; Dabour & LaPointe, 2005), *L. delbrueckii* subsp. *bulgaricus* (Lamothe *et al.*, 2002), *L. rhamnosus* (Peant *et al.*, 2005) and the *L. casei* group (Provencher *et al.*, 2003). The HePS gene clusters appear to be divided into four regions: the first one containing regulation genes, the second region coding for proteins involved in polymerization/chain length determination, the third region containing genes coding for the glycosyltransferase enzymes specifically required for the biosynthesis of the HePS repeating unit, and the fourth one encompassing genes involved in transport and polymerization (Provencher *et al.*, 2003).

Genetic modification of *eps* genes may lead to HePS with different repeating units or with different chain lengths providing an interesting tool for obtaining new HePS with different/specific properties (Jolly *et al.*, 2002). Metabolic and genetic engineering of food-grade LAB strains constitute an interesting strategy to enhance HePS production or to produce HePS with desired specific properties. In general, EPS production may be improved either by optimising the fermentation conditions or by genetic engineering (Årsköld *et al.*, 2007).

HePS Production

One of the main disadvantages concerning HePS-synthesizing LAB is the small amount of polymer produced, usually ranging between 25 to 600 mg/l (De Vuyst & Degeest, 1999; Mozzi *et al.,* 2006). So far, the highest production levels were reported for the mesophilic LAB strains *L. rhamnosus* 9595M (2.8 g/l) and *L. sakei* 0-1 (1.4 g/l) under optimized culture conditions (Degeest & De Vuyst, 2000; Macedo *et al.*, 2002). Temperature, fermentation period, pH and medium composition (carbon and nitrogen sources), dissolved oxygen level and the presence of certain compounds (i.e. citrate, manganese, calcium, adenine) have been described as main factors affecting HePS synthesis (Degeest & De Vuyst, 1999; Font de Valdez *et al.*, 2003; Torino *et al.*, 2005). Culture conditions for optimum HePS production should be optimized for each strain/species to achieve an efficient polymer synthesis. Thus, fermentation strategies such as continuous culture and immobilized cell technology have been applied to improve HePS yields (Grobben *et al.*, 1997; Looijesteijn *et al.*, 2000; Bergmaier *et al.*, 2005; Wellman *et al.*, 2003).

HePS synthesis by thermophilic LAB appears to be biomass associated (Torino *et al.*, 2001); moreover, primary metabolite kinetics has been clearly demonstrated for HePS synthesis by *S. thermophilus* LY03 (De Vuyst & Degeest, 1999). In contrast, HePS production by mesophilic LAB strains is improved under non-optimum growth conditions and may be considered secondary metabolites (De Vuyst & Degeest, 1999).

Degradation of HePS produced by LAB was observed during prolonged fermentation periods, being hydrolysis attributed to the activity of different enzymes (glycohydrolases) responsible for the breakdown of the HePS as observed in strains of *L. rhamnosus* and *S. thermophilus* (Pham *et al.*, 2000; Degeest *et al.*, 2001). Recently, Mozzi *et al.* (2009) have demonstrated partial degradation of the HePS produced by the strains *L. casei* CRL 87 and *S. thermophilus* CRL 1190 after submitting these EPS under the harsh conditions of an *in vitro* gastric system.

Exopolysaccharide From Kefir Grains

Kefir grains, the starter for obtaining the sour fermented milk kefir, are gelatinous irregular particles with a diameter ranging from 2 mm to 4 cm. They resemble cauliflower florets and contain a complex microflora mainly composed of yeasts, lactobacilli, lactococci and acetic acid bacteria (Garrote *et al.*, 1997, 2001; Farnwoth, 2005; Lopitz-Otsoa *et al.*, 2006). The grain is composed of protein, polysaccharide and water. The polysaccharide, produced by microorganisms immersed into kefir grains, is commonly known as kefiran as proposed by La Rivière *et al.* in 1967. At least 24% of the dry material consists of kefiran but this amount changes with the growth media. The polysaccharide content of kefir grain grown in cow's milk is almost twice than that of kefir grain grown in soy milk (Abraham & De Antoni, 1999). This polysaccharide is present in kefir at a concentration of about 250 mg/l (Rimada & Abraham, 2003) and is regarded as safe since kefir milk (containing kefiran) has been consumed for a long time (Koroleva, 1988; Farnworth, 2005).

Exopolysaccharide-Producing Bacteria from Kefir Grains

Several efforts have been directed towards the isolation of polysaccharide-producing LAB from kefir grains; it has been reported that kefiran may be produced by a variety of microorganisms isolated from them (Toba *et al.*, 1987; Mukai *et al.*, 1990; Hosono *et al.*, 1990; Yokoi *et al.*, 1991; Pintado *et al.*, 1996; Micheli *et al.*, 1999).

Since the first report on polysaccharide production by kefir grain microorganisms was published in 1967 (La Rivière *et al.*), the attempt to isolate novel EPS-producing bacteria is yet subject of interest (Santos *et al.*, 2003; Maeda *et al.*, 2004a; Wang *et al.*, 2008). Kefiran biosynthesis was first ascribed to *Lactobacillus brevis* (La Rivière *et al.*, 1967) and later to *Lactobacillus kefir* (Toba *et al.*, 1987). Other studies on EPS-producing *Lactobacillus* strains isolated from kefir showed similar properties to previously reported kefiran (Toba *et al.*, 1987; Yokoi *et al.*, 1990; 1992; Micheli *et al.*, 1999). Frengova *et al.* (2002) isolated the strain *L. delbrueckii* subsp. *bulgaricus* HP1 from kefir grains with ability to produce high amounts of EPS. Whether in fact the diverse bacteria reported in different publications were the same has not been studied nor has any definitive identification been published to fully characterize those bacteria reported as kefiran producers (Farnworth, 2005). Nowadays, kefiran production is associated to *Lactobacillus kefiranofaciens* (Fujisawa *et al.*, 1988; Yokoi *et al.*, 1990) as has been recently reported for the EPS-producing strain *L. kefiranofaciens* ZW3, isolated from Tibet kefir grains (Wang *et al.*, 2008).

Isolation of EPS-producing lactobacilli from kefir has not been very successful probably due to the rigorous growth conditions required by the lactobacilli strains after being transferred in different culture media. Several growth media to isolate EPS-producing lactobacilli from kefir grains were developed; whey-based media have been found to be optimum for kefiran production by these bacteria. Although whey was described to be an essential component of the media, the addition of wine or alcohol improved EPS production (Toba *et al.,* 1986; Yokoi *et al.,* 1990; Yokoi & Watanabe, 1992). Improvement of kefiran yields was also subject of research. Kefiran can be produced in high yields (2.04 g/l) in a modified MRS medium (MRSL) containing lactose, $CaCl_2$ and double amount of nitrogen (Yokoi & Watanabe, 1992). Mitsue *et al.* (1998) obtained the high kefiran-producing strain *L. kefiranofaciens* KF-75 by repeated UV irradiation. They found that kefiran productivity by this strain was successfully optimized by changing the medium composition and culture conditions (Mitsue *et al.*, 1999). Addition of yeast extract and ethanol, aeration of gas containing CO_2, and their combinations promoted kefiran production by *L. kefiranofaciens* in a single culture (Taniguchi *et al.*, 2001). In this way, *L. kefiranofaciens* grows under conditions established by mimicking the action of yeast cells into kefir grain (Taniguchi *et al.*, 2001). The use of gelatinized sago starch was tested as carbon source for improving kefiran production (Yeesang *et al.,* 2008); a mixture of α-amylase and glucoamylase in a 3:2 ratio enhanced the productivity of kefiran and showed the highest kefiran production (Yeesang *et al.,* 2008).

Several attempts to use mixed cultures to enhance kefiran production by *L. kefiranofaciens* were described. During mixed culture growth, lactose is fermented by *L. kefiranofaciens* to kefiran and lactic acid (Cheirsilp *et al.,* 2003). As lactic acid inhibits the growth of *L. kefiranofaciens* (Tada *et al.,* 2007), the consumption of this product by the yeast *Saccharomyces cerevisiae* prevents the accumulation of lactic acid at high concentrations and enhances cell growth and kefiran production by *L. kefiranofaciens* (Cheirsilp *et al.,* 2003).

The best kefiran yields (2380 mg/l), however, have been reported by Mitsue *et al.* (1999) when they combined the kefiran-producing bacterium *L. kefiranofaciens* with the yeast *Torulaspora delbrueckii.*

Kefiran production by whole grains was also reported in milk, whey and soymilk. Kefiran can be obtained from the grains at high yields, being 38% of the total carbohydrate of the grain (Rimada & Abraham, 2006), and kefir biomass can be produced in deproteinized whey (Rimada & Abraham, 2001, 2003).

Kefiran Structure

Kefiran is a water-soluble branched glucogalactan containing approximately equal amounts of D-glucose and D-galactose residues and is fragmented by a β,D (1-6) glucanase giving a pentasaccharide unit and glucose (La Rivière *et al.,* 1967). The structure of the repeating unit, named kefirose, was determined by hydrolysis and methylation analysis. Kefirose is a branched hexa- or heptasaccharide composed of a pentasaccharide lineal unit to which one or two sugar residues are randomly linked (Kooiman, 1968; Micheli *et al.*, 1999). Analysis of kefiran obtained from grains by using different isolation treatments (i.e. solubilization in hot water or acid media, and/or precipitation with ethanol) or from fermented products showed differences in its molecular weight (Toba *et al.,* 1987). Methylation/hydrolysis studies showed that kefiran presents a more complex structure than that described above, indicating the coexistence of molecules with microheterogeneities in the structure (Mukai *et al.*, 1988). Liu *et al.* (2002) demonstrated that kefir grains grown in soy milk produce an EPS that is primarily composed of D-glucose and D-galactose (ratio 1.00: 0.43), with a MW of approximately 1.7 x 10^6 Da.

Several methodologies have been used to verify if the EPS produced by isolated bacterial strains was kefiran. The polysaccharide produced by *Lactobacillus sp.* KPB-167B was found to be similar to kefiran by 13C-NMR and methylation analysis (Yokoi *et al.*, 1991). FT-IR spectroscopy revealed the presence of carboxyl, hydroxyl, and amide groups, in the glucogalactan produced by *L kefiranofaciens* ZW3. A capsular-polysaccharide-producing, slime-forming *Lactobacillus* strain LM-17 was isolated from kefir grains. This EPS was identical to the glucogalactan extracted from kefir grains according to 1H- and 13C-NMR spectral data (Micheli *et al.,* 1999).

Maeda *et al.* (2004a) studied the EPS produced by *L. kefiranofaciens* WT-2BT in a liquid medium containing a rice hydrolysate. Sugar composition analysis showed that the polysaccharide was composed of glucose and galactose in a molar ratio of 1.00:1.05. Methylation analysis, 1H- and 13C-NMR spectroscopy and gel permeation chromatography revealed that the polysaccharide was composed of a hexasaccharide-repeating unit with a weight-average MW about 7.6 x 10^5 g/mol.

Physicochemical and Functional Properties of Kefiran

Physicochemical Properties

Kefiran dissolves slowly in cold water, quickly in hot water and precipitates in two volumes of cold ethanol. These properties are used for their isolation and purification from grains or culture media.

The intrinsic viscosity of a polysaccharide provides an insight of the hydrodynamic volume occupied by a given polymer; this parameter can be determined by diluted solution viscosity measurements. Infinitely diluted solutions of a polysaccharide can be considered as systems in which each individual polysaccharide coil is well separated from one another, free to move independently and each chain occupies a spherical volume. The intrinsic viscosity of kefiran obtained by applying Huggins and Kraemer extrapolations was 6.0 and 5.95 dl/g, respectively (Piermaria *et al.,* 2008) in agreement with Micheli *et al.* (1999), who obtained values of 5.7 and 5.36 dl/g for the same polysaccharide. Lower values for intrinsic viscosity of kefiran were reported by Pintado *et al.* (1996) and Mukai *et al.* (1990), who indicated values of 2.9 and 1.5 dl/g, respectively and the viscosity of the polysaccharide produced by the strain *L. kefiranofaciens* LM-17 was 3.08 dl/g. For random coil polysaccharides radius of gyration (Rg) depends on molar mass and for more expanded polysaccharides Rg is larger at a given molar mass. These observations mean that, to obtain a higher (intrinsic) viscosity, either the molar mass (chain length) or the polysaccharide Rg (stiffness) should be increased (Ruas Madiedo *et al.,* 2008).

Kefiran forms viscous solutions with Newtonian flow behaviour at concentration below 1 g/l. At higher concentrations, a pseudoplastic or shear thinning flow is observed. Experimental data fits the Ostwald de Waele model. The flow index (n) varies from 1.00 to 0.52 when kefiran concentration increases from 1.0 to 27.3 g/l (Piermaria *et al.,* 2008). Carboxymethyl kefiran, at concentration ranging from 1 to 3 g/l, has a higher viscosity than kefiran, as determined by capilar viscometry. The increase in viscosity was attributed to the increase in the net charge of the polymer (Mukai *et al.,* 1990).

Functional Properties

Kefiran also can be used as a food grade additive for fermented products since it enhances the rheological properties of chemically acidified skim milk gels increasing their apparent viscosity and the elastic (G′) and viscous (G′′) modulus of these gels. This phenomenon was strengthened by previous heat treatment usually applied for yogurts manufacture (Rimada & Abraham, 2006).

Kefiran can form weak gels when added to k-carrageenan (1%, 1:4 kefiran/k-carrageenan ratio), which have gelation and melting temperatures similar to those of guar/k-carrageenan gels (Pintado *et al.,* 1996), and can form gels in aqueous solutions containing ethanol. Gel strength prepared with 3% of kefiran in 8% of ethanol was equivalent to the strength of 3% gelatine gel (Mukai *et al.,* 1991).

Freezing kefiran solutions provoke a forced alignment of kefiran molecules that may provide a mechanism for the formation of side-by-side associations, which then remain intact on thawing, forming what is known as cryogel. Kefiran solutions are capable to produce translucent cryogels with high water holding capacity (90.43 ± 1.51% for those obtained from 1% w/v solutions) and sufficiently cohesive to support their own weight. The evaluation of these gels using small deformation mechanical tests shows that after freeze-thaw treatment the rheological behaviour changes from a liquid entangled network system to a gel. The mechanical spectra of frozen samples illustrates that G′ (elastic modulus) is greater than G" (viscous modulus) and both moduli become almost independent on frequency indicating a gel like-behaviour. The G′ value in cryogels at a given frequency (1 Hz) was 35 times higher than G′ value for the solution. The behaviour of kefiran cryogels at about 37°C determines its ability to melt at mouth temperature; this property becomes relevant for its application in food formulations (Piermaria *et al.*, 2008).

Several hydrocolloids are able to stabilize emulsions by increasing the viscosity of the continuous phase. EPS produced by *L. kefiranofaciens* ZW3 was able to stabilize emulsions (hexadecane/water) as well as other commercially available hydrocolloids like xanthan gum, guar gum and locust gum. Besides, this EPS showed similar flocculation stability to xanthan gum but better than guar gum (Wang *et al.*, 2008).

Kefiran is able to produce transparent, flexible, homogeneous and extremely thin edible films. An edible film or coating is defined as a thin layer of edible material, which is used in an independent manner, applied to a food or placed between food components (Piermaria *et al.,* 2009). Edible films and/or coatings must be free of toxic compounds and possess physical, chemical and microbiological stability before, during and after the application to food. In turn, they must have good sensory characteristics and do not have taste or smell because these are key qualities in the acceptability of the final product. These films must have appropriated mechanical and barrier characteristics to the purpose for which they will be used and when possible they should be obtained by simple, low polluting and affordable technologies. In addition, edible films must be compatible with food characteristics and legislation (Debeaufort *et al.*, 2000). Kefiran films observed by scanning electron microscopy (SEM) showed smooth surfaces and compact and homogeneous structures. Kefiran films have good moisture barrier properties but they are brittle and rigid since they show high elastic modulus and tensile strength values and low deformation at break. Incorporation of glycerol as plasticizer decrease resistance to traction and increase elongation (Piermaria *et al.,* 2009). It is noteworthy that films containing glycerol at concentrations greater than 25 g of glycerol every 100 g of kefiran showed very high values of elongation similar to that of high-density polyethylene (Smith, 1986). This mechanical characteristic places the glycerol plasticized kefiran film as superior, according to the classification made by Han & Gennadios (2005), developed on the basis of the compilation of results for edible and synthetic films usually used in food packaging.

BIOLOGICAL FUNCTIONALITY OF KEFIRAN

Due to its chemical structure and type of linkages, kefiran may not be hydrolyzed by enzymes present in the upper gastrointestinal tract, so it may reach the small intestine where it might exert biological effects by: i) stimulating the activity of some probiotic/benefic microorganisms (prebiotic functionality), ii) modulating the cytopathic effect of some intestinal pathogens or iii) modifying specific immune cell populations.

Virulence Factors Antagonism

It has been demonstrated that kefiran is able to antagonize virulence mechanisms of *Bacillus cereus* in an *in vitro* model. Studies performed with human enterocytes in culture (Caco-2 cell line) showed that the presence of kefiran (800 mg/l) was able to diminish cytopathic effects triggered by extracellular factors present in supernatant cultures of *B. cereus* strain B10502 (Medrano *et al.,* 2008) by avoiding cell damage (induced by *B. cereus* extracellular factors) such as morphological alterations, F-actin cytoskeleton disorganization, decrease of mitochondrial activity and cell viability and enterocytes detaching from cell culture plate surface. This antagonism was also demonstrated for cell damage caused by *B. cereus* B10502 vegetative cells interaction with Caco-2 cells (Medrano *et al.,* 2009). In these studies, kefiran was able to prevent bacterial association to enterocytes, avoiding morphological alterations, antagonizing the induction of cell death events and preventing cell detachment from the substrate. Kefiran ability to interact both with bacteria and enterocytes surface could be responsible for the protective effect observed by interfering with some eukaryotic cell membrane domains.

Wound Healing Activity

Antimicrobial and cicatrizing activities of kefiran were studied by Rodrigues *et al.* (2005b). These authors tested kefiran antimicrobial activity *in vitro* (using agar diffusion method) against several bacterial species and *Candida albicans.* The results showed that *Streptococcus pyogenes* was the most sensitive microorganism to kefiran, followed by *Staphylococcus aureus* and *Streptococcus salivarius.* Cicatrizing experiments were carried out with Wistar rats with induced skin lesions infected with *S. aureus*. Topical administration of kefiran (dried grains) showed a protective effect on skin connective tissue and seven days treatment enhanced wound healing.

Anti-Inflammatory Effect

Rodrigues *et al.* (2005a) studied the anti-inflammatory effect of kefiran in an *in vivo* model using a cotton-induced granuloma and paw edema assays in rats and carrageenan, dextran and histamine as stimuli. Kefiran showed significant inhibition in the formation of granuloma tissue and inflammatory process for all the test groups, as compared to the control group.

Kwon *et al.* (2008) evaluated the effect of kefiran on the ovalbumin-induced asthma mouse model in which airway inflammation and airway hyper-responsiveness were found in the lung. BALB/c mice sensitized and challenged to ovalbumin were treated intra-gastrically with kefiran 1 hour before the ovalbumin challenge. These authors found that kefiran suppressed ovalbumin-induced airway hyper-responsiveness to inhaled methacholine. Administration of kefiran significantly inhibited the release of both eosinophils and other inflammatory cells into bronchoalveolar lavage fluid and lung tissue. Interleukin-4 (IL-4) and interleukin-5 (IL-5) cytokines were also reduced to normal levels after administration of kefiran in bronchoalveolar lavage fluid.

Blood Pressure and Cholesterol Levels

Kefiran also demonstrated biological activity by suppressing the increase of blood pressure and reducing serum cholesterol levels *in vitro*. Maeda *et al.* (2004b) isolated kefiran from culture medium of *L. kefiranofaciens* and treated SHRSP rats (Stroke-prone spontaneously hypertensive rats) fed with high cholesterol diet. They found that this EPS was able to diminish blood pressure and serum LDL and total cholesterol and triglycerides levels compared to animals not treated with kefiran. Moreover, the ability of kefiran to diminish blood glucose in KKAy mice (genetically diabetic mice) was demonstrated. In addition, the authors found an improvement in the levels of faecal moisture and wet weight of faeces when it was administrated to SD (Sprague-Dawley) rats with constipation induced by a low-fiber diet.

Antitumoral Activity

Shiomi *et al.* (1982) studied antitumoral activity of kefiran isolated from grains by oral and intraperitoneal administration in mice, finding a decrease of the tumor growth in mice treated with the EPS. These authors related the antitumoral activity to an immune cell mediated response, although the mechanism remains to be elucidated.

Murofushi *et al.* (1983) studied the remission of induced tumors in mice treated with kefiran isolated from grains. These authors observed that oral administration of kefiran to mice was able to reduce tumor growth when it was administered before the carcinoma establishment. On the other hand, they concluded that kefiran was able to induce a cell mediated immune response (Murofushi *et al.*, 1986).

Immunomodulatory Activity

Evidence presented in the above mentioned studies suggest that antitumoral functions of kefiran may be related to its immunomodulatory activity. Furukawa *et al.* (1992; 1996) demonstrated that the soluble fraction of kefir (containing kefiran) could act as immune response modulator. These authors incubated kefiran isolated from grains with cells obtained by disaggregating Peyer`s patches of mice with tumors and found that the obtained supernatant was able to increment B cell proliferation in control mice and to increase the activity of mitogens (lipopolysaccharide and phytohaemaglutinin-A) in proliferation assays with B cells.

Vinderola *et al.* (2006) demonstrated that kefiran (obtained from the supernatant culture of *L. kefiranofaciens*) orally administered to mice induced an immune response on the intestinal mucosa, by maintaining homeostasis and enhancing IgA production. This EPS had an influence on systemic immune response by modifying cytokine balance on serum. Nevertheless, the phenotype of cells as well as lymphoid tissues involved in this response remains unknown.

Conclusions

Traditionally, polysaccharide-producing lactic acid bacteria has been employed in the dairy industry to confer texture properties to fermented product but nowadays, a huge market in healthy compounds has appeared and the application of food grade polysaccharide from lactic acid bacteria is an interesting alternative. A great diversity of heteropolysaccharides from lactic acid bacteria exists with regard to their composition and structure, molecular mass, yield, and functionalities but they are produced in small amount to be used as food additive. Kefiran, the polysaccharide produced by microorganisms immersed into kefir grains, is a water-soluble branched glucogalactan containing approximately equal amounts of D-glucose and D-galactose residues. Kefiran is present into kefir grains and extraction from grains is simple with good yield, besides, deproteinised whey could be utilized as a substrate for kefir grain biomass production.

Kefiran from kefir grain proved to be an interesting potential food grade thickening ingredient in fermented dairy products and is able to form gels at low temperature with interesting properties for applications in food formulations. Kefiran can form edible transparent films. The addition of glycerol allowed obtaining a matrix with low water vapor permeability, and extraordinary flexibility. These characteristics of plasticized kefiran films enhance their potential uses, especially for food industry.

Due to its chemical structure and type of linkages, kefiran may not be hydrolyzed by enzymes present in the upper gastrointestinal tract, so it may reach the small intestine where it might exert biological effects such as antagonism of *B cereus* virulence factors, wound-healing activity, anti-inflammatory effect, decrease of cholesterol levels and antitumoral and immunomodulatory activity. The use of this food grade polymer as an ingredient with health-promoting properties could represent an advantageous alternative for the development of functional foods.

Acknowledgements

The authors thank the financial support of their research work by the projects "Biopolímeros de bacterias lácticas: nuevas aplicaciones en alimentos" Agencia Nacional de Promoción Científica y Tecnológica Proyecto PICTR2004 N°09-20801- CONICET and UNLP.

References

[1] Abraham, A. G. & De Antoni, G. L. (1999). Characterization of kefir grains grown in cow's milk and in soya milk. *Journal of Dairy Research, 66,* 327-333.

[2] Almirón-Roig, E.; Mulholland, F.; Gasson, M. J. & Griffin, A. M. (2000). The complete cps gene cluster from *Streptococcus thermophilus* NCFB 2393 involved in the biosynthesis of a new exopolysaccharide. *Microbiology, 146,* 2793-2802.

[3] Årsköld, E.; Svensson, M.; Grage, H.; Roos, S.; Radström, P. & van Niel, E. W. J. (2007). Environmental influences on exopolysaccharide formation in *Lactobacillus reuteri* ATCC 55730. *International Journal of Food Microbiology, 116,* 159-167.

[4] Bergmaier, D.; Champagne, C. P. & Lacroix, C. (2005). Growth and exopolysaccharide production during free and immobilized cell chemostat culture of *Lactobacillus rhamnosus* RW-9595M. *Journal of Applied Microbiology, 98*, 272-284.

[5] Boels, I. C.; Ramos, A.; Kleerebezem, M. & de Vos, W. M. (2001). Functional Analysis of the *Lactococcus lactis* galU and galE Genes and Their Impact on Sugar Nucleotide and Exopolysaccharide Biosynthesis. *Applied and Environmental Microbiology, 67,* 3033-3040.

[6] Bourgoin, F.; Pluvinet, A.; Gintz, B.; Decaris, B. & Guédon, G. (1999). Are horizontal transfers involved in the evolution of the *Streptococcus thermophilus* exopolysaccharide synthesis loci? *Gene, 233,* 151-161.

[7] Broadbent, J. R.; McMahon, D. J.; Welker, D. L.; Oberg, C. J. & Moineau, S. (2003). Biochemistry, genetics, and applications of exopolysaccharide production in *Streptococcus thermophilus*: a review. *Journal of Dairy Science, 86,* 407-423.

[8] Cheirsilp, B.; Shimizu, H. & Shioya, S. (2003). Enhanced kefiran production by mixed culture of *Lactobacillus kefiranofaciens* and *Saccharomyces cerevisiae*. *Journal of Biotechnology, 100,* 43-53.

[9] Dabour, N. & LaPointe, G. (2005). Identification and molecular characterization of the chromosomal exopolysaccharide biosynthesis gene cluster from *Lactococcus lactis* subsp. *cremoris* SMQ-461. *Applied and Environmental Microbiology, 71,* 7414-7425.

[10] De Vuyst, L. & Degeest, B. (1999). Heteropolysaccharides from lactic acid bacteria. *FEMS Microbiology Reviews, 23,* 153-177.

[11] De Vuyst, L. & Vaningelgem, F. (2004) Developing new polysaccharides from lactic acid bacteria. In: B. M. Mc.Kenna (Ed.), *Texture in Food. Vol. 2: Semi-solid Foods.* (pp. 275-320). Cambridge, Woodhead Publishing Ltd.: Publisher.

[12] De Vuyst, L.; de Vin, F.; Vaningelgem, F. & Degeest, B. (2001). Recent developments in the biosynthesis and applications of heteropolysacharides from lactic acid bacteria. *International Dairy Journal, 11,* 687-707.

[13] Debeaufort, F.; Quezada-Gallo, J. A. & Voilley, A. (2000); Edible barriers: A solution to control water migration in foods. In: S. J. Risch (Ed.), *Food packaging.* (pp. 9-16) Washington, American Chemical Society: Publisher.

[14] Degeest, B. & De Vuyst, L. (1999). Indication that the nitrogen source influences both amount and size of exopolysaccharides produced by *Streptococcus thermophilus* LY03 and modelling of the bacterial growth and exopolysaccharide production in a complex medium. *Applied and Environmental Microbiology, 65,* 2863-2870.

[15] Degeest, B. & De Vuyst, L. (2000). Correlation of activities of the enzymes α-phosphoglucomutase, UDP-galactose 4-epimerase, and UDP-glucose pyrophosphorylase with exopolysaccharide biosynthesis by *Streptococcus thermophilus* LY03. *Applied and Environmental Microbiology, 66,* 3519-3527.

[16] Degeest, B., Vaningelgem, F. & De Vuyst, L. (2001). Microbial physiology, fermentation kinetics, and processing engineering of heteropolysaccharide production by lactic acid bacteria. *International Dairy Journal, 11,* 747-757.

[17] Delcour, J.; Ferain, T.; Deghorain, M.; Palumbo, E. & Hols, P. (1999). The biosynthesis and functionality of the cell-wall of lactic acid bacteria. *Antonie van Leeuwenhoek, 76,* 159-184.

[18] Duboc, P. & Mollet, B. (2001). Applications of exopolysaccharides in the dairy industry. *International Dairy Journal, 11,* 759-768.

[19] Escalante, A.; Villegas, J.; Wacher, C.; García-Garibay, M. & Farrés, A. (2002). Activity of the enzymes involved in the synthesis of exopolysaccharide precursors in an overproducing mutant ropy strain of *Streptococcus thermophilus. FEMS Microbiology Letters, 209,* 289-293.

[20] Faber, E. J.; van Haaster, D. J.; Kamerling, J. P. & Vliegenthart, J. F. (2002). Characterization of the exopolysaccharide produced by *Streptococcus thermophilus* 8S containing an open chain nononic acid. *European Journal of Biochemistry, 269,* 5590-5598.

[21] Farnworth, E. (2005). Kefir - a complex probiotic. *Food Science and Technology Bulletin: Functional foods, 2,* 1-17.

[22] Font de Valdez, G.; Torino, M. I.; De Vuyst, L. & Mozzi, F. (2003). Food-grade heteropolysaccharides: ongoing research and future trends of biopolymers from lactic acid bacteria. *Applied Biotechnology Food Science and Policy, 1,* 223-233.

[23] Frengova, G. I.; Simova, E. D.; Beshkova, D. M. & Simov, Z. I. (2002). Exopolysaccharides Produced by Lactic Acid Bacteria of Kefir Grains. *Zeitschrift für Naturforschung, 57,* 805-810.

[24] Fujisawa, T.; Adachi, S.; Toba, T.; Arimara, K. & Mitsuoka, T. (1988). *Lactobacillus kefiranofaciens* sp. nov. isolated from kefir grains. *International Journal of Systematic Bacteriology, 38,* 12–14.

[25] Furukawa, N.; Iiyama, R.; Takahashi, T. & Yamanaka, Y. (1992). Effect of oral administration of water soluble fraction from kefir grain on antibody production in mice. *Animal Science Technology (Japan), 63,* 428-436.

[26] Furukawa, N.; Takahashi, T. & Yamanaka, Y. (1996). Effects of supernatant of Peyer's Patch cell culture with kefir grain components on the mitogenic response of thymocyte and splenocyte in mice. *Animal Science Technology (Japan), 67,* 153-159.

[27] Garrote, G.; Abraham, A. G. & De Antoni, G. (2001). Chemical and microbiological characterization of kefir grains from different origins. *Journal of Dairy Research, 68,* 639-652.

[28] Garrote, G. L.; Abraham, A. G. & De Antoni, G. L. (1997). Preservation of kefir grain, a comparative study. *Lebensmittel-Wissenchaft und Technologie, 30,* 77- 84.

[29] Grobben, G. J.; Smith, M. R.; Sikkema, J. & de Bont, J. A. M. (1996). Influence of fructose and glucose on the production of exopolysaccharides and the activities of enzymes involved in the sugar metabolism and the synthesis of sugar nucleotides in *Lactobacillus delbrueckii* subsp. *bulgaricus* NCFB 2772. Applied *Microbiology and* Biotechnology, *46,* 279–284.

[30] Grobben, G. J.; van Casteren, W. H .M.; Schols, H. A.; Oosterveld, A.; Sala, G.; Smith, M. R.; Sikkema, J. & de Bont, J. A. M. (1997). Analysis of the exopolysaccharides produced by *Lactobacillus delbrueckii* subsp. *bulgaricus* NCFB 2772 grown in a continuous culture on glucose and fructose. Applied *Microbiology and* Biotechnology, *48,* 516-521.

[31] Han, J. H. & Gennadios (2005) A. Edible films and coatings: a review. In: J. H. Han (Ed.), *Innovations in food packagings.* (pp. 239-262). Amsterdam, Elsevier: Publisher.

[32] Harding, L. P.; Marshall, V. M.; Elvin, M.; Gu, Y. & Laws, A. (2003). Structural characterisation of a perdeuteriomethylated exopolysaccharide by NMR spectroscopy: characterisation of the novel exopolysaccharide produced by *Lactobacillus delbrueckii* subsp. *bulgaricus* EU23. *Carbohydrate Research, 338,* 61-67.

[33] Harding, L. P.; Marshall, V. M.; Hernández, Y.; Gu, Y.; Maqsood, M.; Mclay, N. & Laws, A. P. (2005). Structural characterisation of a highly branched exopolysaccharide produced by *Lactobacillus delbrueckii* subsp. *bulgaricus* NCFB2074. *Carbohydrate Research, 340,* 1107-1111.

[34] Hosono, A.; Tanabe, T. & Otani, H. (1990). Binding properties of lactic acid bacteria isolated from kefir milk with mutagenic amino acid pyrolyzates. *Milchwissenschaft, 45,* 647-651.

[35] Jolly, L. & Stingele, F. (2001). Molecular organization and functionality of exopolysaccharide gene clusters in lactic acid bacteria. *International Dairy Journal, 11,* 733-745.

[36] Jolly, L.; Newell, J.; Porcelli, I.; Vincent, S. J. & Stingele, F. (2002). *Lactobacillus helveticus* glycosyltransferases: from genes to carbohydrate synthesis. *Glycobiology, 12,* 319-327.

[37] Kooiman, P. (1968). The chemical structure of kefiran, the water-soluble polysaccharide of kefir grain. *Carbohydrate Research, 7,* 200-211.

[38] Koroleva, N. S. (1988). Technology of kefir and kumys. *Bulletin of the International Dairy Federation, 227,* 96-100.

[39] Kwon, O. K.; Ahn, K. S.; Lee, M. Y.; Kim, S. Y.; Park, B. Y.; Kim, M. K.; Lee, I. Y.; Oh, S. R. & Lee, H. K. (2008). Inhibitory effect of kefiran on ovalbumin-induced lung

inflammation in a murine model of asthma. *Archives of Pharmacal Research, 31,* 1590-1596.

[40] La Rivière, J. W.; Kooiman, P. & Schmidt, K. (1967). Kefiran, a novel polysaccharide produced in the kefir grain by *Lactobacillus brevis*. *Archives of Microbiology, 59,* 269-278.

[41] Lamothe, G. T.; Jolly, L.; Mollet, B. & Stingele, F. (2002). Genetic and biochemical characterization of exopolysaccharide biosynthesis by *Lactobacillus delbrueckii* subsp. *bulgaricus*. *Archives of Microbiology, 178,* 218-228.

[42] Landerjö, C.; Yang, Z.; Huttunen, E. & Wildmalm, G. (2002). Structural studies of the exopolysaccharide produced by *Lactobacillus rhamnosus* strain GG (ATCC 53103). *Biomacromolecules, 3,* 880-884.

[43] Laws, A. P. & Marshall, V. M. (2001). The relevance of exopolysaccharides to the rheological properties in milk fermented with ropy strains of lactic acid bacteria. *International Dairy Journal, 11,* 709-721.

[44] Lipinski, T.; Jones, C.; Lemercinier, X.; Korseniowska-Kowal, A.; Strus, A.; Rybka, J.; Gamian, A. & Heczo, P. B. (2003). Structural analysis of the *Lactobacillus rhamnosus* strain KL37C exopolysaccharide. *Carbohydrate Research, 338,* 605–609.

[45] Liu, J. R.; Chen, M. J. & Lin, C. W. (2002). Characterization of polysaccharide and volatile compounds produced by kefir grains grown in soymilk. *Journal of Food Science, 67,* 104-108.

[46] Looijesteijn, P.; van Casteren, W. H. M.; Tuinier, R.; Doeswijk-Voragen, C . H. L. & Hugenholtz, J. (2000). Influence of different substrate limitations on the yield, composition and molecular mass of exopolysaccharides produced by *Lactococcus lactis* subsp. *cremoris* in continuous cultures. *Journal of Applied Microbiology, 89,* 116-121.

[47] Lopitz-Otsoa, F.; Rementeria, A.; Elguezabal, N. & Garaizar, J. (2006). Kefir: a symbiotic yeasts-bacteria community with alleged healthy capabilities. *Revista Iberoamericana de Micología, 23,* 67-74.

[48] Low, D.; Ahlgren, J. A.; Horne, D.; McMahon, D. J.; Oberg, C. J. & Broadbent, J. R. (1998). Role of *Streptococcus thermophilus* MR-1C capsular exopolysaccharide in cheese moisture retention. *Applied and Environmental Microbiology, 64,* 2147-2151.

[49] Macedo, M. G.; Laporte, M. F. & Lacroix, C. (2002). Quantification of exopolysaccharide, lactic acid, and lactose concentrations in culture broth by near-infrared spectroscopy. *Journal of Agricultural Food Chemistry, 50,* 1774-1779.

[50] Maeda, H.; Zhu, X.; Omura, K.; Suzuki, S. & Kitamura, S. (2004b). Effects of an exopolysaccharide (kefiran) on lipids, blood pressure, blood glucose, and constipation. *Biofactors, 22,* 197-200.

[51] Maeda, H.; Zhu, X.; Suzuki, S.; Suzuki, K. & Kitamura, S. (2004a). Structural characterization and biological activities of an exopolysaccharide kefiran produced by *Lactobacillus kefiranofaciens* WT-2B(T). *Journal of Agricultural Food Chemistry, 52,* 5533-5538.

[52] Marshall, V. M.; Laws, A. P.; Gu, Y.; Levander, F.; Rådström, P.; De Vuyst, L.; Degeest, B.; Vaningelgem, F.; Dunn, H. & Elvin, M. (2001). Exopolysaccharide-producing strains of thermophilic lactic acid bacteria cluster into groups according to their EPS structure. *Letters in Applied Microbiology, 32,* 433-437.

[53] Medrano, M.; Hamet, M. F.; Abraham, A. G. & Pérez, P. F. (2009). Kefiran Protects Caco-2 Cells from Cytopathic Effects Induced by *Bacillus cereus* Infection. *Antonie*

van Leeuwenhoek International Journal of General and Molecular Microbiology. DOI 10.1007/s10482-009-9366-z.

[54] Medrano, M.; Pérez, P. F. & Abraham, A. G. (2008). Kefiran antagonizes cytopathic effects of *Bacillus cereus* extracellular factors. *International Journal of Food Microbiology, 122,* 1-7.

[55] Micheli, L.; Uccelletti, D.; Palleschi, C. & Crescenzi, V. (1999). Isolation and characterisation of a ropy *Lactobacillus* strain producing the exopolysaccharide kefiran. Applied Microbiology and Biotechnology**,** *53,* 69-74.

[56] Mitsue, T.; Tachibana, K.; Hara, T. & Fujio, Y. (1998). Isolation of kefiran-producing lactic acid bacteria from kefir grain and improvement of kefiran productivity – Note. *Seibutsu-kogaku Kaishi, 76,* 447-450.

[57] Mitsue, T. A.; Tachibana, K. A. & Fujio, Y. B. (1999). Efficient kefiran production by a mixed culture of *Lactobacillus kefiranofaciens* KF-75 and yeast strains – Note. *Seibutsu-kogaku Kaishi, 77,* 99-103.

[58] Monchois, V.; Willemot, R. V. & Monsan, P. (1999). Glucansucrases: mechanism of action and structure-function relationships. *FEMS Microbiology Reviews, 23,* 131-151.

[59] Mozzi F.; Gerbino E.; Font de Valdez G. & Torino M. I. (2009). Functionality of exopolysaccharides produced by lactic acid bacteria in an in vitro gastric system. *Journal of Applied Microbiology, 107,* 56-64.

[60] Mozzi, F.; Rollán, G.; Savoy de Giori, G. & Font de Valdez, G. (2001). Effect of galactose and glucose on the exopolysaccharide production and the activities of biosynthetic enzymes in *Lactobacillus casei* CRL 87. *Journal of Applied Microbiology, 91,* 160-167.

[61] Mozzi, F.; Savoy de Giori, G. & Font de Valdez, G. (2003). UDP-galactose 4-epimerase: a key enzyme in exopolysaccharide formation by *Lactobacillus casei* CRL 87 in controlled pH batch cultures. *Journal of Applied Microbiology, 94*, 175-183.

[62] Mozzi, F. ; Vaningelgem, F. ; Hébert, E. M.; Van der Meulen, R.; Foulquié-Moreno, M. R.; Font de Valdez, G. & De Vuyst, L. (2006). Diversity of heteropolysaccharides-producing lactic acid bacterium strains and their biopolymers. *Applied and Environmental Microbiology, 72,* 4431-4435.

[63] Mukai, T.; Toba, T.; Itoh, T. & Adachi, S. (1988). Structural Microheterogeneity of Kefiran from Kefir Grains. *Japanese Journal of Zootechnical Science, 9,* 167-176.

[64] Mukai, T.; Toba, T.; Itoh, T.; Nimura, T. & Adachi, S. (1990). Carboxymethyl Kefiran: Preparation and Viscosimetric Properties. *Journal of Food Science, 55,* 1483-1484.

[65] Mukai, T.; Watanabe, N.; Toba, T.; Itoh, T. & Adachi, S. (1991). Gel-Forming Characteristics and Rheological Properties of Kefiran. *Journal of Food Science, 56,* 1017-1018, 1026.

[66] Murofushi, M.; Mizuguchi, J.; Aibara, K. & Matuhasi, T. (1986). Immunopotentiative effect of polysaccharide from kefir grain, KGF-C, administered orally in mice. *Immunopharmacology, 12,* 29-35.

[67] Murofushi, M.; Shiomi, M. & Aibara, K. (1983). Effect of orally administered polysaccharide from kefir on delayed-type hypersensitivity and tumor growth in mice. *Japanese Journal of Medical Science & Biology, 36,* 49-53.

[68] Peant, B.; LaPointe, G.; Gilbert, C.; Atlan, D.; Ward, P. & Roy, D. (2005). Comparative analysis of the exopolysaccharide biosynthesis gene clusters from four strains of *Lactobacillus rhamnosus. Microbiology, 151*, 1839-1851.

[69] Pham, P. L.; Dupont, I.; Roy, D.; LaPointe, G. & Cerning, J. (2000). Production of Exopolysaccharide by *Lactobacillus rhamnosus* R and Analysis of Its Enzymatic Degradation during Prolonged Fermentation. *Applied and Environmental Microbiology, 66,* 2302-2310.

[70] Piermaria, J.; Pinotti, A.; García, M. A. & Abraham, A. G. (2009). Films based on kefiran, an exopolysaccharide obtained from kefir grains. Development and characterization. *Food Hydrocolloids, 23,* 684-690.

[71] Piermaria, J. A.; de la Canal, M. L. & Abraham, A. G. (2008). Gelling properties of Kefiran, a food grade polysaccharide obtained from kefir grain. *Food Hydrocolloids, 22,* 1520-1527.

[72] Pintado, M. E.; Lopes Da Silva, J. A.; Fernandes, P. B.; Malcata, F. X. & Hogg, T. A. (1996). Microbiological and rheological studies on Portuguese kefir grains. *International Journal of Food Science & Technology, 31,* 15-26.

[73] Provencher, C.; LaPointe, G.; Sirois, S.; Van Calsteren, M. R. & Roy, D. (2003). Consensus-degenerate hybrid oligonucleotide primers for amplification of priming glycosyltransferase genes of the exopolysaccharide locus in strains of the *Lactobacillus casei* group. *Applied and Environmental Microbiology, 69,* 3299-3307.

[74] Ramos, A.; Boels, I. C.; de Vos, W. M. & Santos, H. (2001). Relationship between Glycolysis and Exopolysaccharide Biosynthesis in *Lactococcus lactis. Applied and Environmental Microbiology, 67,* 33-41.

[75] Rimada, P. S. & Abraham, A. G. (2001). Polysaccharide production by kefir grains during whey fermentation. *Journal of Dairy Research, 68,* 653-661.

[76] Rimada, P. S. & Abraham, A. G. (2003). Comparative study of different methodologies to determine the exopolysaccharide produced by kefir grains in milk and whey. *Lait, 83,* 79-87.

[77] Rimada, P. S. & Abraham, A. G. (2006). Kefiran improves rheological properties of glucono-δ-lactone induced skim milk gels. *International Dairy Journal, 16,* 33-39.

[78] Robijn, G. W.; Thomas, J. R.; Haas, H.; van der Berg, D. J. C.; Kamerling, J. P. & Vliegenthart, J. F. G. (1995). The structure of the exopolysaccharide produced by *Lactobacillus helveticus* 766. *Carbohydrate Research, 276,* 137-154.

[79] Rodrigues, K. L.; Caputo, L. R.; Carvalho, J. C.; Evangelista, J. & Schneedorf, J. M. (2005b). Antimicrobial and healing activity of kefir and kefiran extract. *International Journal of Antimicrobial Agents, 25,* 404-408.

[80] Rodrigues, K. L.; Carvalho, J. C. & Schneedorf, J. M. (2005a). Anti-inflammatory properties of kefir and its polysaccharide extract. *Inflammopharmacology, 13,* 485-492.

[81] Ruas-Madiedo, P.; Abraham, A. G.; Mozzi, F. & de los Reyes-Gavilán, C. G. (2008) Functionality of exopolysaccharides produced by lactic acid bacteria. In: B. Mayo, P. López & G. Pérez-Martínez, (Eds.), *Molecular Aspects of Lactic Acid Bacteria for Traditional and New Applications.*(pp.137-166). Kerala, Research Signpost: Publisher.

[82] Ruas-Madiedo, P.; Hugenholtz, J. & Zoon, P. (2002). An overview of the functionality of exopolysaccharides produced by lactic acid bacteria. *International Dairy Journal, 12,* 163-171.

[83] Sánchez, J. I.; Martínez, B.; Guillén, R.; Jiménez-Díaz, R. & Rodríguez, A. (2006). Culture conditions determine the balance between two different exoplysaccharides produced by *Lactobacillus pentosus* LPS26. *Applied and Environmental Microbiology, 72,* 7495-7502.

[84] Sánchez-Medina, I.; Gerwig, G. I.; Urshev, Z. L. & Kamerling, J. P. (2007). Structural determination of a neutral exopolysaccharide produced by *Lactobacillus delbrueckii* ssp. *bulgaricus* LBB.B332. *Carbohydrate Research, 342,* 2430-2439.

[85] Santos, A.; San Mauro, M.; Sanchez, A.; Torres, J. M. & Marquina, D. (2003). The antimicrobial properties of different strains of *Lactobacillus spp.* isolated from kefir. *Systematic and Applied Microbiology, 26,* 434-437.

[86] Shiomi, M.; Sasaki, K.; Murofushi, M. & Aibara, K. (1982). Antitumor activity in mice of orally administered polysaccharide from Kefir grain. *Japanese Journal of Medical Science & Biology, 35,* 75-80.

[87] Smith, S. A. (1986) Polyethylene, low density. In M. Bakker (Ed.), *The Wiley Encyclopedia of Packaging Technology*. (pp. 514-523). New York, John Wiley & Sons: Publisher.

[88] Stingele, F.; Neeser, J. R. & Mollet, B. (1996). Identification and characterization of the *eps* (exopolysaccharide) gene cluster from *Streptococcus thermophilus* Sfi6. *Journal of Bacteriology, 181,* 6354-6360.

[89] Stingele, F.; Newell, J. W. & Neeser, J. R. (1999). Unraveling the function of glycosyltransferases in *Streptococcus thermophilus* Sfi6. *Journal of Bacteriology, 181,* 6354-6360.

[90] Tada, S.; Katakura, Y.; Ninomiya, K. & Shioya, S. (2007). Fed-Batch Coculture of *Lactobacillus kefiranofaciens* with *Saccheromyces cereviseae* for Effective Production of Kefiran. *Journal of Bioscience and Bioengineering, 103,* 557-562.

[91] Taniguchi, M.; Nomura, M.; Itaya, T. & Tanaka, T. (2001). Kefiran Production by *Lactobacillus kefiranofaciens* under the Culture Conditions Established by Mimicking the Existence and Activities of Yeast in Kefir Grains. *Food Science and Technology Research, 7,* 333–337.

[92] Toba, T.; Abe, S.; Arihara, K. & Adachi, S. (1986). A Medium for the Isolation of Capsular Bacteria from Kefir Grains. *Agricultural and Biological Chemistry, 50,* 2673-2674.

[93] Toba, T.; Arihara, K. & Adachi, S. (1987). Comparative study of polysaccharides from kefir grain, an encapsulated homofermentative *Lactobacillus* species and *Lactobacillus kefir*. *Milchwissenchaft, 42,* 565-568.

[94] Torino, M. I.; Hébert, E. M.; Mozzi, F. & Font de Valdez, G. (2005). Growth and exopolysaccharide production by *Lactobacillus helveticus* ATCC 15807 in an adenine-supplemented chemically defined medium. *Journal of Applied Microbiology, 99,* 1123-1129.

[95] Van Calsteren, M. R.; Pau-Roblet, C.; Bégin, A. & Roy, D. (2002). Comparative analysis of the exopolysaccharide biosynthesis gene clusters from four strains of *Lactobacillus rhamnosus*. *Biochemical Journal, 363,* 7-17.

[96] van Kranenburg, R.; Voss, H. R.; Van Swam, I. I.; Willem, N. J. & De Vos, W. M. (1999). Functional analysis of glycosyltransferases genes from *Lactococcus lactis* and other Gram-positive cocci: complementation, expression, and diversity. *Journal of Bacteriology, 181,* 6347-6353.

[97] Vaningelgem, F.; Zamfir, M.; Mozzi, F.; Adriany, T.; Vancanneyt, M.; Swings, J. & De Vuyst, L. (2004). Biodiversity of exopolysaccharides produced by *Streptococcus thermophilus* strains is reflected in their production and their molecular and functional characteristics. *Applied and Environmental Microbiology, 70,* 900-912.

[98] Vinderola, G.; Perdigón, G.; Duarte, J.; Farnworth, E. & Matar, C. (2006). Effects of the oral administration of the exopolysaccharide produced by *Lactobacillus kefiranofaciens* on the gut mucosal immunity. *Cytokine, 36,* 254-260.

[99] Wang, Y.; Ahmed, Z.; Feng, W.; Li, Ch. & Song, S. (2008). Physicochemical properties of exopolysaccharide produced by *Lactobacillus kefiranofaciens* ZW3 isolated from Tibet kefir. *International Journal of Biological Macromolecules, 43,* 283-288.

[100] Wellman, A.; Maddox, I. & Archer, R. (2003). Exopolysaccharide and extracellular metabolite production by *Lactobacillus delbrueckii* subsp. *bulgaricus*, grown on lactose in continuous culture. *Biotechnology Letters, 25,* 1515-1520.

[101] Yeesang, C.; Chanthachum, S. & Cheirsilp, B. (2008). Sago starch as a low-cost carbon source for exopolysaccharide production by *Lactobacillus kefiranofaciens. World Journal of Microbiology and Biotechnology, 24,* 1195-1201.

[102] Yokoi, H. & Watanabe, T. (1992). Optimum culture conditions for production of Kefiran by *Lactobacillus* sp. KPB-167B isolated from kefir grains. *Journal of Fermentation and Bioengineering, 74,* 327-329.

[103] Yokoi, H.; Fujii, Y.; Mukai, T.; Toba, T. & Adachi, S. (1991). Some taxonomical characteristic of encapsulated *Lactobacillus sp.* KPB-167B isolated from kefir grains and characterisation of its extracellular polysaccharide. *International Journal of Food Microbiology, 13,* 257-264.

[104] Yokoi, H.; Watanabe, T.; Fujii, Y.; Toba, T. & Adachi, S. (1990). Isolation and Characterization of Polysaccharide-Producing Bacteria from Kefir Grains. *Journal of Dairy Science, 73,* 1684-1689.

In: Food Hydrocolloids: Characteristics, Properties... ISBN: 978-1-60876-222-4
Editor: C. S. Hollingworth, pp. 273-289

Chapter 11

A PRELIMINARY STUDY ON ANTIMICROBIAL EDIBLE FILMS FROM PECTIN AND OTHER FOOD HYDROCOLLOIDS BY EXTRUSION METHOD*

LinShu Liu[1†], Tony Jin[1‡], Cheng-Kung Liu[1#], Kevin Hicks[1§], Amar K. Mohanty[2], Rahul Bhardwaj[2], and Manjusri Misra[3]

[1]Eastern Regional Research Center, ARS, U.S. Department of Agriculture, Wyndmoor, PA 19038, USA

[2]The School of Packaging, 130 Packaging Building, Michigan State Universities, East Lansing, MI 48824, USA

[3]Composite Materials and Structure Center, 2100 Engineering Building, Michigan State University, East Lansing, MI 48824, USA

ABSTRACT

Antimicrobial Edible films were prepared from natural fiber of pectin and other food hydrocolloids for food packaging or wrapping by extrusion followed by compression or blown film method. Microscopic analysis revealed a well mixed integrated structure of extruded pellets and an even distribution of the synthetic hydrocolloid in the biopolymers. The resultant composite films possess the mechanical properties that are comparable to films cast from most natural hydrocolloids that consumed as foods or components in processed foods. The inclusion of poly(ethylene oxide) alters the textures of the resultant composite films and therefore, demonstrating a new technique for the modification of film properties. The composite films were produced in mild processing

---

* A version of this chapter was also published in Modern Tendencies in Organic and Bioorganic Chemistry: Today and Tommorrow, edited by Abdulakh Mikitaev, Mukhed Kh. Ligidov and Gennady E. Zaikov, Nova Science Publishers. It was submitted for appropriate modifications in an effort to encourage wider dissemination of research.

† E-mail: linshu.liu@ars.usda.gov

‡ E-mail: tony.jin@ars.usda.gov

\# E-mail: chengkung.liu@ars.usda.gov

§ E-mail: kevin.hicks@ars.usda.gov

conditions, thus, the films are able to protect the bioactivity of the incorporated nisin, as shown by the inhibition of *Listeria monocytogenes* bacterial growth by a liquid incubation method.

Keywords: Pectin, antimicrobial, fiber, films, extrusion

INTRODUCTION

Pectin is a branched heterogeneous polysaccharide, consisting of 11 monosaccharides and their methyl or acetylated derivatives (Varagen 1996). Pectin is produced by extraction from cell walls of higher plants. The isolated pectin is water soluble. In water solutions or as the commercially available form of powders, pectin macromolecules aggregate each other to form fibers or small particles (Fishman, Jen 1986). Figure 1A shows the AFM imagine of pectin fibers in aqueous solution; the fibers with the size ranging from 100 nm to 400 nm are further associated to form a network. Soluble pectin readily reacts with most natural polymers to form hydrogels. Thus, pectin has a long history in the use as a gelling agent and film forming material (Liu et al. 2003). In the industry, the sources of pectin include citrus peel, sugar beet pulp, apple pomace, sunflower heads, or grape skins, etc. In the U.S., more than 300 thousand tons of pectin are available annually from the byproducts of fruit juice and beet sugar processing. However, only less than 2% of the available pectin is produced and they are mainly consumed as food additives. We have been interested in new utilizations of pectin and other agricultural commodities. In the previous studies, we showed that the blends of pectin with glycerol, and high amylose starch or poly(vinyl alcohol) can be used to produce biodegradable films with a wide range of mechanical properties and excellent oxygen barrier activity (Coffin, Fishman 1993; Coffin et al. 1995; Fishman, Coffin 1995; Fishman et al. 1996; Coffin et al. 1996). We also explored the medical applications of pectin-derived composites by incorporating with bioactive substances (Liu et al. 2003; Liu et al. 2004; Liu et al. 2005a; Liu et al. 2005b; Liu et al. 2006a). Recently, we extended our research to active packaging materials (Liu et al. 2006b; Liu et al. 2007a; Liu et al. 2007b). A series of pectin cast films have been prepared with the inclusion of various food proteins and a bacteriocin and tested for food packaging and wrapping applications, showing improved mechanical properties, water resistant and sustained antimicrobial activity.

Extrusion is a cost effective manufacturing process. Extrusion is popularly used in large scale production of food, plastics and composite materials. Most widely used thermoplastics are processed by extrusion method. Many biopolymers and their composite materials with petroleum-based polymers can also be extruded. These include pectin/starch/poly(vinyl alcohol) (Fishman et al. 2004), poly(lactic acid)/sugar beet pulp (Liu et al. 2005c), and starch/poly(hydroxyl ester ether) (Otey et al. 1980), etc. In this study, composite films of pectin, soybean flour protein and an edible synthetic hydrocolloid, poly(ethylene oxide), were extruded using a twin-screw extruder, palletized and then processed into films by compression molding process or blown film extrusion. The films were analyzed for mechanical and structural properties, as well as antimicrobial activity.

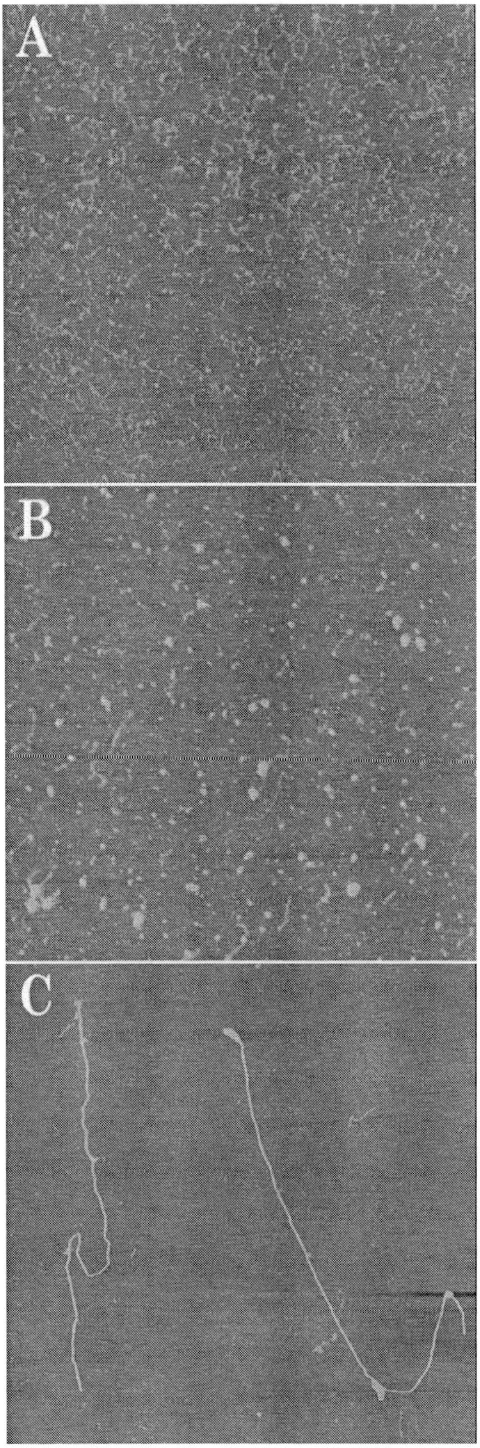

Figure 1. Atomic force microscopic images of pectin (A), soybean flour protein (B) and poly(ethylene oxide) (C). Concentration of solutions: A and B, 100 μg/ml; C 10 μg/ml. Field width: 2.5 μm.

EXPERIMENTAL

Materials

Citrus pectin (MexPec 1400) with the molecular weight (M.W.) of 1.8×10^5 dalton and the degree of esterification of 82 and Nisaplin® (2.5% Nisin) were purchased from Danisco-Cultor (Kansas City, KS). Soybean flour protein (SFP), poly(ethylene oxide) (PEO, M.W. 900,000), and glycerol (reagent grade, < 99.5%) were obtained from Aldrich Chemical Corp. (Milwaukee, WI).

Preparation

Prior to extrusion, dry ingredients of pectin, SFP, glycerol and Nisaplin were blended using a model C-100-T Hobart heavy-duty blender (Hobart Mfg. Co., Troy, OH). The compositions of blends are shown in Table 1. The blends were freeze-dried for 8 h and stored in a plastic bucket covered with lid at 4°C. The premixed pectin, SFP, glycerol were blended with PEO using a Werner & Pfleiderer ZSK-30 twin screw extruder and processed into pellets. This extruder has co-rotating screws (L/D ratio of 30/1) having variable screw profiles throughout its length. The screw speed was 105 rpm, which gave a torque value of 54%. The temperature profile of this extruder from zone 1 to zone 6 was 83, 90, 90, 88, 87, and 86°C. For formulation I, tap water was fed with solids together into the extruder barrel using a series 6300 Digital Feeder at the weight ratio of 19:81, liquid to solid. For formulation II, no water was used. The film blown was processed using a Killion Extruder (Davis Standards Corporation, Troy, OH). The extruder has a bottom fed blown film die. The die has a concentric air opening for film blowing. The air pressure can be controlled with an air valve. The die is also surrounded by an air ring for cooling. The screw rotation was varied between 25-50 rpm. The rotation speed of rollers on the tower was extremely slow to avoid the breaking of the film. The temperature profile of the extruder was also shown in Table 2. Some pellets were ground with the help of Polymer Machinery B.T.P granulator for 5-10 minutes to get powder for film compression processing. A Carver Laboratory press (Fred S. Carver Inc., Summit, NJ) was used for this purpose. The temperatures of upper and lower plates were controlled at 90°C. Composite powders were placed between the 2 metal plates and pressed for 3 minutes under the load of 15000-20000 lbs. Teflon sheet were used between the metal plates to avoid the sticking of the materials to metal plates.

Table 1. Compositions of two formulations*

Formulations	Pectin	SFP	PEO	Glycerol	Nisaplin®
I	50	19	0	30	1
II	35	13	30	21	1

* Date expressed as w% of total solid.

Table 2. Extrusion temperature profiles (°C)

Barrel Zones						
Zone 1	Zone 2	Zone 3	Clamp ring	Adaptor	Die 1	Die 2
1000	100	100	90	85	85	80

All samples thus obtained were stored in a desiccator over $CaCl_2$ at room temperature for structural and mechanical examinations.

Microscopic Analysis

Confocal laser scanning microscopy. A laser scanning confocal microscope system was used to analyze the microstructure of the film components and films. A model IRBE (inverted) microscope (Leica Microsystems, Bannockburn, IL) with a universal stage and 20X objective lens was coupled to a TCS NT/SP scanner head (Leica Microsystems, Exton, PA). Sheets of samples or powders in MatTek dishes (MatTek Corp., Ashland, MA) were illuminated with the 488 nm line of an Argon laser and fluorescence was collected in one channel (520-580 nm) and reflection at 488 nm was collected in another. Fluorescence emission spectra for the sample powders of SFP and Pectin were collected in the range of 500-680 nm.

Atomic force microscope (AFM). Sample solutions at 100 ng/ml or less were cemented onto mica and imaged in a model Nanoscope IIIa scanning probe microscope with TESP cantilevers (Veeco/Digital Instruments, Santa Barbara, CA) operated in the intermittent contact mode on an atomic force microscope.

Scanning electron microscopy. A Quanta 200 FEG scanning electronic microscope (SEM, FEI, Hillsboro, OR) was used to collect images. Prior to examination, samples were mounted to specimen stubs and sputtered with a thin layer of gold. Samples were examined in the high vacuum/secondary electron imaging model at 5,000X and 50,000X.

Dynamic Mechanical Analysis (DMA)

Small deformation dynamic mechanical analysis on compressed or blown films was done using a Rheometrics Scientific RSA II Solids Analyzer. Samples were tested using an initial applied force of 150 grams, an applied strain of 0.1%, and were heated from -100°C to 200°C at 10°C/min. A triplicate set of tests were performed for each samples

Tensile Test and Acoustic Emission (AE)

The mechanical property and AE measurements were performed as previously described (Liu et al. 2005). An upgraded Instron mechanical property tester, model 1122, and Testworks 4 data acquisition software (MTS Systems Corp., Minneapolis, MN) were used throughout this investigation. The strain rate was set at 50 mm/min. Date of tensile strength,

Young's modulus, and fracture energy was collected. The tensile tester was also programmed to perform a cyclic test. Samples were repeatedly stretched to 2 % strain at 50 mm/min and back to 0% strain. A total of 5 cycles were tested and the peak stress was recorded for each cycle.

AE measurement was performed simultaneously with the stress-strain tests. A small piezoelectric transducer (Model R15, Physical Acoustics Corp., Princeton, NJ) was clipped against the samples. AE signals emanating from this transducer when the Instron stretched samples were processed with an upgraded LOCAN-AT acoustic emission analyzer (Physical Acoustics Corp.), which connected to a PC base with enhanced graphing and data acquisition software with all features and options of the SPARTAN 2000.

Bacterial Inhibition Test

Pathogenic bacteria *Listeria monocytogenes* Scott A 724 was obtained from the in house culture collection. Stock cultures of *L. monocytogenes* were maintained at -80ºC in BHI medium (Difco Laboratory, Detroit, MI). Working Cultures of *L. monocytogenes* were maintained on BHI agar at 4 ºC and were sub-cultured bi-weekly and grown aerobically at 37ºC in BHI broth. Prior to inoculation of product the organism was cultured in BHI broth at 37ºC for 16-18 h.

Bacterial inhibition by the antimicrobial films was evaluated using a liquid incubation method, 4 pieces of films (1 cm^2 each) were placed in a glass test tube with 9 ml BHI broth inoculated with 1 ml *L.monocucytogenes* overnight culture. The test tubes then placed in a shaker (Innovas 3100, New Brunswick Sci. Inc., Edison, NJ) at room temperature and shaken at 200 rpm. One ml of inoculated sample was taken at 0, 8, 24, and 48 h. The samples were serially diluted by sterile phosphate buffer (Hardy Diagnostics, Santa Maria, CA), then pour plated onto BHI agar. Plates were incubated at 37ºC for 24 h. A film-free inoculated BHI broth served as a control. Plates with 30-300 colonies were enumerated using a manual colony counter (Bantex Colony Counter 920; Bantex, Burlingame, CA).

Results and Discussion

Characterization of raw materials for films was firstly conducted on the three ingredients and their blends. Figure 1 shows the AFM images of pectin, SFP and PEO in diluted solutions. The three macromolecules were in the form of fibers in the size ranging from 100 nm to 10 μm. The fibers associated each other.

Both the pectin and SFP have intrinsic fluorescence emission (Figure 2A, B) and show a similar profile in their fluorescence spectra (Figure 2C), except that the protein has a narrow maximal intensive peak at around 530 nm, the pectin has a broad peak extended from 530 to 550 nm. Furthermore, the autofluorescence of protein is stronger than that from pectin. PEO doesn't emit fluorescently, PEO can be detected by reflection (Figure 2D). The differences of the ingredients in photoemission were used to probe the microstructure of resultant composite films. Figure 3 reveals the homogeneity of extruded pellets. After processing, we are still able to identify specific regions that consist of pectin or SFP alone. Each of these regions stretched about 10 μm in width; between the two regions of interest there was a well mixing area, which also sized at about 10 μm. The florescent emission from the mixed area reflects the profiles of each spectrum of the two single biopolymers. Figure 4 shows the microscopic images of one of the final products, compression film of formulation II, obtained by confocal reflection and confocal fluorescence in stereo projection. The red area correlated with PEO reflection; the green area correlated with the pectin and protein. The image indicates a well mixed integrated structure, showing an even distribution of the synthetic hydrocolloid in the biopolymers.

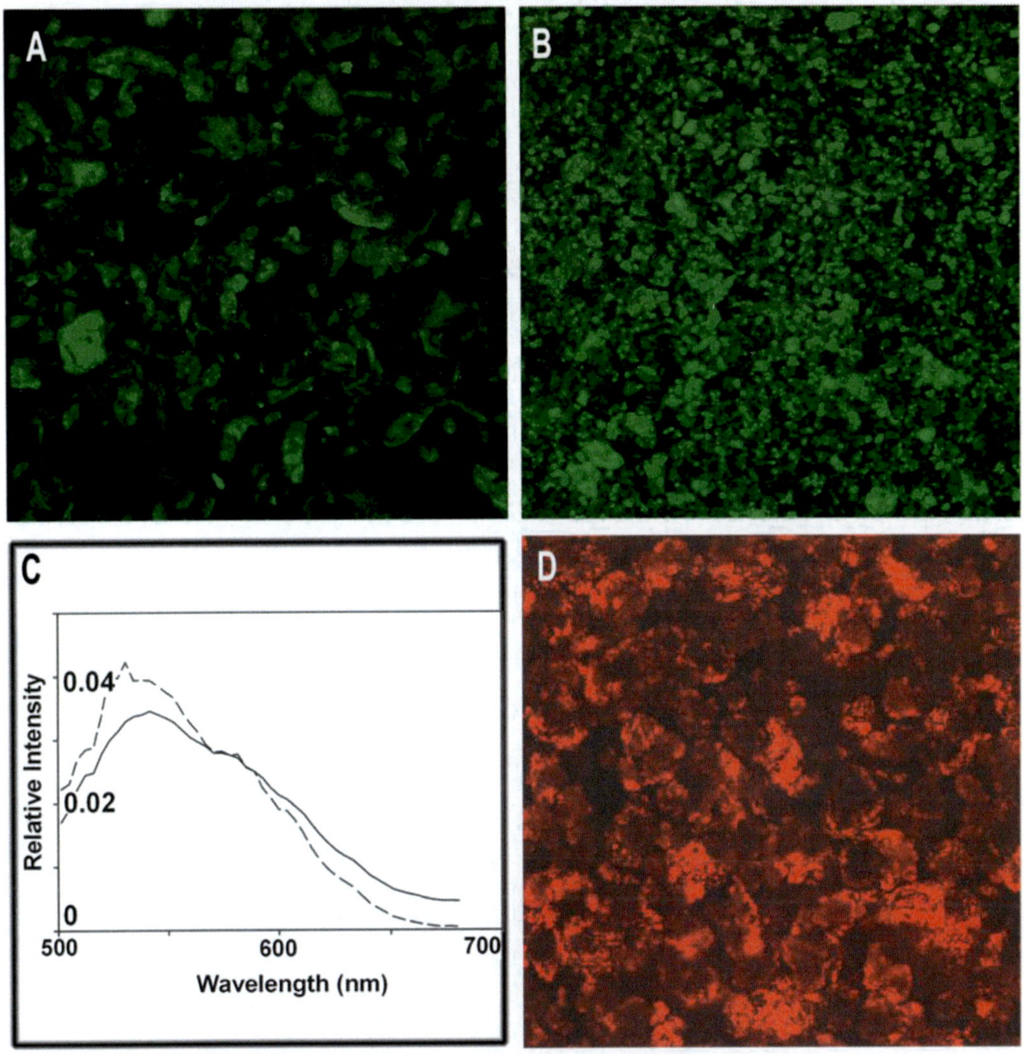

Figure 2. Confocal laser scanning microscopic image: (A) fluorescence of pectin, (B) fluorescence of soybean flour protein, (C) fluorescence spectra of pectin and soybean flour protein, and (D) reflection of poly(ethylene oxide). Field width: A, B and D, 480 μm.

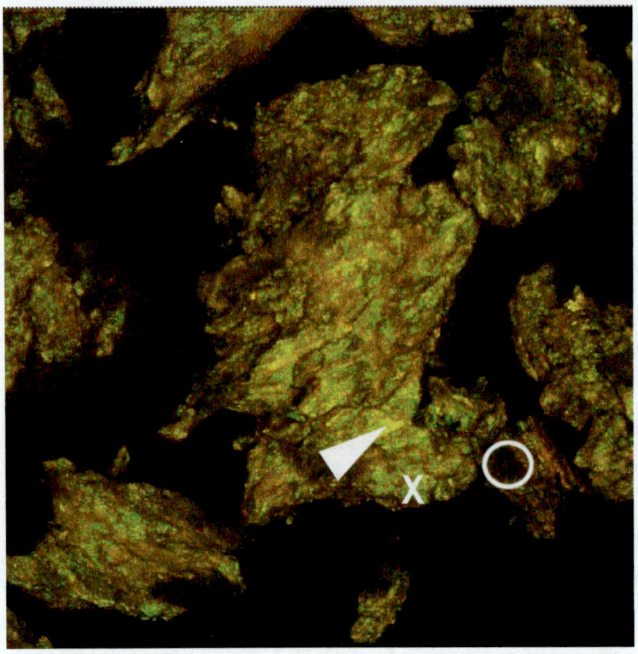

Figure 3. Structure analysis of extruded pellets: Region I (X) colored with green, indicating more pectin component in the area; regions II (Δ) colored with yellow, indicating there is more soybean flour protein; regions III (○) colored with brown, indicating the well mixing areas. Field width, 550 μm.

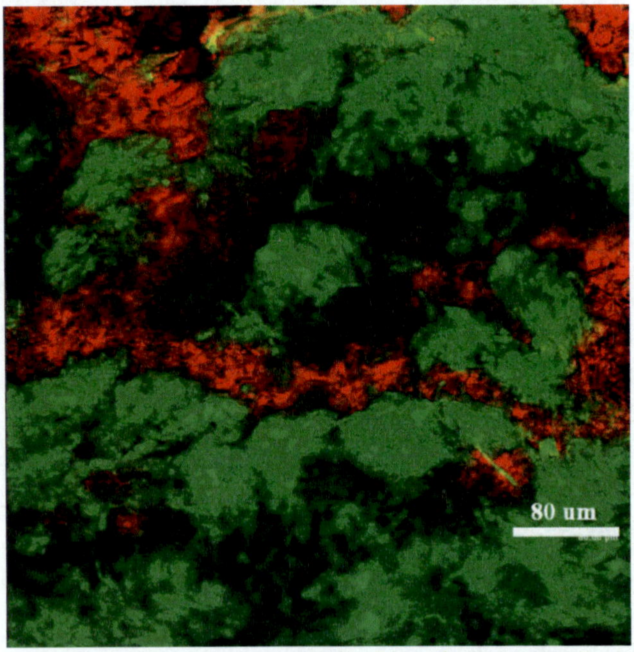

Figure 4. Confocal laser scanning microimages of compression film from formulation II. The organization of the biopolymers were resolved by confocal fluorescence (excitation 484 nm, emission 520-580 nm), the PEO was defined by confocal reflection (633 nm). The micrograph was collected in stereo projection in extended focus images of 20-30 micrometer-thick slabs of the film. Field width, 470 μm.

Small deformation dynamic mechanical analysis is a well known method used as a complement for microscopic examination in material structure study. Figure 5 comprises of DMA curves for samples prepared by compression method from formulations I (PEO-free, Figure 5A), formulation II (PEO included, Figure 5B) and PEO alone (Figure 5C). Two significant differences were seen between the samples. One is the location of the sub-ambient glass transition temperature. This is seen at -55°C in sample of formulation I and at -69°C in sample of formulation II. The other is that samples with PEO shows a melting behavior at about 64°C, while a melting behavior is not seen in the PEO-free samples until about 145°C. These differences are certainly due to the presence of the PEO which has a melting point of 57°C. The PEO-free samples also have a transition at 34°C which is probably related to the protein in the sample.

Mechanical properties of tensile strength, Young's modulus, elongation at break and toughness are important for packaging and wrapping materials. In a variety of end uses, packaging and wrapping materials are often subjected to a force during tensile strain. The materials must be able to resist considerable stress without failing to a fracture at a designed stress. Furthermore, as an edible food wrapping materials, the materials may be taken with foods together either for convenient purpose or to enhance or alter the food texture. In these cases, their mechanical properties directly related to the mouth feeling of accepters, which is an important measurement of food quality. Table 3 shows the mechanical properties of selected samples of compression and blown films. In general, the composite films have mechanical properties that are similar to cast films from most natural hydrocolloids, which consumed in our ordinary life (Liu et al. 2006a). For compression films, the addition of PEO almost doubled the Young's modulus, which is an indication of stiffness; but only had a minor impact on the tensile strength, maximal elongation, and toughness. In comparison of blown film with compression films with same composition, the blown films seemed to be stiffer, but they were not as strong as the films prepared by compression method. Moreover, the replacement of SFP with PSG dramatically enhanced the mechanical properties of the composite films (Table 3). The results of the stress-strain cyclic tests are shown in Figure 6. For the PEO-free samples, the loop created by the first cycle is bigger than those created by following cycles. Among the following cycles, the different in loop size is not significant. For samples containing PEO, the size of loop decreased gradually as cycled, then, became constant in the last two cycles. At the end of the cyclic test, the PEO containing sample expressed a much higher stress than the PEO-free sample. Although the PEO free samples are more resistant to mechanical force and less permanent deformation occurred than the PEO-included sample; the inclusion of PEO strengthens the composite films.

Table 3. Mechanical properties

Samples, processing method	Tensile strength (MPa)	Elongation at break (%)	Young's modulus (MPa)	Toughness (J/cm^3)
I, compression	2.35 ± 0.09	6.14 ± 0.6	77.2 ± 4.49	0.16 ± 0.08
I-b, compression*	8.8 ± 0.3	13.1 ± 2.0	353.4 ± 26.9	0.90 ± 0.12
II, compression	3.1 ± 0.8	7.3 ± 1.5	125.2 ± 25.3	0.20 ± 0.07
II, blown	2.3 ± 0.3	2.8 ± 0.3	151.2 ± 20.8	0.03 ± 0.01

* I-b, PSG was used to replace SFP in this composite film.

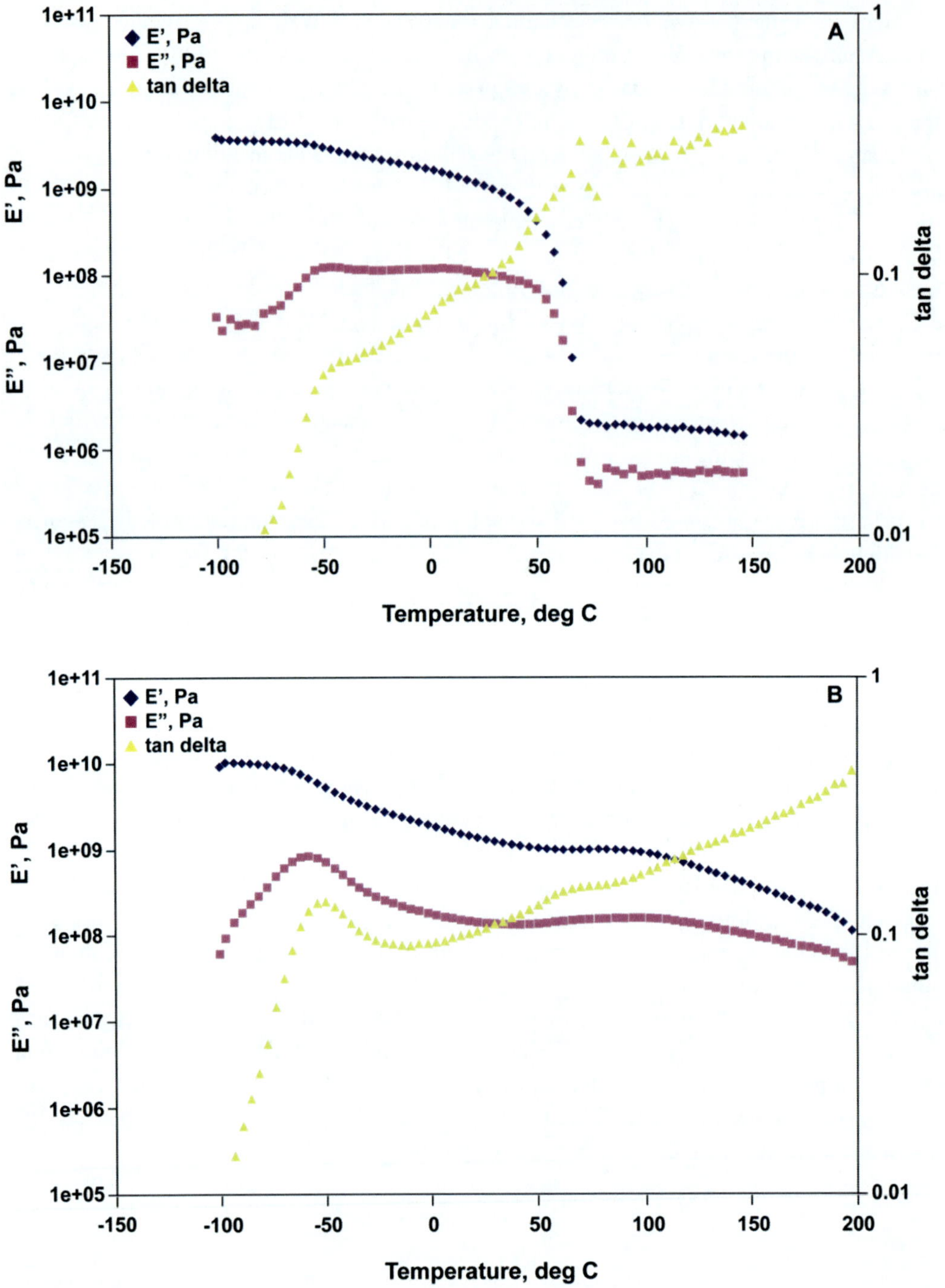

Figure 5. (Conti ued).

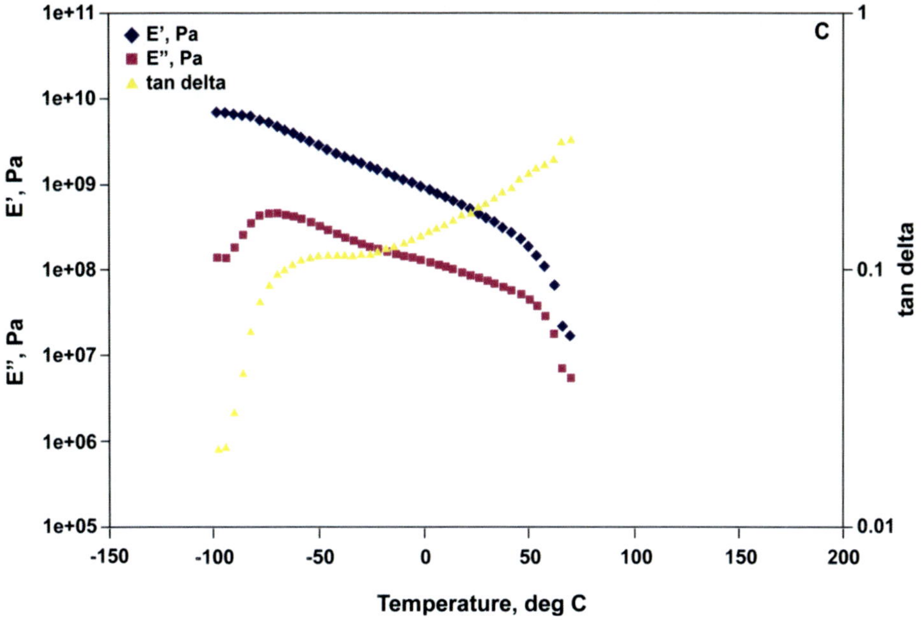

Figure 5. Dynamic Mechanical Analysis of composites containing with PEO (A) or without PEO (B) and PEO alone (C). The PEO-free samples show a melting point at about 145 °C; while, both the PEO alone samples and PEO-included samples a melting behavior at about 64 °C.

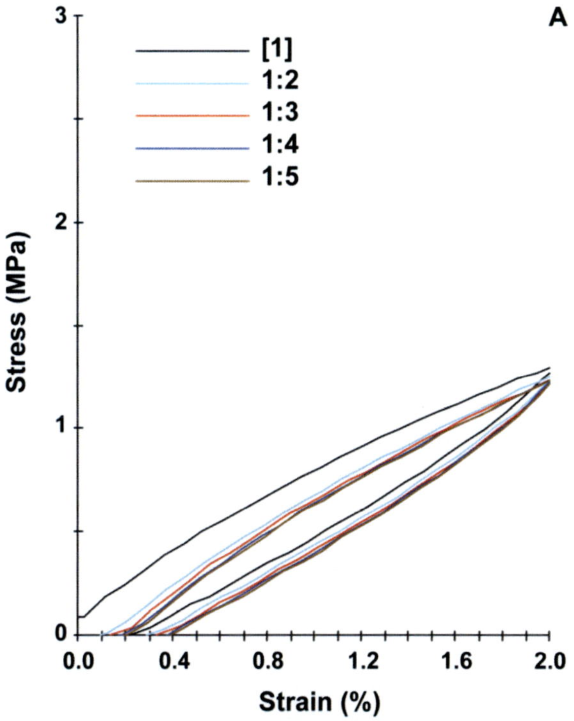

Figure 6. (Continued).

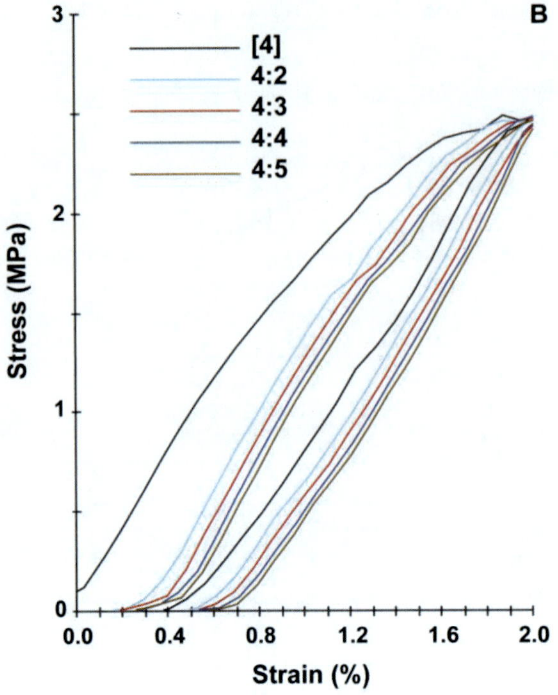

Figure 6. Stress-strain curves obtained from cyclic tensile tests. Samples of soybean flour protein and pectin containing without PEO (A) and with PEO (B). For the PEO free films, the loop created in the first cycle is larger than following cycles. For the PEO included films, the size of loops gradually decreased as cycled, then became constant in the last two cycles.

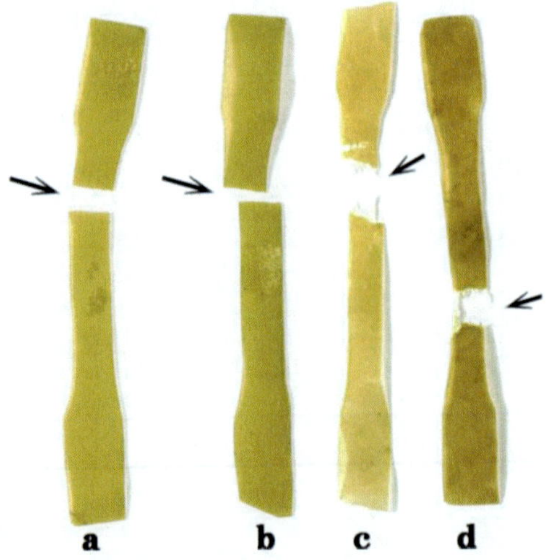

Figure 7. Photographs of fractural surfaces of samples of PEO-Free (a and b) and PEO-included composite films (c and d).

The addition of PEO has an influence on film destruction caused by an external destructive force (Figure 7). Without PEO, the external force created a clear-cut fracture surface, indicating the good adhesion between the two biopolymers. With the inclusion of PEO, the deformation created a fibrous surface. This can be seen more clearly from SEM and fluorescent microscopy. As shown in Figure 8 A and B, fibers were pulled out, extended, and then, broken, but still embedded in the matrix phase. We examine the fibers with confocal reflection and confocal fluorescence in two channels. It confirms that the main component of the fibers is PEO; however, the biopolymers were either inserted or encapsulated within the fibers (Figure 8C).

To probe the structural changes of a composite film that subjected to a destructive force, we measured the AE event simultaneously with the tensile test. Figure 9 shows the correlation between the stress-strain curve and the AE hit pattern. For the PEO-free samples, AE activity detected only at the peak stress, when the samples were completely destructed. This confirms the homogeneity of the composite films. Since the two biopolymers are compatible, they are able to transfer stress evenly. For the samples containing with PEO, the phenomena are similar, the samples emitted sound at the peak stress; however, signals were continually collected as the PEO fiber were pulled and broken, being consistent with the results shown in Figures 7 and 8.

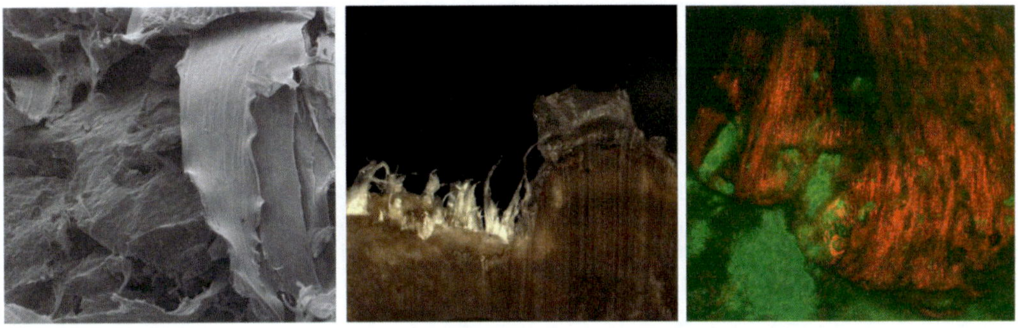

Figure 8. Microscopic images of the fractural surfaces of PEO-included composite samples obtained by scanning electron microscope (left), laser microscope (middle), and confocal laser scanning microscope in confocal fluorescence and confocal reflection two channels (right). Field width: 520 μm (left) and 480 μm (middle and right).

In the colorful family of food packaging and wrapping materials, antimicrobial film is a new member. Besides providing a physical barrel, the films function in prohibition, protection and suppression of microbial migration to or growth in the packages by creating antimicrobial surfaces or releasing antimicrobial substances. The use of antimicrobial materials in food packaging improves food safety and is more convenient to the consumers; therefore, the market for antimicrobial food packaging has a fast growth (Cutter 2006; Ozdemir, Floros 2004). In the present study, we incorporated an antimicrobial polypeptide, nisin, into the film formulation. The antimicrobial activity of resultant films is shown in Figure 10. Nisin alone was firstly tested for stability and the results were compared to that treated in the conditions as same as film compression. No statistically significant differences could be detected. Then, we compared the activity of nisin incorporated in blown films and compression films with that of non-treated nisin. As shown in Figure 9, the non-treated nisin was little more active than that formulated in pectin films at the fist 24 hours; however, the difference disappeared in 48 hours incubation. This result indicates that the extrusion conditions applied in the current experiments are mild and not harmful to nisin. Thus, the resultant composite films are able to retain nisin activity.

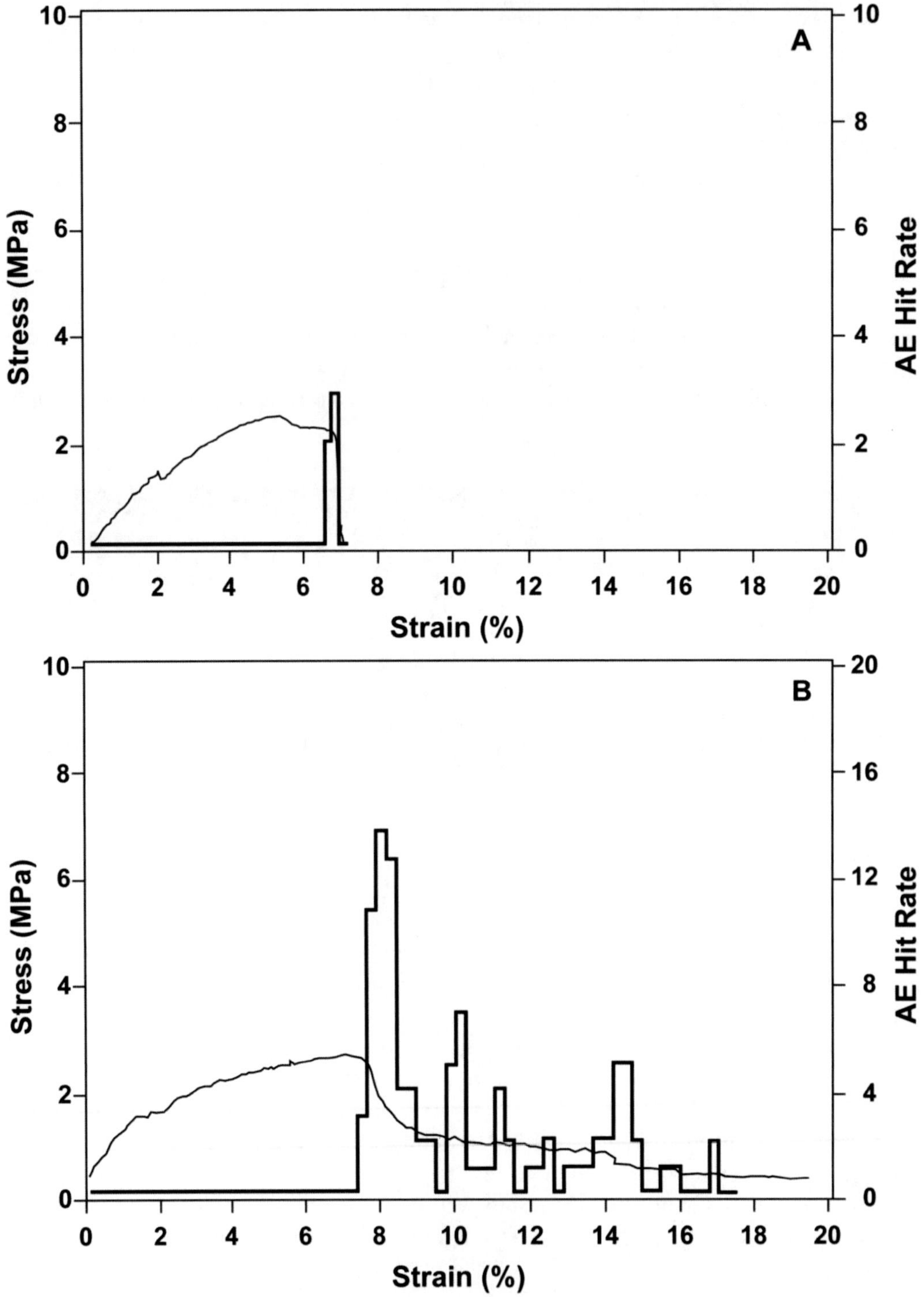

Figure 9. Correlation between Stress-strain curves and AE patterns of PEO-free samples (A) and the samples containing with PEO (B).

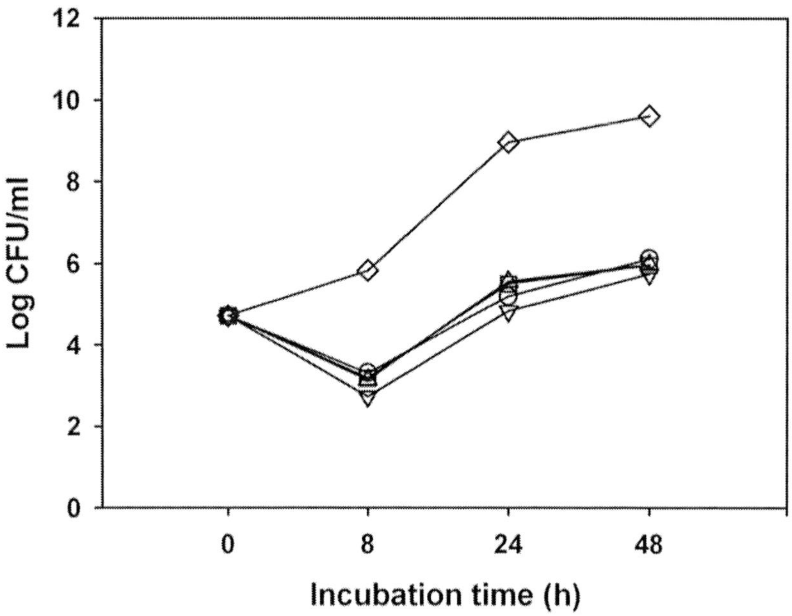

Figure 10. Growth of *Listeria monocytogenes* in BHI broth at 24 °C. Control (diamond), nisaplin prior to extrusion (square), nisaplin post extrusion (circle), nisaplin in compress film (up triangle), and nisaplin in blown film (down triangle).

CONCLUSIONS

The current study provides a new type of edible, antimicrobial food packaging or wrapping films from food-grade natural fibers or hydrocolloid. Besides film casting, the films can also be produced by compression, extrusion blown methods. The inclusion of PEO hydrocolloid in natural fiber formulations makes films tougher and caused less permanent deformation when the films were subjected to an external force. Since the extrusion and compression were performed in mild conditions, nisin can be incorporated into films without diminishing its antimicrobial activity.

ACKNOWLEDGMENTS

Authors acknowledge Dr. Peter H. Cooke, Mr. Nicholas Latona, Ms. Guo-ping Bao, Dr. David R. Coffin and Dr. Vitoria L. Finkensdadt for their technical assistance.

REFERENCES

[1] Coffin, D. R., Fishman, M. L. 1993. Viscoelastic Properties of Pectin/Starch Blends. *Journal of Agricultural and Food Chemistry,* Vol. 41, pp. 1192-1197.

[2] Coffin, D. R., Fishman, M. L. Cooke, P. H. 1995. Mechanical and Microstructural Properties of pectin/Starch Films. *Journal of Agricultural and Food Chemistry,* Vol. 57, pp. 663-670.

[3] Coffin, D. R., Fishman, M. L., Ly, T.V. 1996. Thermomechanical Properties of Blends of Pectin and Poly(vinyl alcohol). *Journal of Applied Polymer Science,* Vol. 61, 663-670.

[4] Cutter, C. N. 2006. Opportunities for Bio-Based Packaging Technologies to Improve the Quality and Safety of Fresh and Further Processed Muscle Foods. *Meat Science,* Vol. 74, 131-142.

[5] Fishman, M. L., Coffin, D. R. 1995. Films Fabricated form Mixtures of Pectin and Starch. US Patent 5,451,673.

[6] Fishman, M. L., Jen, J. J. 1986. Chemistry and Function of Pectins. ACS Series 310, American Chemical Society press, Washington D.C.

[7] Fishman, M. L., Coffin, D. R., Unruh, J. J., Ly, T. 1996. Pectin/Starch/Glycerol Films: Blends or Composites. *Journal of Macromolecular Science: Pure and Applied Chemistry,* Vol. A33, pp. 639-654.

[8] Fishman, M. L., Coffin, D. R., Onwulata, C. I., Konstance, R. P. 2004. Extrusion of Pectin and Glycerol with Various Combinations of Orange Albedo and Starch. *Carbohydrate Polymers,* Vol. 57, pp. 401-413.

[9] Liu, L. S., Fishman, M. L., Kost, J., Hicks, K. B. 2003. Pectin-Based Systems for Colon-Specific Drug Delivery via Oral Route. *Biomaterials,* Vol. 24, 3333-3343.doi:

[10] Liu, L. S.; Won, Y.-J.; Cooke, P. H., Coffin, D. R., Fishman, M. L., Hicks, B. K., Ma, P. X. 2004. Pectin/Poly(lactide-co-glycolide) Composite Matrices for Biomedical Applications. *Biomaterials,* Vol. 25, pp. 3201-3210.

[11] Liu, L. S., Chen, G., Fishman, M. L., Hicks, K. B. 2005a. Pectin Gel Vehicles for Controlled Fragrance Delivery. *Drug Delivery*, Vol. 12, pp. 149-157.

[12] Liu, L. S., Fishman, M. L., Hicks, K. B., Kende, M. 2005b. Interaction of Various pectin Formulations with Porcine Colonic Tissues. *Biomaterials,* Vol. 26, pp. 5907-5916.

[13] Liu, L. S., Fishman, M. L., Hicks, K. B., Liu, C.-K. 2005c. Biodegradable Composites from Sugar Beet Pulp and Poly(lactic acid). *Journal of Agricultural and Food Chemistry,* Vol. 53, No. 23, pp. 9017-9022.doi:10.1021/jf058083w.

[14] Liu, L. S., Fishman, M. L., Hicks, K. B., Kende, M., Ruthel, G. 2006a. Pectin/Zein Beads for Potential Colon-Specific Drug Delivery: *Synthesis and in vitro Evaluation.* 13:417-423.

[15] Liu, L. S., Liu, C.-K., Finkenstadt, V. L., Jin, T. Z., Fishman, M. L., Hicks, K. B. 2006b. Pectin Films for Various Applications. Proceedings of the 35th United States – *Japan Cooperative Program in Natural Resources*. PR31-34.

[16] Liu, L. S., Finkenstadt, V. L., Liu, C.-K., Jin, T. Z., Fishman, M. L., Hicks, K. B. 2007a. Preparation of Poly(Lactic acid) and Pectin Composite Films for Application in Antimicrobial Packaging. *Journal of Applied Polymer Science,* in press.

[17] Liu, L. S., Liu, C.-K., Fishman, M. L., Hicks, K. B. 2007b. Composite Films from Pectin and Fish Skin Gelatin or Soybean Flour Protein. *Journal of Agricultural and Food Chemistry,* 55(6), 2349-2355. doi: 10.1021/jf062612u.

[18] Otey, F. H., Westhoff, R. P., Doane, W. M. 1980. Starch-Based Blown Films. *Industrial Engineering Chemical Products Research and Development,* Vol. 19, pp. 592-595.

[19] Ozdemir, M., Floros, J. H. 2004. Active Food Packaging Technologies Critical Reviews in *Food and Nutrition,* Vol. 44, pp. 185-193.

[20] Voragen, A. G. J. 1996. Pectin and Pectinases. Elsevier Science Publisher, New York, NY.

INDEX

A

B

C

D

E

F

G

H

I

K

L

M

N

O

P

Q

R

S

T

U

V

W

X

Y

Z